# DRAFT HORSES AND MULES

## Storey's Working Animals

# Draft Horses and Mules

## Harnessing Equine Power for Farm & Show

Gail Damerow & Alina Rice

Storey Publishing

The mission of Storey Publishing is to serve our customers by
publishing practical information that encourages
personal independence in harmony with the environment.

Edited by Lisa H. Hiley and Deborah Burns
Art direction by Cynthia N. McFarland
Cover design by Alethea Morrison
Text design by Vicky Vaughn Design
Text production by Jennifer Jepson Smith

Cover and chapter opener illustrations by John MacDonald
Letterpress typography and borders by Yee Haw Industries
Illustrations by Bethany A. Caskey
Maps and infographics pages 42, 43, and 148 by Ilona Sherratt

Back cover photographs by © Dusty Perin, top, and © Joseph Mischka, middle and bottom
Interior photography credits appear on page 261

Indexed by Susan Olason, Indexes & Knowledge Maps

© 2008 by Alina Rice and Gail Damerow

All rights reserved. No part of this book may be reproduced without written permission from the publisher, except by a reviewer who may quote brief passages or reproduce illustrations in a review with appropriate credits; nor may any part of this book be reproduced, stored in a retrieval system, or transmitted in any form or by any means — electronic, mechanical, photocopying, recording, or other — without written permission from the publisher.

The information in this book is true and complete to the best of our knowledge. All recommendations are made without guarantee on the part of the author or Storey Publishing. The author and publisher disclaim any liability in connection with the use of this information. For additional information, please contact Storey Publishing, 210 MASS MoCA Way, North Adams, MA 01247.

Storey books are available for special premium and promotional uses and for customized editions. For further information, please call 1-800-793-9396.

Printed in the United States by Versa Press
10 9 8 7 6 5 4 3 2 1

**LIBRARY OF CONGRESS CATALOGING-IN-PUBLICATION DATA**

Damerow, Gail.
  Draft horses and mules : harnessing equine power for farm and
show / Gail Damerow and Alina Rice.
    p.   cm. — (Storey's working animal series)
  Includes index.
  ISBN 978-1-60342-081-5 (pbk. : alk. paper)
  ISBN 978-1-60342-082-2 (hardcover : alk. paper)
  1. Draft horses. 2. Mules.  I. Rice, Alina. II. Title. III. Series.
SF311.D36 2008
636.1'5—dc22
                                2008004720

To my husband, Dr. Jereld Rice.
Without his support, this book would not have become a reality.
As a veterinarian, Jereld offered careful and thoughtful advice,
as well as professional criticism and review of information.
He also listened to countless hours of reading.
As a spouse, a horseman, and a veterinarian, Jereld has been
an inspiration to me. He completely supports and collaborates
with my ambitions and goals for working draft horses.
I am one fortunate woman to be his wife.

— Alina Rice —

◆

To my husband and best friend, Allan Damerow,
and to all the wonderful people I have met
in the draft horse world over the years.

— Gail Damerow —

# ACKNOWLEDGMENTS

THIS BOOK IS THE RESULT of many experiences coming together, experiences made possible by a number of memorable people and equines. In particular, I wish to acknowledge my mother and father, Christine and Ron Arnold. I also want to mention George Hatley, Otis and Ann Parks, Gail Damerow, and the great horses and mules I have worked, ridden, and driven in my short life. All have influenced the creation of this book.

Without my parents' interest in and pursuit of draft horse ownership and use, I would not have the knowledge or practical experience necessary to write this book. Although they continue to scratch their heads over this crazy, wide-open daughter of theirs, I appreciate their support.

George Hatley continues to inspire me with his knowledge of and dedication to the preservation of equine history. After I moved away from my parents' Kentucky farm, George kindly let me get my draft horse fix whenever I needed it, with the use of his two beautiful Belgian geldings.

Otis and Ann Parks will always hold a special place in my heart. They have inspired me with their hardworking approach to life, with their dedication to horses, and with their interest in so many people's lives, mine included.

Gail Damerow was my fearless and fierce collaborator. She is responsible for the success of this book, for she took the time to educate me in the ways of publishing, to review my work, to help me understand the beauty of ruthless organization, and to patiently answer my infinite questions.

I have been fortunate to have the opportunity to learn and improve my skills as a teamster by driving great draft horses and mules — from the many horses I drove on my parents' Kentucky farm to the beautiful Southern German Drafts I drove on a farm in Germany. The learning continues. Thank you to all.

— Alina Rice

# CONTENTS

Foreword viii

Preface ix

**1 WHY OWN A DRAFT HORSE OR MULE?** 2

**2 LEARNING THE BASICS** 20

**3 FINDING THE IDEAL EQUINE** 30

**4 HORSES: SELECTING A BREED** 64

**5 MULES: SOMETHING SPECIAL** 79

**6 EQUINE COMMUNICATION** 89

**7 HOUSING YOUR HORSE OR MULE** 108

**8 FEEDING YOUR HORSE OR MULE** 128

**9 ROUTINE HEALTH CARE** 152

**10 FIRST AID AND ILLNESS** 173

**11 HARNESSING AND DRIVING** 188

**12 WORKING WITH DRAFT POWER** 219

Glossary 243

Resources 249

Index 252

# FOREWORD

**Just go anywhere** experienced teamsters are working with their draft horses or mules and you will usually see them. It might be a field day where a regional draft horse and mule club demonstrates plowing, tilling, and planting with horse-drawn equipment. Or it might be at a county fair where horsemen and women put their teams and larger hitches to show wagons to compete with one another before a crowded grandstand. Or it might be in a small town holding a local parade. Or it might just be on the side of the road where a team is working in a field.

Go to any of these places and you will see them there: men and women wishing they had horses or mules of their own to work in a field, show at a fair, or drive in a parade. A few of them will make this dream a reality.

And it's not long after they get their first horse that they call somebody like me, looking for a book of information and advice. The first-time draft horse or mule owner wants a reference book that covers all the basics of caring for and working with their animal, from selecting the proper feed to correctly fitting a collar. Alina Rice and Gail Damerow have put together just such a resource, and you are holding it now.

Drawing on a number of expert sources, Alina and Gail begin by covering the basics of equine ownership and what it really means to be responsible for a horse or mule of your own. They discuss what to consider when buying your first horse: what to look for at the seller's farm, whether to buy a mare or gelding, how to identify personality traits in the prospective animal, and how to assess the horse's physical condition.

The authors move on to talk about the many decisions a horse owner must make when preparing the stable and pasture for the new horse. Bedding, feed and tack storage, and animal space requirements are all taken into account. How much pasture is enough? What kind of fencing is best? All the questions a new horse owner might have are answered, and even more are covered that the reader may not have thought to ask.

A full chapter is devoted to feeding your horse. New information regarding feed requirements and avoidable health hazards are presented in a complete and easy to understand fashion. Balancing your horse's diet between fiber, oils, and proteins is discussed, and the reader is warned of what can happen when that balance goes awry.

Great care is given to the topic of equine health. Knowing how to identify health problems and what to do about them takes a lot of the worry and guesswork out of owning a horse. From top to bottom, the entire animal is presented in terms of its health requirements and, like the rest of the book, each discussion is accompanied by clear and informative illustrations.

Once your new horse is healthy at home in its stable and being fed correctly, Alina and Gail walk you carefully through the mechanics of harnessing, hooking, and driving your animal. What type of equipment do you need and where can you get it? What kinds of harness are available? What exactly can you do with your new horse and how can you safely learn how to do it?

Finally, the detailed glossary and generous index of additional resources including books, magazines, and draft associations finish the job of making this volume as complete a reference tool as I've seen for the new draft horse and mule owner as well as those who have enjoyed draft equine ownership for years. And what makes the book even more valuable is the inclusion of many real-life stories offered by people who work with horses and mules every day. These anecdotal sidebars offer priceless advice from genuine experts and effectively illustrate the topics covered in the book.

—JOE MISCHKA, Publisher
*Rural Heritage* Magazine

# PREFACE

**When my husband, Allan,** and I moved to our Tennessee farm so many years ago, the cornfield across the road was worked with harnessed mules. While other areas of the country were experiencing a resurgence of interest in draft animals, this area had never fully entered the tractor era. Our closest neighbor, Willie, was an old-time horse trader who always had a horse or mule ready to put to work.

In those early days we paid our mortgage by thinning our woodlot and selling firewood. Our operation was powered by a chain saw, four human legs, and an old pickup truck that, hitched to a cable and pulley, served as a log skidder. Our logging operation wasn't as lucrative as it might have been, because a lot of our earnings went to periodically replacing the truck's transmission.

I don't remember how many transmissions we financed before Willie showed up one day with his stout horse, Bob, and began skidding logs for us. No longer did Allan have to run up and down the hills, positioning the cable and hollering instructions that I strained to hear, as I operated the pickup and tried my best not to burn out yet another transmission. With real live horse power bringing out the logs, our operation was not only more efficient, but also more profitable.

Thereafter, whenever Willie heard Allan start his chain saw, he'd show up with big, patient Bob, who knew quite a bit more about log skidding than his human handler. Whenever Bob got stuck pulling a log up the trail, an exasperated Willie would throw down the lines, whereupon Bob invariably turned in the opposite direction to free the log and trot off to the landing, with Willie huffing and puffing to catch up.

When we decided to purchase 10 acres adjoining our farm, we paid the mortgage in 18 months by sorting saw logs, pallet logs, hickory logs for ax handles, and firewood — all logged by horse. Unlike neighboring woodlots that were being clear-cut with monster machines, when our woodlot had been selectively harvested, you could scarcely tell the land had been logged.

On the days when the weather wasn't conducive to logging, I worked as a freelance writer. Eventually, I became the editor of *Rural Heritage,* a bimonthly journal for proponents of real horse power.

Alina Rice (then Arnold) burst onto the draft horse scene with the publication of her home-school essay, written at age 11, about her family's horse-powered farm in Kentucky, not far from our Tennessee farm. Through her teenage years she continued to occasionally publish her horse-farming adventures, and when she started college we invited her to spend a summer with us as an intern in agricultural journalism. That summer she spent three days a week living on our farm and working in our office; the rest of the week she went back to Kentucky to work on her family's farm.

During that summer Alina decided that a career in agricultural journalism was not for her — she'd rather be outdoors. But she enjoyed writing, especially about her favorite subject — draft horses — and so became a regular *Rural Heritage* columnist. Her column, "Farm Fresh," first appeared in the Holiday 2000 issue, and in it she introduced herself as "a total horse power enthusiast." As promised, her columns described incidents that occurred on her family's farm.

Before long our office began receiving mail, e-mail, and phone calls from Alina's fans. She inspired other young people to get involved with draft horses, and she instilled in other women the notion that working horses isn't just for men. But the most surprising and unexpected fan mail came from fellows, who were encouraged by the notion that if this intrepid 5-foot 3-inch, 100-pound gal could drive two, three, and four heavy horses — and not only drive them, but plow and cultivate and make hay — why, then, anyone with a mind to could do the same. One fellow was so enraptured he suggested marriage, sight unseen, but Alina was waiting for Mr. Right.

She went off to graduate school in Idaho, earned her master's degree in soil science, met and married Jereld Rice, and all the while continued to inspire novice and experienced teamsters alike with the knowledge and enthusiasm expressed in her column.

When Deb Burns at Storey, publisher of many of my books on farming, asked me to write a book on draft horses and mules, I wasn't sure I could get it done in a timely manner while fulfilling my job as editor of *Rural Heritage*. I also felt the book would be written better with the energy and enthusiasm of someone like . . . like . . . Alina! But just as I had trepidations about finding sufficient time to write yet another book, Alina had trepidations about being ready to write her first book. So this volume came about through collaboration, and is imbued with Alina's fresh outlook and unique ability to energize readers with her passion about draft horse and mule power.

— *Gail Damerow*

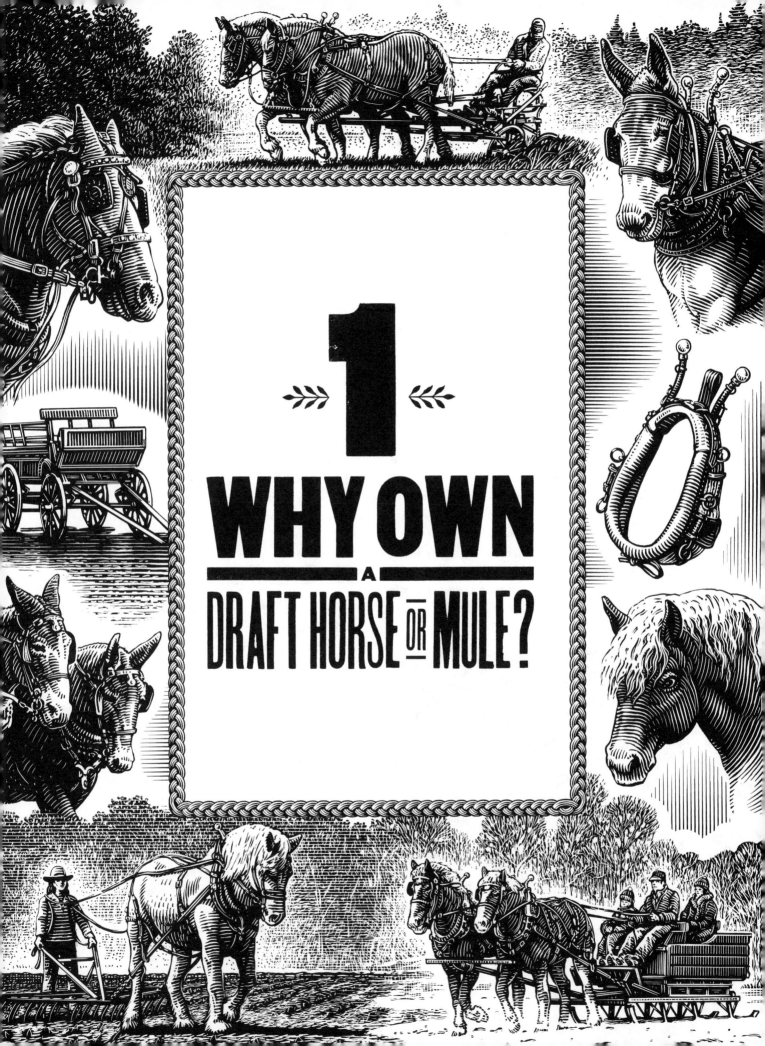

Growing up on a horse-powered farm, I observed my fair share of dreamers. I could spot them a mile away as they slowed down to look at a teenage girl guiding three enormous Belgian mares up and down furrow after furrow, turning the rich, fertile soil. They would wait for us to plod to the end of the field and then bend my ear until I clucked to the team, eager to move on with my work. These people were curious about how I arrived at that moment, sitting on the plow. They wanted to understand the training of horse and teamster, the handling of the equipment, and the care and management of the horses. What they were seeing was the end product of much learning (and exasperation on the part of my father), many bone-tired days, and numerous costly mistakes.

Before you begin actively searching for your dream draft horse or mule, it is critical to understand exactly what you are taking on. You may have a good idea, but until you have directly experienced the good and the bad of equine ownership you will not be able to comment with authority. Once you embark on this adventure, you will have a lifetime of learning and growing ahead of you as you develop your teamster skills. You have two options for pursuing this learning curve: you can dive in headfirst, only to realize that you don't know how to swim, or you can educate yourself thoroughly on the draft equine world and make your decisions with knowledge and confidence.

Experienced teamsters often express the sentiment that the successful teamster has a small ego. Horses and mules will always find ways to test you and make you look stupid; you might as well accept this with grace. An effective teamster has an avid interest in learning from other teamsters, even if it means eating humble pie from time to time. You will also have many opportunities to learn from your equine partners as you watch them closely and learn each animal's rhythms. By learning from other people's mistakes, observing your own animals, and building on previous experience, you will significantly improve your efficacy as a teamster, as well as minimize the sting of mistakes caused by handler error.

## Your Commitment

Weighing the advantages and disadvantages of equine ownership is an important step on your journey toward committing to ownership. Even more important is a critical assessment of your own skill level. Are you completely ready for this leap into horse or mule ownership? An honest assessment is crucial in making a commitment to a lifestyle involving draft horses or mules. You could put the lives of you, your family, your friends, and your animals at risk if you overestimate your skill level or your ability to take care of equines.

### YOUR SKILLS

The skills amassed by successful teamsters are many and varied. They are also subject to sarcasm, dispute, and exaggeration. Teamsters and muleteers are a unique breed of people who are, in general, fiercely independent and sure of their own personal meth-

No matter how young you are when you start working with draft equines, you will always be learning from your animals.

ods. In time, you will develop techniques that you'll feel are best suited to your needs and completely bulletproof, but at this point, honestly answering a few questions will help you to gauge your level of experience with equines and your skill in handling them. If you are completely new to horse or mule ownership, you won't be able to answer some of the following questions, but they present issues that you need to think about carefully.

**Have you ever ridden or driven a horse or mule?** If you have some experience with riding or light driving horses, then you are familiar with how equines react to stimuli, how they learn, and how they protect themselves (for example, a horse is more likely to startle or shy at an unexpected sight, while a mule tends to stop and study the situation).

**Have you ever bridled, collared, and harnessed a horse?** If you plan on working your draft horse or mule, you will need to know how to properly fit and put on a harness.

**Do equines respond well to you?** Do you feel completely in control when haltering or leading, or do you feel out of your depth? An often-used excuse by novices who lack control is "Oh, Bob just likes to be the boss." As an equine owner, you have no room for any situation where you are not the leader.

**Are you comfortable working in close quarters with equines?** Draft horses and mules are large animals, and tight spots such as box stalls and trailers can prove intimidating for the novice handler.

**Are you comfortable walking into a herd to remove a single animal?** If you keep your animals in a large pasture, you will need to catch them; if they escape, you will be responsible for their retrieval.

**Are you comfortable handling equine legs and feet?** While you may be able to rely on your farrier to do most of the trimming and shoeing, you will need to clean hooves on a regular basis.

**Do you understand the nutritional needs and veterinary care required by horses and mules?** Although equine health care is a lifelong learning experience, at the outset you must understand the basics, including vaccinations, parasite control, the prevention of colic and laminitis, and being prepared for emergencies.

**Do you react calmly in emergencies?** When a horse gets himself into trouble, his first reaction is usually to run, and run fast. The mule, thanks to his donkey ancestors, is far more cerebral about the situation and more likely to carefully assess the options before making any fast moves. Panicking along with your equine won't help the situation at all.

**And finally: Do you have any practical skills in the fields of carpentry and tack repair?** Equine ownership invariably includes unexpected barn, fence, and harness/tack repairs. Are you prepared to handle them?

If you answered positively to most of the above questions, then you are ready to acquire draft horses and mules. If your current comfort level with horses and mules is good, any remaining issues may be resolved with time and mentorship. You might commit to equine ownership, with a further commitment to work closely with a teamster who will help you to learn more about the intricacies of working with draft animals under harness. If you are unable to answer at least half of these questions positively, you are not yet ready for equine ownership. Commit to working with a teamster and his or her well-trained animals until you become confident of your own skills.

## WORKING WITH A PROFESSIONAL

If you have never had any experience with drafts, you will need thorough instruction from a skilled equine handler. This type of training costs time and money, but is not optional — it is critical for your successful journey toward having a draft horse or mule of your own.

If you have had some experience with handling horses or mules, such as working with saddle horses, but are new to draft animals, consider hiring a well-reputed teamster to critically evaluate your skill level in handling draft horses or mules. This evaluation will probably show you areas where you need to learn more and give rise to questions you hadn't thought to ask, but is also likely to increase your confidence and enthusiasm.

## The Right Team for You

Once you have determined to buy a team of your own, set some realistic goals for you and your animals. These goals should be based on what you want your team to do; they will help you to narrow your search. Keep in mind that everything your animals do will involve you, so set your goals based on your physical capabilities. Whatever your goals, you should be looking for thoroughly trained animals with years in the harness and the temperament to put up with the mistakes you are bound to make. For the draft horse, you are looking at an animal at least 8 to 10 years old, for the mule, 10 to 12 or older.

Setting goals for you and your team can be difficult if you don't yet know what draft horses and mules are capable of doing. Ask yourself some of the questions below and you will begin to discover where your interests meet the capabilities of draft animals. If you have an endless supply of money and land, you may not be as particular as those who wish to incorporate a team as part of their livelihood. Whatever your situation, figuring out why you want to own draft horses or mules will help you to find your ideal team.

- Do you enjoy learning how things were done a hundred years ago?
- Do you wish to work with draft equines on your own land?
- Do you need a powerful team to snake logs out of the woods?
- Do you want animals that will walk in an arrow-straight line along a plow furrow?
- Do you hope to incorporate draft animals into a living history museum or other public-interest program?
- Do you dream of driving a fancy carriage behind a high-stepping team?
- Do you want animals that will trot down country roads gently and quietly?
- Do you want a companion animal that will love you and nicker whenever you are around?

In short, what exactly do you want your team to do for and with you? Don't limit yourself when setting initial ideals for your future draft animals. In the beginning of your search, the sky (and your pocketbook, of course) is the limit. As you get into the heart of your search, your goals will morph into a realistic set of priorities. Determining your priorities will keep you realistic about your potential purchase. Your equines can always learn new skills, but to begin with, you need an animal or a team already well educated in the skills that primarily interest you. The single most important factor for the novice draft equine owner is the purchase of an animal or set of animals that have been trained well, worked often, and handled responsibly by a seasoned teamster.

### SETTING GOALS

You may decide to purchase draft animals with the goal of utilizing their energy for agricultural functions and/or transportation. You might start your search wanting an animal or a team trained for carriage-business work as well as farm cultivation and fieldwork. As you continue your search, you might decide you would rather have a team with an excellent reputation purchased directly from a carriage business in a busy city, rather than a team that has won plowing competitions. If your priority is driving passengers, having animals that are safe for hauling customers in a carriage becomes more important than producing an arrow-straight furrow.

Another important consideration when setting goals is to remember that whatever your drafts do will intimately involve you. You will not be jumping in and out of a sheltered and climate-controlled cab. Instead, you will be slogging through the April barnyard mud to feed your animals, sweating in the July heat as you and your team put up hay, and freezing your fingers while trying to unbuckle harnesses during a February cold snap. Be realistic about your own level of fitness, energy, and strength when evaluating your needs and desires.

Only you can decide what is most important to you, what your team's strengths should be, and what weaknesses you can tolerate. Draft horses and mules are intelligent and can be easily taught new skills, but you cannot expect them to be equally

Decide what you want to do with your horse, and then set about achieving your goal whether it is pleasure driving, showing, or farming.

talented at every skill. Like people, they have their strengths and weaknesses. Starting out with a set of priorities will help you to find your ideal team. As you become better acquainted with your horses or mules, you will most likely discover that you want to try new and different things.

**MAKING THE BIG DECISION**

A commitment to owning any equine should come after careful consideration of a lifestyle that can support and maintain these animals, and in the case of draft horses or mules, animals that eat and drink a lot and require large living spaces. The seed of interest in draft equine ownership is the first step in the decision process. Once you have considered the general questions above, the decision to go ahead with your purchase immediately generates a cascade of more specific questions:

- What are some of the advantages and challenges of draft horse and mule ownership?
- How will you know what to look for in a well-trained draft horse or mule?
- Where should you look?
- Do you have enough land on which to keep a draft horse or mule exercised, busy, and happy?
- Is your land zoned to allow draft equines?
- Where will you house your animals in inclement weather?
- How will you manage your team to ensure optimal health and strength?
- What routine health care does an equine require and what does it cost?
- Who will provide health care for your draft animals?
- What will be expected of *you* in this relationship?
- Who will help you to train your draft horses or mules to acquire new skills?
- Who will mentor you on your journey of equine ownership?
- Where will you acquire harnesses and equipment for your team?
- What is your budget for the initial expense of a team with harness?

For a first-time equine owner, these questions may seem daunting. Indeed, a horse or mule requires diligent care, responsibility, and wisdom. Your decision to own a team will no doubt cause you frustration at times, but if you are serious about your decision and persevere, you will be amply rewarded.

## Safety, Safety, Safety

It can not be said too many times that working with horses and mules is dangerous business, no matter how well trained the animals may be. Accidents are always waiting to happen in the form of equipment failure, weather conditions, other people making bad decisions, or any number of unexpected events. You will have your own possible errors to deal with, and your animals will make less-than-optimal choices from time to time.

One thing is sure: hurrying will usually land you in trouble. You will forget something important or overlook something critical, and the next thing you know, you're on the verge of a big wreck, with only yourself to blame. Each and every time that you harness your team, you have a recipe for potential disaster if the correct ingredients come together. As a responsible teamster, you must do your best to avoid accidents by checking and rechecking your harness for faults or missed buckles, by knowing your equines' potential weaknesses and how they may be influenced by outside variables, and by being thoroughly familiar with your equipment.

It's also important to train yourself to observe everything: the weather, your animals, the soil, your surroundings, yourself. When you first meet your team, you will not know them intimately. As you work with them and come to know each animal individually, you will begin to notice body language that can indicate warning signs. You will learn how your animals think and react, which helps you to work in harmony with a minimum of scary incidents. Your observations can mean the difference between an accident and a crisis averted.

Accidents can be extremely costly, not only in terms of money, but in the loss of trust your animals have in you. As a teamster, you put yourself in the position of leader. You call the shots, give the commands, and provide safety. Should you let your animals down, they may never fully regain the trust that they had in your ability as a competent leader.

## Learn from Others

While it is true that some folks are born with a knack for equine handling, it is far more accurate to say that many folks start with an interest in all things equine and then choose to learn from the experience of other owners. By taking an active interest in learning from others, you will not only increase your skill level, but you will also significantly reduce the frequency of mistakes and accidents on your own journey to successful equine ownership.

Take the following examples of two greenhorn owners with similar skill levels. One family chose to learn from the crises that occurred while the other chose to learn ahead of the game.

### CRISIS MANAGEMENT

The Jones family was full of girls, and what girl does not at some time long for a horse? Although completely inexperienced as horse owners, the parents were interested in using the animals for work around their small farm to harvest hay and to plow. A teamster friend of the Joneses' advised them to buy older, more experienced animals, but they decided it would take too much time and money to find such animals. Without doing any research, asking questions, or seeking advice, the family went to the nearest auction and purchased a green-broke 1,800-pound (820 kg) Belgian gelding and a green-broke 1,700-pound (770 kg) Percheron mare. The horses that the Joneses purchased were quite a bit less expensive than some of the better-trained teams at the auction, but the family was not in the financial shape to make a purchase that would have suited their needs more closely.

The Joneses brought their horses home and put them in a rickety pen they had slapped together the day before; fencing of the pasture was still underway. A neighbor had strongly recommended completing the fence before bringing the horses home (he knew what would happen to his yard and crops that bordered this pasture). That night the phone rang and the family learned that their new horses were happily grazing in the neighbor's yard — an occurrence that would become far too frequent. They caught the horses and tied them to the fence for the rest of the night. By morning one horse had suffered severe rope burns.

The fencing project was completed the next day and the horses were content until the hay supply ran short and they began to search for more forage. From then on the pair exploited every opportunity to escape into the neighbor's yard, corn crop, or hay fields.

The Joneses experienced significant frustration as they struggled to work with their green horses, and they incurred serious injury to themselves and the animals. In fact, one of the girls was hurt when the team ran away with her while pulling a harrow and another girl was kicked and sent to the emergency room. One mare colicked badly and the family chose not to call the veterinarian and instead tried some home remedies that they had heard of. By the time they realized they should call the vet, it was too late and the mare had to be euthanized. It was a sad situation that the vet pointed out could have been avoided with earlier treatment.

The Jones family chalked up their unfortunate experiences to a run of bad luck, but the simple truth is that they chose not to take advice or make an effort to learn. Without any prior experience, the Joneses jumped into equine ownership without much forethought. They turned a deaf ear to advice, did not seek proper medical care, and refused to invest the time or money necessary to make a safe, secure, and comfortable environment for their horses to live and work. Unfortunately, they and their horses paid dearly for their failure to make the effort to learn.

## AHEAD OF THE GAME

The Smith family was also full of girls who wanted horses. After much discussion, the father realized that he would like to have horses, too. Exposure to Amish culture had shown this family that horses could be used in a practical way, and they were eager to try what they had seen.

The family had little experience with horses, so the parents delegated research to the girls, asking them to gather information on breeds, capabilities, housing requirements, health care issues, and the like. By the time the data-gathering process was complete, one daughter commented, "This project looks like an awful lot of work!" But they all still wanted horses, and after much time and thought,

Learning from other people who have lived their lives with draft animals is the best way to start your draft power education.

the family purchased two Haflingers, a breed of small draft horses suitable for light garden work.

Before finalizing the purchase, the family had a veterinarian check the animals. They ensured that the animals were vaccinated and they learned about critical health-care issues. Before bringing the team home, they erected stout, safe fences and built a solid barn. They bought hay and made water available. Preparing for the arrival of their animals required quite a bit of time and money, but the family was committed to providing for their animals to the best of their ability. The breeders of the two horses spent hours with the family, answering questions and educating them about Haflinger ownership.

The family brought their two well-trained Haflinger mares home and carefully began to get acquainted. Their caution enabled them to successfully catch a colic in the early stages, notice a weak spot in the fence before the horses did, react quickly and appropriately when one of the mares spooked while driving, and promptly call the vet to treat a case of lameness when they noticed symptoms.

This family also chose to learn from other horse owners' mistakes and tried to incorporate such advice as "Keep your training creative," "Don't make the mares pull more than they can handle," and "Call the vet before the horse is down." In short, this family had a successful introduction to draft horses and went on to enjoy a number of years of pleasant and productive gardening, haying, riding, and driving.

## Advantages of Owning Drafts

For many people, the primary reason to own draft animals is the fact that they can provide a somewhat self-sustaining source of energy. For others, it is the appeal of the partnership between teamster and team. Draft animals have been trained for generations to work for and with man. Why not put them to use in your own operation? In addition to their work capability, draft animals can provide calm and gentle companionship and transportation. Today, many draft horse and mule owners do not use their animals in a full-time operation; instead, they work with them around the constraints of other employment and commitments. Keeping draft animals is an expensive undertaking, so if you are a hobby farmer, you need to have pretty deep pockets. Below are just a few of the specific reasons folks choose to own draft horses and mules.

### FARMING WITH REAL HORSE POWER

The ability of draft animals to utilize feed grown with their own energy offers a remarkable way to tighten the circle of agricultural self-sufficiency. These animals make fertilizer in the form of manure, which you may use to add necessary nutrients to your soil. In addition to providing energy for your agricultural tasks, a mare can produce foals that you can sell, adding to your profitability.

Draft horses and mules can help to mow, plow, plant, harvest, rake and collect hay, and provide transportation. The *small* in *small farmer* needs emphasis, because the horse farmer will never make money managing a monster monoculture business with a team of draft horses. The draft horse or mule owner with agriculture in mind must be dedicated to a diverse, dynamic, and vibrant approach to farming.

In assessing your need for equine energy, consider how much land you intend to cultivate. Are you planning on farming five acres or a hundred?

An equine will be more content with a companion, even of another species.

# LEARNING TO LEARN

**Rachel Seemar**
**Kennebunk, Maine**

Type of Farm: 65 acres (26 ha); 4 under cultivation with organic produce; some hay, woods

Animals Owned: Two Belgian mares

Rachel Seemar came to love and appreciate draft horses slowly and through an unconventional route. Rachel grew up in New Jersey and attended college in Boston, majoring in social work. Although she became disillusioned with social work, she particularly enjoyed the opportunity to study community and social development through agriculture. She credits this focus for helping to plant the seed that yielded her current lifestyle.

While managing a vegetable farm in Maine, Rachel further developed her opinions about sustainability, energy, and petroleum use. During the winter, she found work in local riding stables where she developed an appreciation for horses. One year she signed on with John Plowden, a logger in Maine who takes apprentices. Working with John, she learned a great deal about driving, maneuvering, and working draft horses.

After her internship, she wanted to find a way to incorporate draft horses with working in an agricultural setting, so she apprenticed with Buckwheat Blossom Farm in Wiscasset, Maine, a CSA (community-supported agriculture) farm that raises vegetables and meat. While working there, Rachel acquired two older well-trained Belgian mares, Queen and Queenie.

These two mares have taught Rachel many things. Using well-trained horses allows the novice to have a more productive, enjoyable, and safer work experience. For Rachel, this was very important. "They are intelligent, intuitive creatures, and a joy to work with," she says.

Rachel set out to farm on her own in October 2006 when she leased a 65-acre farm. With the help of two friends (one working full-time, the other part-time) and one apprentice, Rachel manages a CSA that provides 60 families with fresh produce for 20 weeks of the year. She also sells eggs from her flock of laying hens and takes produce and flowers to a farmers' market in Portsmouth, New Hampshire.

Rachel stresses that acquiring draft horses that were already well trained and accustomed to agricultural work was one of the best things she ever did. These horses not only help her each day with her work, but they are also training her to become a better teamster.

Rachel enjoys her horses and believes with all her heart that what she is doing is making a better earth. Rachel has pursued opportunities to learn, applied them to her life, and is now passing her knowledge on to others. And the horses, of course: "They are a beautiful, mystical, and practical source of energy."

## Public Interaction

If you enjoy having the public share in the joy of your team, you will need to train your animals to endure such experiences as hauling 25 shouting preschoolers on a hayride. Many horse and mule owners diversify their animals' use by providing transportation services for weddings, funerals, and other occasional events. The draft horse or mule called upon at such times must be of excellent character, and the driver must be prepared for all types of potential problems.

In general, the public has no perception of how an equine thinks or is likely to act. If an inexperienced horse or mule is approached rapidly by a screaming child, his flight instinct may kick in and a wild ride could ensue. If an overeager mother just wants a picture with her "darling" perched on Bob, and you don't find out about it until the darling is in a heap on the ground, you could be involved in some unfortunate legal action. If a new team taking a bride to her wedding suddenly encounters a shrieking fire engine weaving in and out of traffic, their flight instinct must be sufficiently in check for them to stand quietly out of the way until the commotion has subsided.

---

If you plan on managing only a few acres, chances are that the sustainability of your operation will be severely undermined if your animals are so large they can barely turn around in the field.

Draft horses and mules come in many sizes and shapes. Maybe a team of draft ponies would better suit your needs than a team of heavy horses. Perhaps a single horse or mule will do everything you need. At any rate, starting with one equine will allow you to do many things, while incurring half the cost of owning a team. Start small and add as your knowledge, time, and funds allow. If you purchase a single horse or mule, however, keep in mind that they are herd animals and are happiest with company. You may want to consider a goat or pony as a companion animal.

### RUNNING A TRANSPORTATION SERVICE

Commercial transportation today has little to no use for the draft horse or mule. One area that remains open, however, is the carriage or tour business, and some draft horses and mules earn their daily hay by hauling carriages of tourists past downtown sights. By incorporating equine power into a market often served by fume-spewing buses, the draft horse or mule provides green energy, a quiet ride, and a look back in time for tourists interested in moving more slowly around a city.

Carriage work typically involves busy streets and commotion. To have a successful business in which you cater to the public, both you and your draft equine must have advanced skills. The advantages of using draft animals for public transportation will be obvious as your passengers comment about the quiet ride, the beauty of your animals, and the enjoyable pace. Some carriage businesses in large cities operate full-time, seven days a week with horses that rotate through so that each gets a certain amount of rest. Others provide services for weddings, funerals, proms, and other special events as needed.

### DRIVING FOR PLEASURE

Although draft horses and mules have been used for personal transportation for centuries, today the lighter breeds dominate the driving scene. The heavier draft horse, in particular, has a significant drawback when it comes to driving. The animal's size makes him more difficult to hitch and maneuver than the lighter breeds. Still, a dedicated group of people in the United States has maintained draft breeds specifically for hitching and driving, resulting in an animal that is leggier and quicker than his more agriculturally suited ancestor. The draft mule is lighter than the draft horse, and therefore more suited for driving, although he, too, lacks speed and agility compared to lighter-bred mules.

Draft horses and mules are most commonly driven as pairs or in large hitches, and are generally seen on the show circuit. In some rural areas, however, small farmers continue to be part of the driving scene as they transport hay, produce, animals, and other farm goods with sturdy, strong, calm, and reliable draft horses or mules.

## ATTENDING PUBLIC EVENTS

Plowing and threshing bees and wagon drives attract large crowds of curious people and are excellent places to find equine mentors, swap equipment secrets, and shoot the breeze with other teamsters. These community events are the highlight of the year for some folks who keep their animals in shape specifically for showing their skills on these days.

The sheer number of horses, mules, and humans mixing in a limited area, however, necessitates reliable teams. To be welcome at these events, your draft animals must be obedient, calm, and ready to work. Events such as these often proceed slowly, and bottlenecks do occur. Your horses must be willing to wait patiently until they are asked to move out.

A significant advantage to these old-time days is the interest shown by an older generation, a generation that once used draft horses and mules because they had to. I have spent many hours visiting with older folks leaning on canes or sitting quietly, watching enthralled as the animals and their drivers move. The opportunity to interact with a generation of people who may feel left behind by the technical age is intensely satisfying for any draft horse or mule owner.

## LOGGING WITH HORSE POWER

Horses and mules have been used for years to log and are still employed to remove logs from ecologically sensitive areas, particularly in the American Northeast and Northwest. The people running these operations range from environmentalists to old-timers

With draft power, you can relive history as you experience a day on the trail in an old-style wagon train.

determined not to give in to the mechanical skidder. Compared to machinery, draft equines offer the advantage of dramatically reduced soil compaction and disturbance to the forest floor. While you may not plan on joining a horse-logging crew, perhaps you have a few acres of timber on your land that you would like to manage with horses.

Even with an experienced team, logging is dangerous work, so make sure that you, as well as your animals, are prepared. Horses that pull logs should be capable of starting heavy, deadweight loads without giving up and must be comfortable with the noise of chain saws and falling trees. Though demanding, logging with horse power is rewarding work, and it's combined with the majesty of the deep, dense, beautiful forest; little can beat it for a daily workplace.

### ENTERING PULLING COMPETITIONS

Pulling is a sport in which draft horses and mules compete to see which team can move the most weight a certain distance. Classes are based on the team's weight. The sport originated when one teamster would challenge another to prove which team could pull more; historically, tests involved pulling heavily loaded sleds or enormous logs. Today's teams pull blocks of concrete or sometimes pull against a dynamometer. This device was developed at the University of Iowa in the early 1900s as a way to study horsepower. Today, it is used as a method of determining exactly how much weight a horse (or mule) can pull.

To watch a team of horses or mules pulling for all they are worth is truly awe-inspiring. The horses throw themselves into their harness with bulging muscles and snorting nostrils while the teamster urges them on. Unfortunately, the extreme exertion of this event can lead to tendon and ligament injuries; pulling horses and mules have the highest rates of leg injury seen in draft horses and mules.

Some competitors travel thousands of miles to compete in pulls. Thirty-one states have pulling events, many occurring at local and state fairs. A number of organizations and associations for pullers keep official results and other data.

### RIDING YOUR DRAFT ANIMAL

If you enjoy doing splits, you may consider the riding potential of your draft animals. Draft horses make wonderful, calm, steady riding partners — as long as your legs can do the gymnastics. Draft mules and draft horse crosses, with their lesser bulk and athletic ability, make excellent mounts for trail riding, general pleasure, and show. Draft crosses are often used as dressage horses and hunter/jumper eventers.

A pulling competition offers an amazing display of the draft equine's strength and beauty.

Packing with sure-footed draft mules is a great way to access backcountry areas.

Crossing a draft with a Thoroughbred produces an animal of amazing balance, grace, and power that is generally adaptable and easy to train.

Draft mules and horses tend to be gentler and more forgiving than lighter saddle animals, and less inclined to startle or shy. Riding your workhorse is generally not a problem, but not all horses or mules know how to handle a rider. On our farm, we could ride all our draft horses, though it was a bone-jarring experience, for a draft gait is anything but smooth. We often rode them back to the barn after a day of work, when they were covered in sweat. Aside from these minor negatives, few things quite equal the majesty and sheer power of jolting along aboard an 1,800-pound animal, the thick mane flowing straight back, huge feet thudding like thunder.

Once you determine your animal is amenable to being ridden and you decide to use a saddle, you'll need to find a saddle cinch large enough for his belly. You also need to make sure the tree of your saddle is wide enough for your animal, as a narrow tree can create or exacerbate problems with the withers. A number of resources available on the Internet offer saddles specifically (and often custom) made for draft horses.

## PACKING AND CAMPING

Using animals to transport goods and equipment on their backs has long been a method of moving freight over difficult terrain and remote country. Packing is a favorite activity of western horsemen and their draft mules and draft-cross horses. In the mountainous western states, the United States Forest Service used pack animals for years as part of its early fire suppression scheme. For the entire fire season large strings of pack mules were stationed at trailheads — their sole purpose in life being to supply remote fire service personnel manning remote lookouts. A string of mules, often numbering as many as 22 animals, would regularly wind its way over treacherous passes, steep talus slopes, and tricky river crossings to supply the lookouts.

A full-size draft horse may be too bulky on a trail, but draft mules and draft crosses, although larger than their smaller saddle counterparts and thus somewhat harder to maneuver down narrow trails, have substantially more carrying capacity. When you are packing out a bull moose from a remote hunting camp, you want all the carrying capacity you can get.

Mules are renowned for their agility on the trail. On packing trips horse riders are often forced to walk their animals over steep, precipitous talus slopes, while the pack mules are turned loose to find their own way. Mules know where their feet are and when to step carefully.

Many a packer rides a horse and packs mules. Because of the unique bond a mule forms with a horse, the packer will often tie his horse and let the mules loose around camp. Rarely will the mules leave the horse.

## THE JOY OF COMPANIONSHIP

The equine-human relationship is distinctly different today than it was a hundred years ago. Back then, the horse or mule was viewed more as a car or tractor might be now — an item necessary for easing the workload of life, a tool. Today, many people view the equine-human relationship more as a working partnership, with each individual understanding his own role.

### ◀ CAUTIONARY TALE ▶
### Know What Your Team Can Do

A man once bought a team of beautiful and well-trained Belgian mules that had worked most of their lives together in the woods. These mules were in excellent shape and accustomed to pulling heavy loads. The new owner wanted to use them in the woods, but decided they could further pay for their oats by spreading their own barn manure. He purchased a brand-new manure spreader, hooked this bright, alert, magnificently strong team to it, and filled the spreader. Once the spreader was filled, he felt it would be an effortless job to drive out to the field and watch the manure fly.

As it turns out, this journey with the manure spreader was the maiden voyage not only for the spreader, but also for the mules and their new owner. These mules were used to starting heavy, dead-weight logs and pulling for all they were worth to get the logs to the deck. Do you think they altered this approach when hooked to the spreader? Why should they? They had never known anything but pulling logs out of the woods.

So when the mules started that load of manure, they slammed into their collars just as if they were pulling a great, big, stout oak log. The owner was anything but prepared and grabbed the nearest steadying object: the gear levers. This action threw the beaters and chains into motion — it is a noise the mules had never before heard. The resulting wreck included two traumatized mules, an injured owner, a splintered manure spreader, and harness parts spread all over the field. It was not a pretty sight.

Why did this happen?

The new owner did not take into account that these well-trained mules may never have experienced a manure spreader. He expected them to work as well on the spreader as they did hooked to a log. But to the mules, a spreader was alien. For these mules trained in the woods, the command to pull meant "pull as hard as you can and keep pulling," not "pull steadily until the load starts and then slack off as the wheels begin to roll." The urge to hit the collars hard at the first feel of the load may never completely leave their brains, because they have done it for many years.

Lesson learned: When you look for your team, know what you want them to do and what they are capable of doing. Given time, patience, and training, most horses and mules can be taught to do what you want them to do. But until you become an accomplished trainer, you're better off starting with a team that is already trained and experienced at the work you will be doing with them.

Many people take this partnership a step farther and elevate the animal to the level of a companion that provides friendship, entertainment, and diversion. Because the draft horse or mule is a calm, steady animal that easily fills the role of companion animal, he is often left in the pasture to be only petted and brushed. Even though the draft temperament allows him to be a quiet companion, a draft animal is happiest when he does what he was bred for — working for his hay.

## Challenges of Ownership

Along with the many advantages to owning draft equines, you need to consider the challenges before making a commitment to ownership. Horses and mules are complicated creatures with specific needs for shelter and care, and in even the most carefully planned operation, equines can find themselves in trouble at the blink of an eye. They have a predilection for running up vet bills. They are huge hay burners, always wanting to eat more. These challenges are just some of the many you will encounter as the owner of a draft horse or mule.

### YOUR TIME COMMITMENT

As you think about purchasing and housing equines, you will become increasingly aware of the enormous commitment that is involved. These animals need attention and care at least twice a day, every day, preferably at the same times each day.

You will be hauling heavy bales of hay, filling water containers, measuring out feed, checking for injuries and signs of illness in all sorts of weather conditions, and no matter how you are feeling that day. When you can't attend to them you'll need to find a suitably capable and responsible individual to take your place.

As the owner of draft animals you must spend time educating yourself on equine nutrition, health and diseases, fitness, training, harness and tack, equipment acquisition and maintenance, and a host of other factors. Time is a precious commodity to begin with, and equine ownership will add to the list of things you *must* do.

### YOUR FINANCIAL COMMITMENT

Quality horses and mules cost money, and lots of it. The money you pay to purchase your team is only the beginning, and will be followed by regular feed and routine care bills. Additional funds will go toward fence repair, barn upkeep, harness maintenance and repair, equipment, and hoof trimming and shoeing.

Veterinarians don't set out to charge you enormous sums of money, but medications, surgical equipment, treatments, and farm calls are all expensive and add up fast. An unexpected health crisis can be overwhelmingly costly; surgically treating colic, for example, can cost at least $5,000. Are you willing to pay the money required to keep your team in good health?

If you plan to take your equines to shows, fairs, plowing bees, or other events, you also need to factor in the cost of buying a suitable trailer and a truck to pull it, or hiring someone to haul your animals.

### DEALING WITH UNEXPECTED EVENTS

A seasoned teamster can often handle unexpected events without breaking stride. She knows her animals, knows how they will react to stimuli, and can anticipate an event before it becomes an accident. In addition, her relationship with her animals is built on their trust in her leadership. The novice teamster, on the other hand, is still building a solid relationship with her animals and lacks the experience to anticipate an event before it turns into an accident. The animals may wonder who is leading whom.

Every teamster inevitably encounters a thunderstorm, a loud truck, or a fence wire wrapped around an animal's leg. The relationship of trust, the capability of leadership, and the ability to keep your wits firmly in tow will help you respond to these unexpected events.

Although drafts are calmer and gentler than light breeds, they are also larger. A team with a combined weight of two tons has enough strength to do considerable damage to you, others, buildings, vehicles, and equipment. Preventing accidents, rather than doing crisis management, is the best approach to draft equine handling.

## FIGURING OUT SPACE REQUIREMENTS

Draft horses and mules require a lot of living space. If you have not considered the amount of land required to keep a draft horse or mule, now is the time. If you will be using a structure built for lighter horses, you must account for the sheer size and weight of draft equines by making doorways higher, reinforcing walls, and enlarging box stalls.

A single draft horse should never be confined to fewer than two acres (0.8 hectares, or ha), which is in no way enough to provide all the forage that he will eat. Depending on climate, an 1,800-pound (820 kg) draft horse will consume around 13,000 pounds (6,000 kg) of forage per year — that's a *lot* of biomass. The amount of pasture required to fulfill this need depends on soil type, climate, vegetation, and fertilization. The more pasture you have, the less hay you will have to put up or buy. You should, however, plan on supplementing your pasture. Even in a moderate climate, you cannot expect a horse or mule to survive on pasture alone during all seasons. When fresh forage is unavailable, hay must be substituted.

Horses and mules need enough space to kick up their heels and roll in the dirt. Like any animal kept in confinement, your draft equine will eventually initiate a full-blown rebellion if he does not have the opportunity to exercise at his own leisure. To some animals, a good roll in the dirt after a long day of work is as necessary as eating and drinking. Horses and mules need enjoyment and relaxation along with all their hard work.

## YOU MUST BE PHYSICALLY FIT

You need not be a marathon athlete to own draft animals, but your ability to maneuver carefully, quickly, and with strength may mean the difference between a disaster and a close call. A harness, especially a good-quality leather harness, can weigh 50 to 70 pounds (23–32 kg), and that doesn't include the collar. The stronger you are, the more easily you will be able to handle heavy gear and wrestle with large pieces of equipment. Height makes a difference, too. For shorter folks, lifting a heavy-duty work harness onto a big horse is a challenge.

In addition to harnessing, you will need strength for many other activities, particularly if you plan on using your animals for cultivating the land. While the horses or mules may do the bulk of the grunt work, you will be along every step of the way, walking behind the plow or the harrow, rolling logs into position with the peavey, groaning under the weight of a pitchfork full of hay — and you will sweat. Working with draft equines produces strong,

Trusting your animals is an important part of the relationship: even a child can work with well-trained, reliable animals.

muscular bodies — the opposite of the large bellies and flabby arms sported by so many mechanized agriculture producers. Livestock husbandry is not a job for the faint of heart or the slow of step.

## LOOKING AT THE ANIMAL-HUMAN RELATIONSHIP

How do you rate your relationship with animals? Some people have a natural affinity for and confidence around animals. Dogs love them, cats rub on their legs, and horses seem to know that they mean business. Other people, however, are met with barking, growling dogs, hissing cats, and nervous, jittery horses.

If you were not raised around animals, learning how to approach and interact with them can be a challenge. Mules, especially, are able to read human body language exceptionally well. Many a mule person will quote the old saying, "To work a mule, you must be smarter or meaner." Understanding where you fall in this relationship with animals will help you assess the reception you will be given by your potential new team.

The animal-human relationship is a complex mixture of love, respect, obedience, and tolerance. For an animal to obey a human one-tenth his weight, his response must come from one of two reactions — the animal must either fear the human enough to bow to commands, or respect the human enough to willingly do as requested. Both systems achieve results. The significant difference is in the level of the human's satisfaction and the animal's willingness to perform.

When a relationship is based on fear, one party will be bitter and resistant to the demands of the other. A relationship based on trust, however, can lead to mutual satisfaction and willingness from the subservient party to give more than requested. Your horse or mule must always know that you are the boss, the leader of the herd, but you don't have to be a bully to establish that relationship.

While you may not be aware at first of the importance of your position, you won't need too many trips to the pasture to figure it out. Does your horse or mule challenge your authority by being pushy or threatening? If so, you might consider enlisting a trainer to help you reestablish your authority. Having authority means commanding respect by assuring your animals that you are a competent leader worthy of their respect.

## YOUR RESPONSIBILITY AND LIABILITY

As an equine owner, you have a legal liability for the safety of anyone visiting your farm. Draft animals are large, appealing creatures with giant feet and powerful kicks. You are responsible for ensuring that your guests are informed, cautious, and obedient to your wishes regarding your animals. You must also be aware of your legal obligations in the event of an accident; your homeowner's insurance may not cover everything you thought.

Draft horses and mules usually have wonderful personalities and gentle natures, but are massive, strong, and often strong-willed creatures that, like any equine, may behave unpredictably. Any animal, if not properly trained or if handled irresponsibly, can be a serious liability and danger to a human unprepared to understand and respond to him.

# Realizing the Dream

Although the dream of owning a powerful, fluid horse or mule is compelling, think long and hard before making your first purchase. If you make your decision quickly and then ride the wave of enthusiasm directly through to purchase, you will be putting the cart before the horse. Just because you decide

---

**◀ CAUTION ▶**

### Equine Activity Liability

Many states have liability laws relating to the equine industry. Each state has a slightly different view on the subject, so learn what you, in your state, are liable for. No state has a zero-liability law that exempts owners from responsibility for what their animals do. Most laws put the responsibility for animals and their equipment squarely on the owner, who must use due diligence to avoid being accused of negligence.

you are ready to acquire a draft animal, don't be in a rush to purchase one.

At no time during this acquisition process do you want to be hurried, flustered, or confused. Your family and friends may have already called your sanity into question after hearing you proclaim your intentions of owning draft equines. You must prove that all your decisions are based on the maximum amount of information you can obtain, on logic and rationale, and with a promise of economical consideration.

In spite of the preparation and hard work involved, the reward of communicating with those enormous, powerful, willing animals far outweighs the amount of work they create. Whether you choose to cultivate the land, to haul tourists around a metropolitan area, pack far into the wilderness, or drive your team around the countryside for fun, you will revel in the intimate connection between animal and human. Feeling a kinship with animals — particularly beautiful and intelligent animals like horses and mules — is a remarkable experience. You will likely get bruised, humbled, and sweaty, but you will be proud of who you are: a person in the modern age, coming to terms with the twenty-first century from behind the lines of draft power.

Draft equine use is still going strong in the twenty-first century.

# 2
# LEARNING THE BASICS

Much of my knowledge about horses has come from people whom I respect and who were mentors for me, even if they were not aware of their roles. For a good number of my teenage years I lived near an old-order Mennonite community where all of the power was generated by wind, horse, or human energy. I spent many hours talking with, watching, and attempting to emulate the farmers' prowess with horses and equipment. In a society known for valuing humility and hard work, the wide range of approaches to animal handling was shocking. I soon realized that even among people who farm with horses and drive buggies daily, some individuals have brutal tempers and horrible work ethics.

I am fortunate to have parents who chose to incorporate draft horses into their paradigm of a simple life. While I have done and will continue to do many things differently from my parents when it comes to equine husbandry, I cannot thank them enough for giving me the opportunity to work with draft horses daily, to see that draft animals may be used in a practical way to increase the productivity and economics of a farm, and to learn the many things *not* to do. The years I spent on my family's horse-powered farm were critical to my present skill and my appreciation and passion for draft horses.

## The Essentials of Learning

**I**n order to start your journey, you need to line up the following four elements:
- Older animals that are accustomed to and appropriately trained for a beginning handler
- Instruction from an experienced teamster or muleteer
- A safe environment in which to work your animals
- Safe equipment and an understanding of how to operate it

As a novice, you have serious groundwork to do before climbing aboard the horse or mule ownership wagon. At this point, you know you want to own drafts, you think your skill level can handle them, but you haven't yet had the experience of actually using them. How will you make this quantum leap? How will you know what equipment you need? How do you know the equipment is safe? How do you *find* it? How do you know your animals are prepared? How do you know the environment is safe for you and the animals to work? How do you know if a wreck is about to happen? In short, how do you know anything, except what you have read in a book?

No formula, cookie-cutter recipe, or book will tell you everything you need to know as you prepare to put your horses or mules to work. So where do you begin to learn how to work your team? You basically have two options — teach yourself or find an expert to teach you. Having someone teach you how to safely and successfully work your draft animals is by far the better option. Take advantage of someone else's lessons learned, techniques perfected, and experience gained. Doing so will likely cost some money and will definitely cost time and possibly some injured pride. But think of what it could save you — large vet bills for injured animals, large repair bills due to wrecked equipment or harness, large hospital bills for personal injury. You will come out far ahead if you take time to learn from others before you begin operating on your own.

## Opportunities for Mentorship

Opportunities for learning how to practically, safely, and successfully work draft horses or mules are available in many forms. Start with books and videos as part of the mix. You can ask an experienced teamster to mentor you, arrange an apprenticeship with a farm or ranch, take workshops and clinics where intensive, hands-on training is conducted, and, of course, constantly question knowledgeable people. Once you start working with your own well-trained horses or mules, don't overlook them as invaluable instructors. They can teach you skills of character, such as patience, courage, and endurance, as well as practical skills, if you observe their subtle communication cues.

Workshops and clinics are important for intensive education about a particular skill and for continuing education once you get started. Events such as plowing and threshing bees and pulling contests expose you to different styles of equipment, animal handling, and harness. But finding a mentor willing to work closely with you is the best option. An apprenticeship is an excellent way to learn to use draft animals in a real working situation.

Such an arrangement typically involves being away from home for an extended period, but nothing beats having an expert all to yourself — a teacher who understands horses and mules, who has owned, trained, worked, appreciated, admired, and nursed many equines. What you *don't* need is someone who *thinks* he knows everything. Far too many people think they are *the* authority on all things equine. A true mentor knows and willingly accepts his limits.

## FINDING A GOOD MENTOR

Finding a qualified mentor is as important as choosing your horse or mule. The relationship must be developed through trust, experience, and friendship. Your mentor may need to give you critiques that you don't want to hear. You must respect his expertise and be willing to listen.

But how do you locate a mentor when you don't know a soul who owns draft horses? The process is similar to finding an attorney, banker, doctor, or veterinarian. Like these professionals, your mentor should be an expert at what he does, respected by his peers and community, and interested in helping you. Don't hesitate to ask for recommendations.

Clinic participants pay rapt attention to a log-loading demonstration.

A good place to start looking for a mentor is in draft horse publications listed at the back of this book. Through these journals you may find someone in your area willing to help you learn. Online forums let you communicate on message boards and chat rooms with people from all over who care about draft power.

Be persistent in seeking someone willing to undertake your education as a project. If you are humble enough to ask for help, you will likely get more than you anticipate. The draft horse world is quite small, intimate, and connected. We want people to learn about drafts and to join us; this gives you an excellent chance of finding someone to assist you on your journey of discovery.

A mentor can be many things: friend, guide, teacher, counselor, motivator, coach, adviser, role model, and door opener. The mentor(s) you choose to work with in your education of draft power and usage must have all these characteristics and more. A mentor must be honest with you and with himself in order for a constructive relationship to develop among the mentor, you, and your draft animals.

## THE QUALITIES OF A GOOD MENTOR

More than one person will probably fulfill your mentoring needs. People communicate in different ways. With each mentor you must discover how best to draw out the information vital to your journey with draft power. As you search for a mentor who is willing to work with your skill level, consider the following traits:

**Skill in communicating.** A good mentor will clearly and calmly communicate to both you and your team exactly what he feels you need to know. If you feel intimidated by, uncomfortable with, or downright scared of your potential mentor, look elsewhere. Your emotions will be relayed to the animals that you wish to understand, and the result could be a series of unfortunate mistakes.

A quiet, unassuming person can be the trade's best-kept secret. If you persist (without being pushy) and cultivate a friendship, you may hit a gold mine of information. You may have to listen a little more carefully and watch a little more closely to understand what you're being told, but many old-timers are willing to teach you the secrets of their lifetimes with horses or mules — secrets that all too often are otherwise carried to the grave.

**Commitment to teaching.** A good mentor *wants* to teach you how to work with horses and mules. A good teacher takes advantage of every opportunity to connect you to your draft animals and encourages you to think and to be curious. He treats you and the animals with respect and consistency.

### Evaluating a Mentor

**W**hen selecting someone to teach you, whether one-on-one or in a group setting, you must determine that the person is the genuine article. Before accepting anyone's advice or teaching:

- View any claim of skill or expertise skeptically, until you have hard evidence.
- Never trust someone who claims (typically quite verbosely) to know everything.
- Don't believe anyone claiming the ability to make you an expert teamster in a few days or weeks.
- Take careful notice of the condition of the person's animals, equipment, and property, and his interest in safety.
- Don't become discouraged by someone who appears to handle animals and equipment well but is hesitant to teach you. The best mentor is often the person who is reticent about blowing his own horn. Get to know him on a personal level, and then ask for help.

Working with a mentor is an invaluable way to learn about draft animals. A good mentor is happy to share his experience with you.

**Dedication to learning.** Find a mentor who is secure in his position in the horse world. He will not know everything equine (no one person can), but always continues learning and developing his own skills. A good mentor seeks to increase his own skill level with each interaction with his animal, whether it involves plowing a straighter furrow or managing a new dressage move.

**Ability to motivate.** A strong mentor not only helps other people achieve their goals, but also motivates them to succeed and excel. A good mentor expresses encouragement and excitement about your discoveries and the achievement of your own goals. A good mentor inspires you to learn and excel and does not demean your sense of pride.

### YOUR RESPONSIBILITY AS A STUDENT

What is your responsibility in this mentoring relationship? You will find yourself facing the unknown, relying heavily on the advice and instruction of your mentor, sometimes afraid to step out of your comfort zone, and occasionally just plain intimidated. How do you learn to use your mentor as a ladder, not a crutch?

Here are some important considerations for you as the student:
- View your education as a long-term journey and adventure.
- Ask questions and listen to the answers.
- Accept your mentor's assessment of your abilities.
- Be willing to accept criticism.

## Exploring Other Mentoring Possibilities

In the unfortunate event you are unable to find a mentor in close enough proximity to help you learn about your horses or mules on a regular basis, don't despair. Many novice teamsters are in the same position these days because too few capable and learned horse folks are still around to teach the many new people eager to learn about working draft animals.

To fill the draft equine education gap, a number of workshops, clinics, apprenticeships, and other opportunities have been developed across the United States and Canada that specifically offer education

An apprenticeship is an excellent way to gain firsthand experience before you take the step of purchasing your own animal.

in draft animal skills. Due to the dynamic nature of these opportunities, finding a program that will help you on your journey requires searching through the draft animal publications and perusing the Internet.

## LEARNING THROUGH AN APPRENTICESHIP

Unlike a mentor-mentee situation, which can last a lifetime, an apprenticeship requires a finite time commitment, which may be for one season or for a year or more. An apprenticeship or internship offers an excellent opportunity to hone your equine and equipment-operating skills under the tutelage of an expert teamster, but consider signing up only after you have gained some experience by working with a mentor and/or attending workshops and clinics.

Having at least minimal experience gives you something to offer the farmer, rancher, or logger whose apprenticeship you wish to sign up for. Working a farm, ranch, or woodlot is grueling enough without having to stop constantly and explain things. An apprenticeship therefore involves a trade-off in which the intern pitches in to help with the work both to gain firsthand experience and to make up for the time the teamster uses for teaching.

## PASSIONATE MENTORS

**Rob Borsato and Cathie Allen**
**Williams Lake, British Columbia**

Type of Farm: 105 acres (43 ha), mostly forested; 5-acre certified organic market garden; chickens and pigs

Animals Owned: Four Percherons

For Rob Borsato and Cathie Allen, the draft horse choice was initially one of economics; should they buy a small tractor or draft horses? They chose one old mare, and from that first single horse their vision grew into what it is today, more than 20 years later: a piece of land where they continually seek to tread lightly, live off the proceeds of their land, and educate young folks on organic farming. They still do not own a tractor. Initially they generated income from a horse logging business Rob managed in the winter. Five years ago they went full-time into agriculture, selling to the local farmer's market and to a robust CSA clientele.

Of their preference for horse power, Rob says, "On a small farm like ours, using horses for all the heavy work is manageable, and we don't experience the soil compaction problems that can be associated with tractors. Horse manure, combined with spoiled hay and garden residue, provides just enough compostable material (about 50 tons) to meet our garden's annual fertility requirements. And people sort of understand when you talk to your horses, but might put you away if they caught you talking to a tractor!"

Rob and Cathie have a deep commitment to mentoring people interested in their way of life. They view apprentices as a reaffirmation of their own dreams of wanting to carve out a life on the land and they want to play a role in educating people and speeding up the learning process that took them so long. They offer a full introduction to managing a small-scale farm, including the operational aspects as well as economic considerations and environmental concerns.

Rob and Cathie have hosted apprentices since beginning their program in 1990. They advise trainees to spend an entire season, April through October, to understand not only the labor involved with soil preparation, planting, and harvesting, but also the planning involved

Unless you can live without an income for a while, an apprenticeship is probably not for you, as the usual arrangement entails engaging in hard labor in exchange for the experience gained. When a stipend is offered, it is typically enough to cover only incidental or personal expenses. Room and board are frequently included as part of the apprenticeship, and where meals and lodging are not offered, a larger stipend may be given to help to defray living expenses. Although an internship may seem like slave labor, it offers a unique opportunity to learn in an intensive, hands-on manner, and you will probably deal with equines to some extent each day. More than likely, however, not all activities will involve horses or mules, so you will be learning the broader scope of a draft animal farming or logging operation.

An apprenticeship is sure to strip away any over-idealized romance of draft power, leaving you to either appreciate the sweat and responsibility, or loathe the tedious and methodical ways of the slower pace. If you have the opportunity and your finances allow, an apprenticeship is an excellent way to learn about both yourself and draft animals.

regarding what to plant, how much seed to purchase, how to market, and what profit to expect. In October, at season's end, the entire fiscal process is explained.

A good apprentice, Rob feels, is someone who is genuinely interested in draft horses and agriculture. Not all apprentices are interested in using drafts for their own operation; some are interested strictly in the agriculture side of the operation. Others are more interested in the fiscal/marketing part of the operation. Some folks arrive with a pipe dream. Rob chuckles when he mentions how easy a pipe dream is broken after two to three days of tedious weeding.

A good mentor, in Rob's opinion, must be sensitive to a trainee's interest and learning style. When folks arrive with a specific goal in mind, the mentor should provide that education (when feasible, of course). Some folks learn better hands-on, while others, accustomed to the structure of an academic setting, want more textbook involvement.

Most apprentices who arrive on Rob and Cathie's place have never handled horses and have had minimal gardening experience. When they leave after a season, they may not be prepared to conduct their own operations (unless they are highly motivated), but they do have the ability to prepare the soil, plant seed, and harvest crops. They have an understanding of marketing and the fiscal side of the operation. They have had a general introduction to market gardening that allows them to decide whether they want to continue this pursuit. The trainees gain experience working four draft horses and operating equipment such as a plow, disc, manure spreader, and potato digger. Those who are interested also get logging experience.

An apprentice arriving at Rob and Cathie's farm with a genuine interest will be well prepared for the hard work, productive learning, and accomplished mentors who will meet them. Rob and Cathie's main goal is to help trainees realize their agriculture dreams.

### Finding an Apprenticeship

Search for the right apprenticeship using the same methods used to find an individual instructor, for indeed, the apprenticeship will probably involve one-on-one mentoring — though some are group apprenticeships. If you must make a time commitment, make sure the match will be satisfactory. Spending time on the same farm with the same people for days on end can lead to relationship breakdowns if expectations on either side are unclear.

Resources for finding draft animal apprenticeships can be few and far between. Start your search on the Internet, in draft animal magazines, and with organic farmer membership organizations. *Rural Heritage* sponsors the Good Farming Apprenticeship Network, an online up-to-date list of draft animal apprenticeships offered throughout the United States and Canada.

## ATTENDING WORKSHOPS AND CLINICS

Workshops and clinics can be found in many different choices of format, setting, and cost. Although the terms are often used interchangeably, a clinic tends to last a matter of hours, while a workshop lasts a day or more. Before settling on a workshop or clinic, carefully examine the instructor's credentials and then determine whether the program being offered is what you wish to learn.

> ### Learning from Evaluation
>
> **Self-taught teamsters** often struggle with determining their personal skill level, as well as that of their draft animals. How do you know where you and your animals stand now? How do you know what else you are capable of doing?
>
> Attending a hands-on workshop in which you and/or your draft animals are evaluated can be a nerve-racking but eye-opening experience. You will be told with little reserve where you currently are and what you need to do to reach the next skill level. These sessions are a great way to learn what you're doing right and where you need to do more work.

Most workshops and clinics involve an instructor presenting information to paying attendees who simply watch. Sometimes the instructor will take a problem animal and show the audience how to work through his issues. Equipment may be presented, displayed, and offered for sale.

The presentation-style group-learning event is better than no instruction, but not nearly as good

A community plowing bee is a good opportunity to expose your team to a new setting, to learn from others, and to have fun.

as a hands-on format, in which you learn under the direction of the instructor, have your hands on the driving lines, are evaluated on your progress, and can ask as many questions as you like. Most of these workshops furnish the animals for learning purposes, but in some cases you are allowed to bring your own animals. The hands-on workshop may be difficult to find and will cost more than the spectator clinic, but provides a better learning experience because it gives you the opportunity to be personally helped and evaluated by a professional.

**ATTENDING PUBLIC EVENTS**

Events such as covered-wagon trail drives and plowing or threshing bees offer an excellent opportunity to become involved in equine draft power. Although such events are not as intensive or teacher-student oriented as purposeful learning experiences such as workshops or clinics, they attract people from many generations who can give you a good idea of the general nature of the equine draft circle, as well as the opportunity to get your hands on the lines. Don't be pushy, but if you see a particularly well-mannered team and driver, ask if that driver is willing to show you a thing or two. Most drivers would be flattered by your attention.

Plowings, threshing bees, and wagon trail drives occur in many locations throughout the year. A plowing event typically involves teamsters coming together with their teams (some as large as 12 horses in a hitch) to plow a field for a future planting, for competition, or just for fun. A threshing bee incorporates many of these same people and their animals, but the focus is on harvesting grain. Steam engines are often used at threshing bees to run the old threshing machines. A wagon drive may last a day, a month, or even longer and may commemorate a historical trail, route, or event.

Finding one of these events in your area may be a challenge. *Rural Heritage* puts out the annual *Evener Directory* that includes events occurring throughout the year, and also maintains an online calendar of events. Since many draft animal events are publicized only locally, your best bet is to find someone in the loop and put the tentative dates on your calendar, and then keep checking for the final date. Weather conditions often dictate changes of plan.

### Join a Draft Club

In addition to sponsoring public events such as plowing, threshing, and planting bees, many draft clubs offer great opportunities for novice draft owners to broaden their equine education. Draft clubs may sponsor lectures, vaccination clinics, and other educational events. At these meetings you will meet experienced teamsters as well as veterinarians interested in draft animal care (not all are).

Members of these groups are sincerely devoted to draft animal power and often drive long distances to participate in events or meetings. People in attendance typically love nothing more than to talk about their draft horses or mules to anyone who will listen. Joining such a club is a good way to be around experienced teamsters and possibly find a mentor.

### Go to Horse Progress Days

Since 1994 a group of dedicated teamsters and horse-equipment manufacturers have gathered during the Fourth of July weekend for a draft horse farming trade show called Horse Progress Days. This event, which takes place in a different state every year, attracts teamsters from all over the world and has evolved to offer many opportunities to many people. Some people attend to see different breeds in action. Others go to see the latest innovations in horse farming or logging equipment. Still others attend to network and share ideas with fellow teamsters.

Horse Progress Days includes field demonstrations of equipment hitched to teams of horses and mules in a variety of configurations. You will find vendors offering every sort of draft animal equipment and paraphernalia, and most of them are willing to talk to you and answer your questions. The teamsters visiting the vendor booths are always eager to enter into discussions involving horse or mule power. This event also offers short clinics that might cover harnessing, collar fitting, horse training, health care, or hoof care.

Horse Progress Days is an excellent place to soak up knowledge about all things pertaining to working horses and mules. Information on dates and locations may be found online or in any of the draft horse magazines.

## Your Four-Legged Mentors

As a teamster, you must teach yourself to watch your animals constantly. Watch their ears, their tails, their backs, their legs, their heads, their mouths — watch everything about them. A good, quiet, responsive team will teach you about driving line tension, endurance, equipment operation, harness adjustment, and many other skills, if you will only observe them, respond to their communication cues, and correct yourself accordingly. No matter how much you know about them, you can continue learning from your animals throughout your life as a teamster. Never consider yourself to be above learning from animals; they are often quite a bit smarter than we humans.

## Dealing with Your Mistakes

Part of the responsibility of working with animals of any type is a willingness to accept the circumstances in which your choices put you. One of the worst ways you can handle a mistake is to take it out on your draft animal. Far too often people who claim to be excellent horsemen lose their tempers at an animal. Anger is directed at an animal usually because the teamster feels that the animal is not listening. In reality, just the opposite is true. The teamster is not listening to the horse or mule. The teamster does not understand or notice the animal's apprehension, lack of experience, or unusual pep that day.

Horses and mules are living creatures with quick, intelligent minds. Sometimes they like to make their own choices. If you are neither listening to them nor watching them, why should they listen to you? This lack of communication takes a varied route with an almost universal result: the equine receives a tongue lashing, bit jerking, butt slapping, or worse; the teamster feels ticked off for the rest of the day; tension and unrest hum along the driving lines, and no one is remotely happy.

To gain the respect of your draft animals and of the people who watch you interact with them, you must take responsibility for your decisions, choices, and results. While you learn and your skills improve, you may sometimes have to eat a lot of crow. If you have to eat crow, eat it fresh and get it over with.

> ### The Novice and the Mule
>
> If you are considering mules as your first draft animals, please think about this choice very carefully. A horse is willing to work with you and help you learn; a mule is an intelligent and intuitive creature that is more likely to try to exploit your weaknesses or become frustrated with your lack of competence and irritated with you for disrupting its schedule. In general, a horse is relatively forgiving of mistakes, while a mule will never forgive or forget. Even if your mistake was unintentional, a mule will eventually get even with you. Only a determined, careful, and persistent novice with excellent mentorship can make a success story out of a first mule.

Communication with your equine is vital for success.

# 3
# FINDING THE IDEAL EQUINE

Our furrow horse Maxine was a grade Belgian mare and anything but a beauty. She did not have the curvy body that many mares exhibit. She was angular, lean, and muscular — more like a hardworking gelding than a mare. Maxine's head was too large for her body and her ears were too long for her Belgian heritage.

What Maxine lacked in looks, the seller assured us, she made up for in brains. He couldn't have been more correct. Out in the pasture, Maxine never elicited the praise that some of our better-built mares did. But hitch her to a plow or riding cultivator and all she received was admiration. With her muscled shoulders bulging and her big head bowed, Maxine put everything she had into the plow. She was our furrow horse for ten years and not once did we dream of replacing her. When she and a teammate were pulling the riding cultivator, she seemed to have antennae on her feet and knew exactly where each plant was; rarely did she take out any corn.

So what, exactly, makes the ideal horse or mule? As Maxine proves, the question has no definitive answer, because the concept of ideal is like the concept of beauty — it's in the eye of the beholder. You can start, however, with a list of pointers to create your own checklist of what to look for in your ideal horse or mule. You are unlikely to find an animal that meets all your criteria, but having a standard is a start. In addition to the critical factors of temperament and personality of each animal, consider general aspects of symmetry and conformation, age, and ability, and then add to your list from there.

## Choosing Your First Team

Before moving forward with your purchase you must fully understand the intensive physical and health-care requirements of draft horse or mule management, housing and feeding requirements,

Choose your first draft equine wisely, considering many different factors such as conformation, ability, and beauty.

and other important factors involved with successfully owning draft equines. First, though, let's have some fun and go window shopping for the beautiful animal you would like to see in your pasture — an animal that will work with and for you, meet your goals and expectations, and help you learn how to be a teamster; in short, an animal that will fulfill your dream.

You have already thought, in a broad sense, about some general skills you would like your draft horse or mule to have. You know you need older, well-trained, experienced animals. It is extremely important for the novice handler to also look for calm temperament, even disposition, willing attitude, and a high level of tolerance for handler error. As you narrow the scope of your search a bit more, consider some further refinements of the draft equine resume:

- Do you prefer horses or mules?
- Do you want an equine with pep or one that is more laid back?
- Is breed preservation important to you?
- Do you prefer the breeding potential of a mare or the dependability of a gelding?
- Are you sure you need a team, or will one horse do the job?

In addition to gender and breed differences, size must be factored into your equation. Depending on the size of your operation, the amount of hay needed to feed through the winter, and the thickness of your pocketbook, you might do best with smaller animals. Do you need an 18-hand-tall, 3,800-pound (1,700 kg) team of Belgians, or will a 15-hand-tall, 2,000-pound (900 kg) team of Haflingers get the same job done? Your decision should be based on a rational assessment of your intended lifestyle, your skill level, and the general breed characteristics of the animal.

A final consideration for your dream draft equine should be its aesthetic value. For some folks, the way an animal looks is not as important as his soundness, health, and training. Personally, if I am going to be intimately involved with these animals each day, I want to be proud of how they look as well as of their ability to perform. For some folks, the aesthetic consideration would immediately remove mules from consideration, while others gravitate toward mules. For those who prefer the looks of a horse, the difference between the practical chunkiness of the Suffolk and the athleticism of the Percheron could be an important deciding factor. Whatever your decision, never forget your main goal: to acquire an older, well-trained animal that has worked for many years under a caring, responsible teamster or muleteer.

Don't let someone talk you out of your preference just because he has an animal he wants to sell. As you begin your search for your ideal draft equine, the most important favor you will do yourself is to listen to a seasoned teamster or muleteer. Your heart, your eyes, your mind, and, not least of all, your budget will also help you narrow down your purchase options.

## How Many Equines Do You Need?

As the potential owner of horses or mules, chances are you love the sight of a pasture full of beautiful animals. Before charging full-steam into multi-animal ownership, however, consider that many horsemen call themselves horse poor. Being horse poor, in the simplest of terms, is the ever-present conundrum of horse owners — which one to sell when you want to buy or produce more animals?

Another consideration is your age and physical condition. Do you want to haul hay to 12 hungry horses every morning and evening in the dead of winter? I know far too many old horsemen who forget they are old until it comes to winter feeding, when they wonder why on earth they have all those horses.

If you are looking for a pleasure animal, one draft horse or mule is all you will ever need to jog around the countryside, provide entertainment, keep you up at night, and haul to shows. However, equines need companionship and you will find that a pair working together will be more content than a single horse or mule.

A team of draft horses or mules kept strictly for pleasure will likely begin to alarm you with the amount of hay they eat. If you intend to keep draft equines primarily for pleasure purposes, consider showing them at least a few times a year or using them for driving events. Draft horses and mules were bred to work, and are happiest when they perform some function. If you intend to regularly do heavy work, such as plowing, mowing hay, or moving logs, a team will do well for you. A third animal is nice as a backup for times when one animal becomes ill or foals.

If you are looking for a team, you should buy them as a unit, with the animals having worked together for at least a couple of years before you purchase them. The experienced teamster can easily match two horses or mules together and make the team work well; as a novice, you will save yourself much headache if your animals already know each other, work comfortably together, and get along together in their pasture.

## HOW MUCH LAND DO YOU HAVE?

A major consideration for how many drafts you can keep is how much land you have for them, both in terms of the work they can do on your property and the amount of pasture they require. If you have 40 acres (16 ha) or less, one stout and willing horse or mule is likely all you need, though a smaller team, such as Haflingers or Fjords, would be an excellent choice as well. A single large horse or a smaller team will eat less of your pasture and require putting up less hay.

An ideal situation for one horse or a team of smaller horses on 40 acres (16 ha) would be something like 5 to 8 acres of excellent-quality hay, cut two to three times per year, 10 to 15 acres of pasture, 2 acres of garden, and 20 acres in a nice woodlot.

One team, if in good shape, should be able to easily manage between 50 and 75 acres (20–30 ha).

A single draft animal may provide enough horse power for a small woodlot or farm.

A fit team can handle a farm with 20 to 30 acres of pasture, 10 acres of cultivated land (produce, garden, and so forth), 10 to 15 acres of hay, and 20 acres of woodlot.

If you are managing 20 acres (8 ha) or more of cultivated land, you will probably want two or more teams for the heavy spring fieldwork, such as plowing and seedbed preparation. On some of the larger Amish and Mennonite farms, you'll see 8 or 10 horses slowly making their way across a field, pulling an enormous disc. Such larger hitches are also used during harvesttime, when 10 or 12 animals are teamed up to turn the system that powers the threshing machine. When horses are the sole form of agriculture power, quite a number of them may be needed to get the job done.

Most equipment available today for draft equines is manufactured for a team, but equipment may easily be modified for one horse. Any piece of equipment with a tongue — such as a wagon, forecart, or mower — must be modified to have shafts for your single horse.

### EVOLVE WITH YOUR NEEDS

When our family moved to a larger farm, Dad decided it was time to get a larger team of draft horses capable of doing all the work. Prince was the smaller of our two all-purpose horses, and while he had immense heart and power for his size, he couldn't pull the same load as Royal or our new team of draft mares. And so, practicality being the rule on Dad's farm, he sold Prince. This pragmatic approach to equine ownership was hard on me growing up since I became attached to all the animals. I learned with time, however, that to improve your herd, the quality of your animals, and the economics of your operation, selling and buying horses is sometimes necessary.

As the years went by, many more horses came and went on our farm. The animals we sold had fulfilled their purpose; the animals we bought were expected to produce foals and energy to further our family's agrarian interest. At the peak of production, our farm had 16 horses and mules, each with its own job and talents. We did not own one team, or even one horse, that was perfect (although some came close), but they all served their roles. The equines we owned were not all registered, beautiful, or meant for reproduction, but they all had potential as quality animals that would increase our farm's productivity.

## Choosing a Mare or a Gelding

In deciding what you want from an animal, one of your decisions will be reproductive potential. Since a stallion is never an option for the inexperienced horse owner, a mare is your choice if you want to produce foals. Breeding mares, monitoring pregnancies, overseeing the foaling, and then raising the foals is a complicated process that you should consider only if you have bred and raised foals from lighter animals and therefore know the risks involved. If you are not experienced but still determined to breed your own foal, you should be willing to spend the money to turn over the responsibility of later gestation and foaling to a professional on a breeding farm.

◀ **CAUTION** ▶

### Stay Away from Stallions!

The wild stallion's role is that of protector and breeder, and he fights aggressively to defend his rights. A domestic stallion retains the testosterone level necessary to be violently aggressive, dominant, and protective, *even if he is well trained and obedient.* Stallions should be handled *only* by experienced, cautious, and knowledgeable people willing to assume the liability of an unpredictable animal. This is not to say that stallions are incapable of work. A number of small farmers work a well-mannered stallion beside a mare in harness. But there are plenty of horror stories involving inexperienced horse handlers and stallions. As a beginner/intermediate equine handler, steer clear of stallions for now and concentrate on choosing between a mare and a gelding.

## WORKING WITH MARES

A mare can be a much higher-maintenance option than a gelding. Even if you do not plan on breeding, you may find the regular hormonal cycling of a mare annoying. Some mares show little sign of cycling, while others may act royally ornery — kicking, biting, and squealing. Mares experience estrous cycles from early spring to late summer/early fall and ovulate (estrus) every 18 to 22 days. They generally do not cycle in the winter.

If you have a mare that you do not plan to breed but that displays significantly difficult behavior while in season, spaying (ovariectomy) will halt the mare's cycling, as well as any distracting or unpleasant behavior she displays during estrus. This option is expensive and carries some risk. Other options, such as bead implantation into the uterus, can trick your mare's body into thinking she is pregnant, which halts cycling but does not permanently end reproductive capability.

On our farm in Kentucky we worked mainly mares. Each mare foaled nearly every spring. A big, but not insurmountable, challenge was to manage the foaling and then handle the foals throughout the spring and summer. For us, the advantages to working reproducing mares outweighed the challenges. Handling mares almost ready to foal, that had foaled, or that were ready to go back to the field required extra effort, but was the best way we found for our horses to pay their own way.

Before committing to purchasing a mare for breeding, educate yourself on the many challenges associated with successful reproduction. If you plan on using reproducing mares in an agricultural setting, you will find that foaling usually occurs smack in the middle of field preparation and haying. While a mare may be used — with care — up to the day she foals, she should be given at least a two-week break after foaling. Once the foal is delivered, you will also have the new baby to deal with.

## WORKING WITH GELDINGS

If you are interested in accomplishing work with an animal that will not be affected by hormones, consider the gelding. Plenty of horsemen will not have a working mare on their property. They claim that mares are moody and unpredictable. Accordingly, geldings are the working horses of choice for many ranches and farms.

Like the historical eunuchs who worked in royal palaces, the gelding has no distracting interest in the opposite gender. His job is to eat, work, and be happy, making him a great choice for the small farmer who does not want to deal with the problems of reproduction. A gelding makes an excellent, steady year-round worker.

## AVOID THE STUDDY GELDING

If you are surprised by the stallionlike behavior of a gelding, don't let the owner pass it off as a bad day for the horse. A gelding exhibiting stallionlike behavior may be a cryptorchid or a false rig.

A *cryptorchid* is a male horse or mule with one testicle that has not descended into the scrotum. The retained testicle does not produce fertile sperm, but does produce plenty of testosterone, hence the stallionlike behavior. Like a stallion, this so-called gelding can exhibit unsafe and unpredictable behavior.

You cannot be certain of the presence of a retained testicle unless your veterinarian performs exploratory (also risky and expensive) surgery. If found, of course, the offending organ may be removed and your gelding should truly be a gelding, but he could still act like a false rig due to long-established behavior.

A *false rig* is a gelding that has had both testicles removed, but continues to exhibit stallionlike behavior. A common term used to describe these geldings is *proud cut*, a condition often attributed to retention of part of the epidydimis — a long, narrow tube in the reproductive system. The epidydimis, however, does not produce testosterone, so masculine behavior cannot be attributed to its retention. More likely the behavior is learned or innate, though early castration often reduces it. Some people claim that early castration also reduces muscling and size, but research has disproved this.

Bottom line: If a gelding that you are considering acts like an aggressive stallion, look for a different horse to purchase.

# Assessing Personality and Manners

Earlier you asked yourself what you wanted your horse or mule to do. In narrowing your search, you should have located animals that you believe match your expectations, such as experience with logging, plowing, carriage work, or whatever. Before you watch an animal show off his major talents, however, you need to know that he is capable of performing menial tasks in a dignified and correct manner. If you focus on the big things, such as how well the team performs on the plow, the cultivator, the carriage, or the cart, you will probably miss something just as important, such as the animal that kicks when frightened, refuses to be caught in the pasture, or is jumpy when harnessed.

Remain objective and keep your radar tuned exceptionally high for weaknesses in prospective animals. Taking along an experienced horse person to consult with is invaluable in assessing behavior. The points mentioned below are qualities any well-trained mule or draft horse *should* or *should not* be expected to exhibit:

- When you ask a horse or mule to step over in the stall, he should do so.
- A horse or mule should stand quietly for the veterinarian, to be groomed, or simply because you ask him to.
- Under no circumstances do you want an animal that fights a farrier. The animal should stand quietly, lift the requested leg promptly, and hold it up.
- When you lead the animal, he should not lag behind or pull ahead.
- He should not rub on you, push you around, or in any way intimidate you.

## CATCHING AND HALTERING

Before you arrive at the seller's farm, ask that the animal be left in the pasture or paddock until you arrive. By having the animal brought to you from its free-roaming home, you can observe how easy he is to catch and halter. You may hear every excuse in the book, but accept absolutely *no* rationalization for an animal that eludes capture and haltering. Not every horse or mule will come when called, but neither should one run when approached. An animal that is already caught and tied when you get there should send up a big red flag.

## ALERTNESS AND SENSITIVITY

The horse or mule should be alert, but not hypersensitive. An equine that jumps when you slam a door or pat him on the shoulder goes beyond alertness. Some horses are inherently jumpy and some never settle down even after years in harness. An experienced teamster can deal with such a horse but he is not a good choice for the novice. A horse or mule that never outgrows his jumpiness is unsafe for a beginner.

The animal should watch you and be aware of your movements. He should listen to your voice and anticipate your activities. While the animal should not be overly sensitive, neither should he be relaxed to the point of dozing while you examine or work around him.

## THE IMPORTANCE OF TEMPERAMENT

Once you have spent a few moments letting the animal learn to trust you, he should allow you to examine his ears, eyes, legs, feet, and hindquarters without showing *any* concern. If you notice ear pinning, feet stamping, or other signs of unease, head on down the road. Don't let the owner tell you the animal is having a bad day. If he's having a bad day today, he will likely have another bad day tomorrow. Look for an animal that will respect you and not increase your level of anxiety through his own nervousness.

The animal should accept the harness without concern. The crupper (if used) should be buckled under a loose tail, not one that is tightly clamped to the hindquarters. The head should drop into the bridle. When asked to step over with pressure to the shoulder or hip, the animal should do so willingly. You should feel completely comfortable fastening the quarter straps under the belly. Any head tossing, eye rolling, or jumpiness during harnessing often indicates that the equine is green-broke at best. A well-mannered, well-trained horse or mule will accept harnessing the way you accept putting on your own clothes and shoes.

### BEHAVIOR UNDER HARNESS

Examine each prospective horse or mule both alone and with a teammate. Alone, he should willingly leave the barn once harnessed. He should stand quietly while being hitched. *Do not accept any excuses for animals that are hitched when you arrive.* You must watch and be part of the hitching, which means you should insist ahead of time that the animals not be hitched before you get there.

When looking at a prospective team, ask the owner to drive the animals around some, then switch sides on the tongue and drive them some more. If the owner tries to tell you, "Daisy knows how to work only on the off side," these animals are not for you. A well-trained animal is capable of working on both sides of the tongue, without regard to the noise of the equipment, the extra weight requirement, or any other excuse the owner may come up with. After you start using your animals you will likely always hitch them the same way, just for the ease of harnessing (we do things better, it seems, where repetition is involved), but your prospective draft animals should not be limited from the get-go as to where they can or cannot work.

### WORKING WITH TEAMMATES

Notice how the horse or mule treats his teammate. Do you hear contented nickering or do you see ear pinning and teeth baring? A grumpy horse is a liability to you as well as to his teammate and should not be tolerated. Some mares act cranky during their estrous cycle, but crankiness should not be a personality trait. As with a single horse, the team should stand quietly while being hitched to an implement. They should not have to be told to whoa more than once. Although circumstances may require a reinforcement to stand still, the owner should feel comfortable leaving the lines loose while fastening the traces. One person should be able to harness and fasten the traces and neck yoke without trouble or anxiety.

When buying a team, be sure that they are truly a team, well used to each other and able to work together. You do not want two animals unaccustomed to each other, which makes three strangers that must become acquainted, instead of you getting to know two animals as one unit. So watch the animals carefully as they interact. Do they pull well together? Do they seem to work as one unit? Does one animal appear to be intimidated by the other? A lack of trust in each other may mean that the animals have not worked together for long. As a novice, you want two horses or mules that understand each other and will work together to help you accomplish your goals.

### TAKE A TEST DRIVE

Once the team (or single) is hitched, drive around a paddock or other contained space before taking a trip down the road. They should not shy at traffic, bridges, or such scary objects as mailboxes. An animal that shies at everyday objects likely has many other fears as well. Some horses and mules never lose their fear of traffic. Nobody knows what makes some animals completely comfortable around a roaring 18-wheeler and others practically kill themselves and you trying to get out of the way. Some animals are just plain unsafe around traffic, no matter how many times you expose them to it. You do not want an animal that is safe in every situation *except* traffic; at some point in your draft horse experience you will encounter vehicles, guaranteed.

After one trip down the road, switch the side of the tongue the animals are hitched on and make another road trip. You want to make sure the seller is not covering up for a traffic-shy animal by hitching him on the off side, away from traffic.

On the return trip, the team should *not* speed up in anticipation of returning to the barn. They should maintain a steady pace and accelerate or decelerate as requested. The trip should be calm and controlled, not a nerve-wracking *yee-ha* experience. If you ever have the feeling of being out of control, you are in the wrong place at the wrong time with the wrong team. Horses and particularly mules can sense your lack of experience, as well as your fear. If you can't control the animals, *do not buy them*.

### TRY OUT THE TRAILER

Before leaving the prospective seller's farm, ask to have the animal(s) loaded into a trailer. Few things are less agreeable than passing the greenbacks, only to spend the next two hours trying to get your new team loaded. The well-trained equine will load easily and calmly. He should also back out of the trailer in a collected manner, rather than staging a pell-mell retreat that leaves you with a rope-burned hand.

An animal that accepts loading and unloading well will make your life far easier. During your years of equine ownership, you will make visits to the veterinarian and to various events where a balky animal is not only embarrassing, but also a hassle and a danger.

## Thinking about Age

For a safe experience, the novice should have an older team of well-trained horses or mules. Horses can live well into their twenties, so a 10- or 12-year-old draft horse that has experienced consistent, meaningful work with a skilled teamster beginning at around three years of age should be mature and wise enough to understand and work with a novice's mistakes.

A mule, on the other hand, takes longer to mature. Although the mule will never accept a novice's mistakes, he will have a good idea of what is expected of him by the age of 12 or so — assuming, of course, that he has been worked since he was young under a careful and responsible muleteer who has treated the mule with respect.

Most people worry about getting cheated on how old an animal really is; for the novice, the older the better, as long as the animal is in good health and body condition. Senior horses and mules lose body condition, and you should therefore avoid horses over 20 and mules over 30 years of age. How can you, the neophyte, know the age of a horse or mule? Chances are you won't be able to tell exactly, but certain factors can point you in the right direction.

The number and conformation of the teeth are two of these factors. Equine teeth have unique eruption and wear patterns that can be used to gauge an animal's age. Body condition is another key to aging. Although examining these general aspects will not give you a precise birth date, they will at least help you determine a ballpark age.

### The Old Young Equine

**Discrepancies often occur** between the registration-paper age and the age that the teeth tell. No, the registration papers are not lying. Rather, the animal has probably been eating pasture or hay forage containing a high amount of sand or other abrasive material that unnaturally wears down the teeth. Or he may be a habitual wood chewer. Even though the animal's true age may be younger than the teeth indicate, an equine that cannot chew is not much good. Consider any animal to be as old as his teeth.

### LOOK AT THE TEETH

You can tell the age of a young equine by examining the eruption patterns of his teeth. Young horses have milk teeth — temporary teeth that start to be replaced by permanent adult teeth at around age two. At three years of age, the two central incisors are fully erupted. At three and a half, the middle incisors appear. At age four and a half, the corner incisors begin to erupt and are fully erupted at age five, thus completing the equine's 24 permanent premolars and molars.

Some equines develop wolf teeth, a second set of premolars just in front of the main premolars. These teeth usually appear by age two and are generally found in males, but occasionally in females; they may or may not appear on both sides of the jaw. These teeth should be removed since they may interfere with biting and chewing, as well as causing the bit to sit uncomfortably in the animal's mouth. A horse with painful teeth may throw his head around or shake it frequently and might seem irritable or easily distracted.

At age four, the canine teeth, also called *tushes*, erupt behind the corner incisors, which in turn erupt at age four and a half. Canine teeth appear almost exclusively in male horses; when they do appear in female horses, they are quite small. They function as a defense mechanism and can be dangerous, so are often reduced in length (by an equine veterinary dentist) to avoid injury to other livestock and to humans. Canine teeth that are not deliberately reduced in size may interfere with the set of the bit, creating pain and resulting in behavioral problems.

After age five, you can estimate equine age by the wear on the incisor cups (pits in the center of the teeth), the shape of the incisors, and a groove that appears in the upper corners of the incisors. The cups of the central incisors disappear at age six, of the middle incisors at age seven, and of the corner incisors at age eight.

The shape of the teeth also changes. At age 12 the chewing surfaces of the central incisors become round and slant slightly forward when viewed from the side. At age 17 the chewing surfaces of the incisors are round and the teeth slant forward even more. At age 18 the chewing surfaces of the central incisors have become triangular again and continue to slant forward. By age 23 the incisors regain an oval shape (now from front to back rather than side to side), the Galvayne's groove begins to disappear, and the teeth continue to slant forward.

The Galvayne's groove, a line first appearing at the gum line of the upper corner incisors when a horse is about 10 years old, is another indicator commonly used to determine equine age. With age, this groove becomes stained yellow and extends downward. At age 20, the groove will extend down to the surface of the tooth; by age 30, the groove will be gone. Many horse folks consider Galvayne's groove to be the least reliable sign of a horse's age after 10 years.

The teeth of a horse or mule grow continuously. If a tooth breaks off or is lost — perhaps from a kick or other injury — the opposing tooth will grow without interruption and, if not rasped off, may interfere with chewing or prevent the mouth from closing properly.

At age 12, the central incisors become round.

At age 17, all incisors are round.

At age 18, the central incisors are triangular.

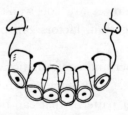

By age 23, all incisors are oval.

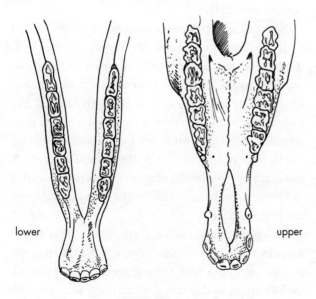

**THE EQUINE JAW**

# WORKING GOOD HORSES AND MULES EVERY DAY

**Dave Feltenberger**
**New Braunfels, Texas**

Type of Farm: 60 acres (24 ha); mostly hay with 15 acres in tillage (oats)

Animals Owned: Six Percheron mares, a Belgian mare, a Mammoth jack, three Belgian mules, and two Percheron mules

Talking about his appreciation for horses and mules is second nature to Dave Feltenberger. Dave was exposed to the concept of working horses while living with his grandfather as a boy. This experience benefited him later in life when, after a career in the military, Dave sought a quieter life raising quality draft mules — if donkeys, horses, and mules may be called quiet.

He started out the right way by purchasing a dead-broke team of Belgian mares. These horses provided all of his hands-on education. No one in his area knew anything about draft animals or hitching, let alone farming. The part of his education not provided by his venerable Belgian team came mostly from reading and watching videos. In the 15 years since that initial purchase, he estimates that his farm has seen 150 horses and mules come and go, many of them trained by Dave himself.

Dave uses his horses and mules as supplemental power on his farm to make hay and till the soil. He also operates a busy carriage business, which he started with Belgians but switched to Percherons because many brides want white horses. His horses are equally comfortable pulling a mower, a disc harrow, a spike-tooth harrow, a single shovel left-hand plow, a four-section harrow, a forecart, a wooden training cart, a 12-passenger wagonette, and a cabriolet carriage.

In Dave's view, any rookie who purchases a mule as a first animal is making a big mistake. "A horse," he says, "is much harder to screw up than a mule." The mule is an incredibly intelligent and careful animal that will turn sour once he figures out you don't have a clue what you are doing. Dave has seen a well-trained mule turned into an outlaw in two weeks by novice hands.

Horses, on the other hand, are generally willing to help you to learn along the way. Dave notes that the only way you will keep a well-trained horse team well trained is always to be consistent and correct. You may not have to do any training, but if you handle your animals incorrectly or irresponsibly, no matter how well trained they are, they will learn bad and potentially dangerous habits.

A good teamster regularly works his animals, knows his animals, and cares for his animals. He is willing to admit their faults and doesn't describe every animal as "grandma or kid broke." Dave encourages the novice to take along an admired and respected teamster when looking at a potential purchase. If you are buying at an auction, go only to a reputable one, such as the Amish auctions in Ohio and Indiana, where owners provide detailed information about each animal and his training.

Dave emphasizes the need for forming a relationship with a knowledgeable teamster who can help you on the challenging road to successful draft horse and mule ownership. He also emphasizes the need to find gentle and well-trained horses willing to help train a greenhorn. "I am living my dream. I spent my adult life working toward retirement. Now I am there and I play with horses and mules every day."

## EVALUATE BODY CONDITION

Although teeth are the most accurate indicator of equine age, you must also consider the animal's body condition. (See pages 132–135.) Some key factors that flag a senior equine include:

- Sway back, particularly in an older brood mare with a long back
- Coat lacking luster and shine
- Recessed temporal fossa (rounded hollow above the eye)
- History of difficulty maintaining weight

As in most situations, exceptions occur. Many old horses age gracefully and remain in great condition. Some 25-year-olds look not a day over 10 until you open their mouths. A team of "senior citizens" can come out of winter rest in perfect body condition and step right into another season of plowing without a backward glance.

## Assessing a Draft Equine

When you look at a horse or mule that interests you, the first step is to make sure the animal is sound and healthy. Being a good judge of horseflesh takes years of experience and the examination of many horses, so don't feel overwhelmed. If at all possible, take along an experienced horse or mule person. Even if he or she is not familiar with draft animals, you will get a nonbiased view and critical assessment.

You can assess a potential draft horse or mule in two easy ways. The first is a quick-and-dirty check for overall symmetry of general appearance, to decide if the animal deserves a closer look. If the animal passes, move on to a closer examination of his conformation. Correct conformation is critical to the animal's health and future.

**PARTS OF THE EQUINE BODY**

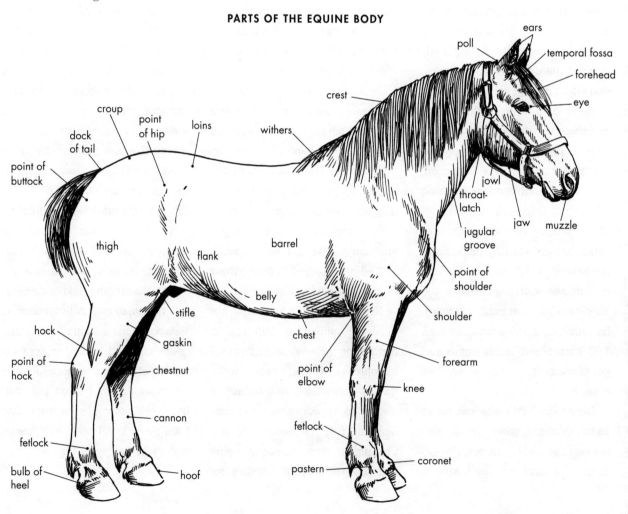

## SYMMETRY OF GENERAL APPEARANCE

Symmetry, or the equine's balanced proportions, is one quick way to determine whether the animal deserves a closer look. Obvious inconsistencies in symmetry may indicate serious problems. Reviewing an animal's general appearance is as simple as asking yourself: Does the horse or mule look the same on both sides? Look carefully from both the front and the rear while the animal is standing still. He should stand square (with weight equally on all four feet) and at ease. Do the shoulders, barrel, and croup all look symmetrical? Is the head too small? Does the neck look too long?

The legs of a horse or mule must carry great weight, often over unforgiving surfaces. Do the two front legs look symmetrical to each other? How about the back legs? Do you observe any differences in the size or structure of one leg compared to the other? Does one leg have swellings that make it look different from the other?

### Proper Muscling

An equine, particularly a draft animal, must have good muscling for the work it is expected to perform. Ensure that your potential animal will be able to do the job you require by examining his muscling and inquiring about his history. Are the shoulders well rounded and symmetrical? Are the pectoral muscles hard and even or soft and flabby? Are the hindquarter muscles symmetrical? Asymmetry of muscling may indicate an injury or disease.

One muscle disease seen in heavily built drafts and occasionally in draft mules is equine polysaccharide storage myopathy (EPSM). This condition, caused by an inability to metabolize carbohydrates, results in symmetrical atrophy of muscle groups, such as both left and right croup muscles or both left and right thigh muscles. EPSM often causes more pronounced atrophy in the hindquarters than the front. Although EPSM may be controlled through diet change, an animal with advanced signs may never fully recover enough to work. EPSM is not always visible to the casual observer, but muscle atrophy, trembling muscles after exercise, and an abnormal gait in the hind limbs should all be considered red flags.

Another condition, known as *sweeney*, is evidenced by atrophied shoulder muscles. Sweeney is caused by the crushing of the nerve against the shoulder muscle, often because of a poorly fitting collar. It may also be caused by injury, such as a well-placed kick. Today's draft horse is often valued more and worked less strenuously than horses 100 years ago, thus reducing the risk of finding a horse with sweeney.

### Proper Movement

Symmetry of movement means that each hoof receives the same amount of weight and each stride is the same length. The animal does not have any aberrations such as stiff, short strides or head bobbing. Don't confuse the head bob of a lame horse with the rhythmic up-and-down motion of the sound horse at work. I remember well the comfortable sight of watching my dad drive a team of horses cultivating corn, and watching their heads nod up and down to the rhythm of their walk. The head bob of the lame horse is also rhythmic, but can be more pronounced, and occurs only when the affected limb takes weight.

When you test symmetry of movement, have the horse walk in both a straight line and a circle. Have the animal move away from you, come toward you, and turn in a circle to the right and to the left. Watch carefully. Subtle lameness may be hard to detect. Although the normal draft gait is a walk, trotting the animal can bring out faults not noticeable at the walk.

## THE IMPORTANCE OF GOOD CONFORMATION

Conformation of the draft horse is slightly different from conformation of either the average light horse or the mule. The draft horse generally has a wide, deep chest, massive neck, great, rounded barrel, deep belly and flanks, well-muscled croup, powerful, well-muscled foreleg and hind leg, and huge feet. He is far stouter and more muscular than the athletic and fleet light sport horse. He is also quite a bit heavier than the draft mule (discussed in chapter 5).

# Is Your Horse *Really* Lame?

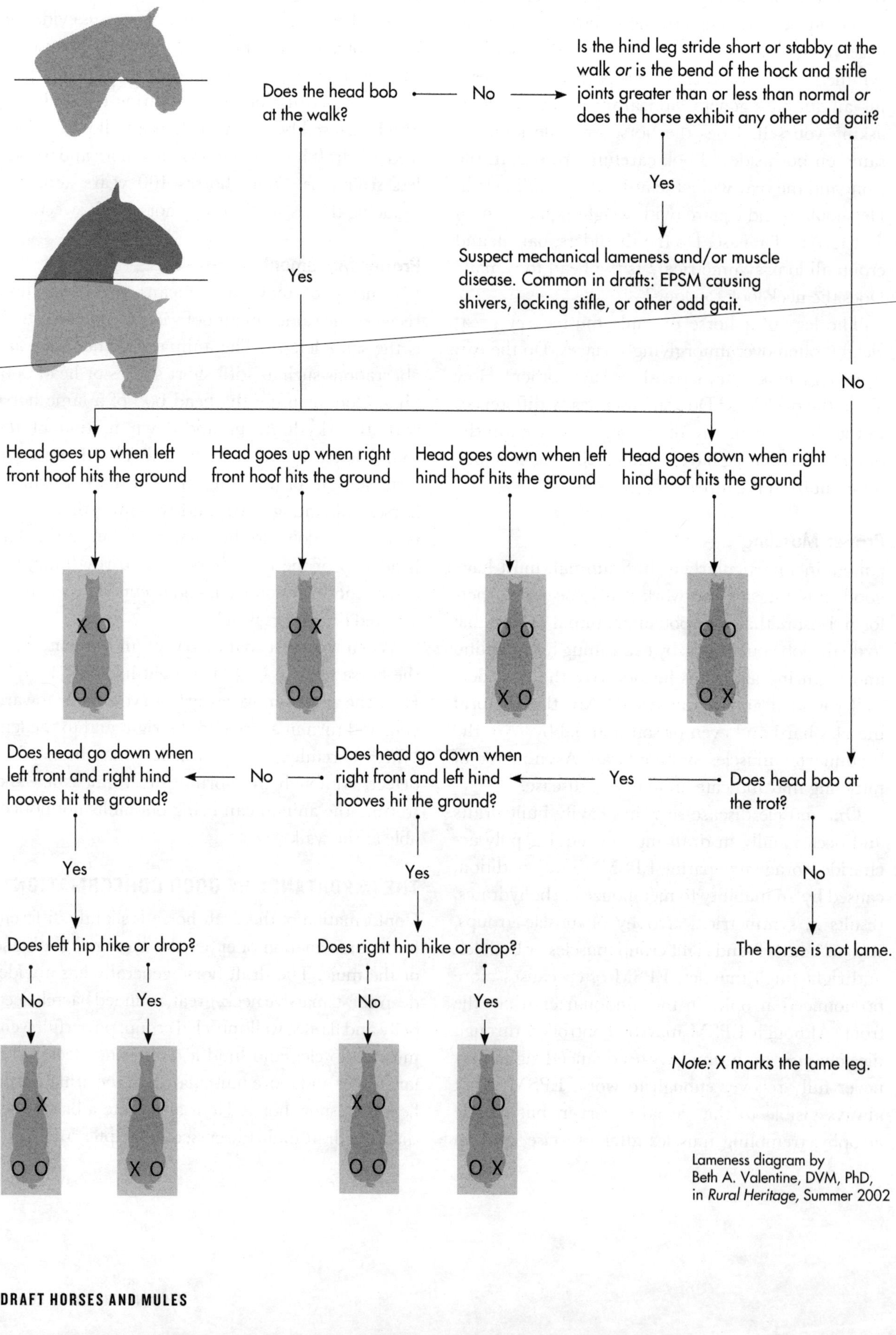

Does the head bob at the walk? — **No** → Is the hind leg stride short or stabby at the walk *or* is the bend of the hock and stifle joints greater than or less than normal *or* does the horse exhibit any other odd gait?

↓ **Yes**

Suspect mechanical lameness and/or muscle disease. Common in drafts: EPSM causing shivers, locking stifle, or other odd gait.

**Yes** ↓ (from head bob question)

- Head goes up when left front hoof hits the ground
  - X O
  - O O
- Head goes up when right front hoof hits the ground
  - O X
  - O O
- Head goes down when left hind hoof hits the ground
  - O O
  - X O
- Head goes down when right hind hoof hits the ground
  - O O
  - O X

Does head go down when left front and right hind hooves hit the ground? ← **No** — Does head go down when right front and left hind hooves hit the ground? ← **Yes** — Does head bob at the trot?

↓ **Yes** ↓ **Yes** ↓ **No**

Does left hip hike or drop? / Does right hip hike or drop? / The horse is not lame.

- **No**: O X / O O
- **Yes**: O O / X O
- **No**: X O / O O
- **Yes**: O O / O X

*Note:* X marks the lame leg.

Lameness diagram by Beth A. Valentine, DVM, PhD, in *Rural Heritage,* Summer 2002

## START WITH THE HEAD

A horse's head should be proportional to his body. Each breed has its own distinctive head, but whether it is plain or beautiful is less important than its proportion to the body. A Roman nose, in which the profile bulges slightly outward from the eye down, is desirable in many draft breeds. The features of the equine head — the eyes, the ears, the jaw, and the muzzle — are important for an equine's survival and, to the trained eye, tell much about intelligence, disposition, and temperament. Each of these features should be in excellent working order.

**Look at the eyes.** The eyes supposedly show intelligence if they are large, bright, clear, and alert. While intelligent horses can be hard workers, they can also be tricksters. A mischievous horse is not necessarily a bad horse; he just may require a bit more patience.

Some breeds are known for their gentle eye. Draft horse eyes are often referred to as gentle, particularly in contrast with a hotter breed such as the Arabian. Good eyes are widely spaced, allowing for proper range of vision. If an abnormal amount of white shows around the edges or the eyes are hard or suspicious, keep looking.

The equine has an almost 360-degree range of vision, since each eye operates independently of the other. Each eye sees a different picture, with a slight overlap of picture in the 60- to 70-degree range. The equine has a blind spot directly behind and in front, for approximately four feet. To see anything in either of these locations, the animal must turn his head.

**Assess the jaw and muzzle.** The jaw and muzzle should be sufficiently wide for the heavy breathing that takes place with exertion, and the throatlatch should be broad. The jaw should not have an underbite or overbite. Stallions typically have larger jaws than mares. The muzzle should include large nostrils for excellent airflow. Lips should be firm and muscular. Floppy lips can indicate disinterest or illness, though many horses relax their lips when at rest.

## CHECK OUT THE NECK

The neck should be long for flexibility and grace, with the topline short and smooth for strength. The underline of the neck should include length for freedom of movement and should be straight. The neck of a draft horse should have good crest, although too much crest may mean the horse is overweight. A

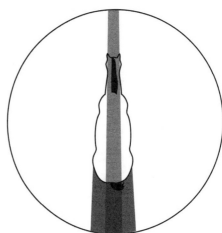

Equine range of vision: shaded areas represent blind spots.

cresty neck can also be a sign that a horse has foundered or might be prone to foundering. Necks may have one of three faults:

- A short neck can lead to overcresting. This fault is often seen in stout breeds such as Fjord, and can considerably reduce the horse's maneuverability. A horse with a short neck often has a choppy stride.
- A long neck may earn the animal the nickname of Snake or Giraffe. The neck should be no longer than 1½ times the length of the back. A neck that is too long can reduce the animal's agility.
- A ewe neck is characterized by downward arching of both the upper and lower sides of the neck, so the horse looks like he is stargazing. This condition causes the horse to carry his head high, interfering with vision and bit contact.

## CHECK OUT THE CHEST

The neck, back, and shoulders meet at the withers. The peak, or highest point, of the withers should be the same height or higher than the highest point of the croup. Withers should be smooth and without irregularities. A draft horse can develop recurring sores known as *fistula of the withers* if the collar fits badly. Prominent, high withers or an extreme lack of weight can cause an animal to be susceptible to wither lesions. A draft horse's chest should be broad, with a substantial gap between the legs. Excessive muscling or fat deposits can hide a poorly formed chest. From the horse's side, the pectoral muscles should be well defined. From the front, you should see a clear V on the chest.

Being thinner and leaner than a horse, the mule will not have the same broadness in the chest. His muscles, however, should still be defined.

## EXAMINE THE FORELEGS

The equine's foreleg encompasses the shoulder, upper arm, elbow, forearm, knee, cannon, fetlock, and pastern. Sixty-five percent of the body weight is carried on the forelegs, making them more susceptible to injury. Poor conformation increases this susceptibility. While the draft horse is not typically associated with the injuries common to lighter sport horses, he still must have excellent foreleg conformation. The legs of a draft horse are shorter and stouter than those of a lighter horse, giving him immense power to pull loads greater than his own body weight. A good foreleg is therefore a critical component of soundness.

When you examine the foreleg for conformation, observe it from all angles. From the front, the legs should appear neither *toed in* nor *toed out*, conditions caused by crookedness in the bone somewhere

Cresty neck

Ewe neck

Correctly proportioned neck

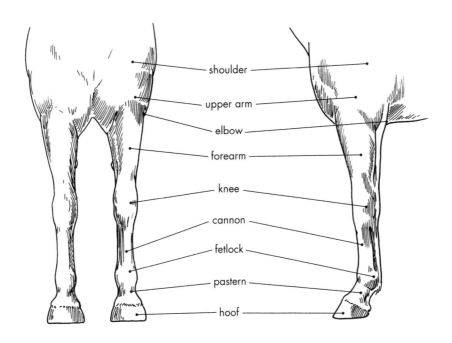

**PARTS OF THE FORELEG**

between the elbow and the fetlock. Viewed from the side, the foreleg should be perpendicular to the ground from the elbow to the fetlock.

**The shoulder** varies among draft breeds, although the ideal shoulder for all breeds is heavily muscled. The shoulder from the peak of the withers to the point of the shoulders should be almost as long as the neck from the poll to the front of the withers. A straighter shoulder is more acceptable in draft horse breeds than in light horse breeds. The

## Measuring Equine Height

**A** hand is the unit used to measure equine height. One hand equals four inches (10 cm). A designation such as 16 hh means the horse stands 16 hands high; 16-2 hh means 16 hands plus two inches (5 cm). The term *hand* originates from the approximate distance across the width of a man's hand, as a ready way to measure equines. Height is measured in a straight, vertical line from the top of the withers down to the bottom of the hoof.

To accurately measure the height of your horse or mule, use a measuring stick purchased from a livestock supply company. This tool has a vertical bar attached to a horizontal bar with a bubble level, allowing you to make sure that you have an accurate vertical measurement.

Have your horse or mule stand squarely with all four feet on a hard, level surface such as a concrete carport or sidewalk. Place the measuring stick crossbar squarely on the highest point of the animal's withers. Since the withers are behind the foreleg, make sure that the vertical part of the measuring stick is positioned behind the foreleg at the withers. Once the bubble in the crossbar is in the center, read the height indicated on the stick.

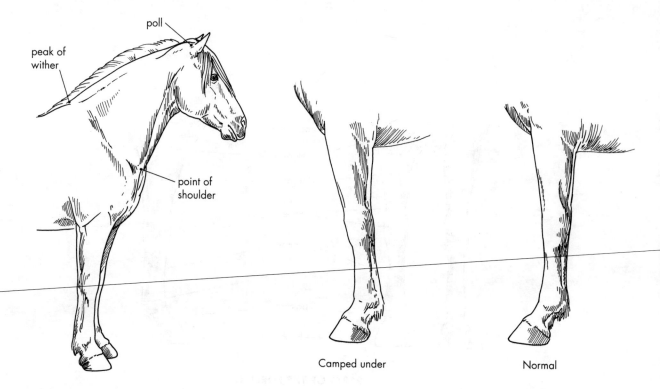

**CONFORMATION OF THE SHOULDER AND WITHERS**

**CONFORMATION OF THE FOREARM**

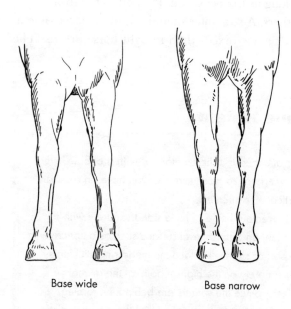

**POOR CONFORMATION OF THE ELBOW**

mule also has a fairly straight shoulder. The straightness of the shoulder may help provide pulling power, though it is not critical.

**The elbow** should be well defined and balanced with the rest of the body. An enlarged or swollen elbow may be the result of injury or being clipped with a shoe when lying down. The elbow is the point of connection for the upper arm and the forearm, and so is subject to many conformational issues. Improper position of the elbow can make a horse base wide or base narrow — respectively, having greater or less distance between the forearms than between the feet.

**The forearm** connects the elbow to the knee. The forearm should be in a straight line with the knee and the cannon from all views. A wide forearm is optimal for the attachment of muscles for strong propulsion. Muscles should be distinctly defined. Little fat is deposited in this area, making it a good indicator of the animal's true muscling. A short upper arm can cause a choppy gait, as well as

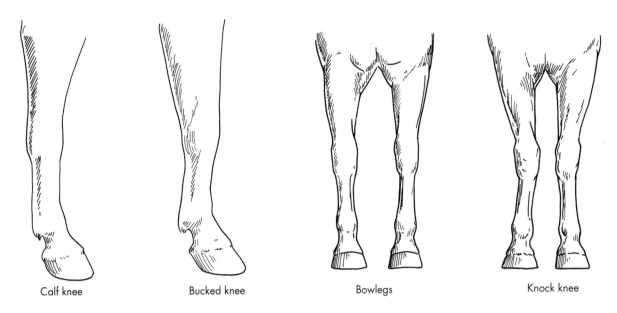

**POOR CONFORMATION OF THE KNEE**

a camped-under look in which the forelegs are set too far under the body.

**The knee** joins the forearm to the cannon and should be straight and square. When viewed from the front, the horse should not be *in at the knee* (knock-kneed) or *out at the knee* (bowlegged). *Offset knees,* in which the cannon bone is offset to the outside of the knee, are particularly problematic; a horse with this condition could easily break down under exertion. Offset knees create stress on the inside splint bone, often leading to the painful condition known as *splints*. From the side, the leg should appear neither bent back at the knee (calf knee) nor arched forward at the knee (bucked knee).

**The cannon** lies between the knee and the fetlock. In some draft horses, the rear of the cannon and fetlock is obscured by feathering. The cannon should be short and straight, giving the horse good stride, and should be narrow from the front and wide from the side.

**The fetlock** should be smooth and wide for good shock absorption and weight bearing without puffiness or swelling. Edema (a liquid-filled swelling) in the fetlock may indicate or accompany lameness and should be carefully investigated.

> ### Splints: Lameness or Blemish?
>
> **Lameness due to excess stress** on or injury to the inside splint bones is known as *splints*. Splints are typically seen in animals between the ages of two and five that have sustained severe concussive action on a hard surface or generally overworked their immature bones and ligaments. Splints may also be a result of offset knees, a conformational defect. Splints can heal and be expressed in older horses merely as bumps, typically seen on the inside of the forelimb cannon bone, but may also indicate ongoing unsoundness.

## Weight-to-Bone Ratio

A horse's **weight-to-bone ratio** helps to determine the optimal length of his cannon bone, but because accurate weights can be difficult to ascertain, the height-to-bone ratio is more often used to help determine an animal's potential long-term soundness. A ratio of 5.0 between the horse's height and the length of the cannon bone is ideal for heavy draft animals. A higher ratio (which is typical in a light horse or mule) means that the bone is too short to support the heavy horse's massive frame, making him prone to lameness.

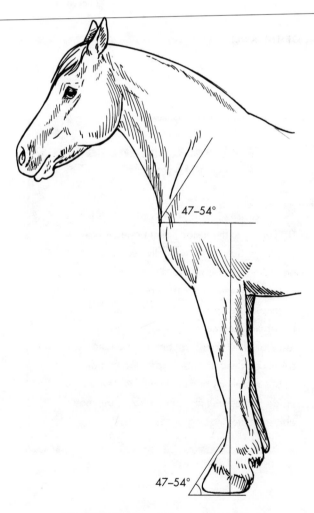

**ANGLE OF THE CANNON AND FETLOCK**

**The pastern and the foot** should be on the same axis, running on a straight imaginary line through the core of the pastern bones. If this line is broken or crooked, the bones, joints, ligaments, and tendons will be compromised by stress, often leading to lameness.

The angle of the pastern should be almost exactly equal to that of the shoulder. The correct angle (slope) of the pastern and foot in most horses is 47 to 54 degrees in the foreleg and 49 to 56 degrees in the hind leg. Hoof trimming and shoeing can greatly influence this angle, for better or worse. The pastern is responsible for absorbing and dissipating much of the force of concussion received when the animal is moving.

The draft horse, because of his great muscle and weight, is meant to walk more than to travel at higher speeds, thus reducing the excessive concussive force many light horses experience while trotting or galloping on hard surfaces. The greater weight of the draft horse, however, still generates impressive stress on the pastern, which must be well conformed to preserve soundness.

## LOOK AT THE BODY

The horse's body consists of the barrel and the back. The barrel houses the abdomen and chest cavity, while the back refers to the area between the end of the withers and the point of the croup.

## CONFORMATION OF THE BACK

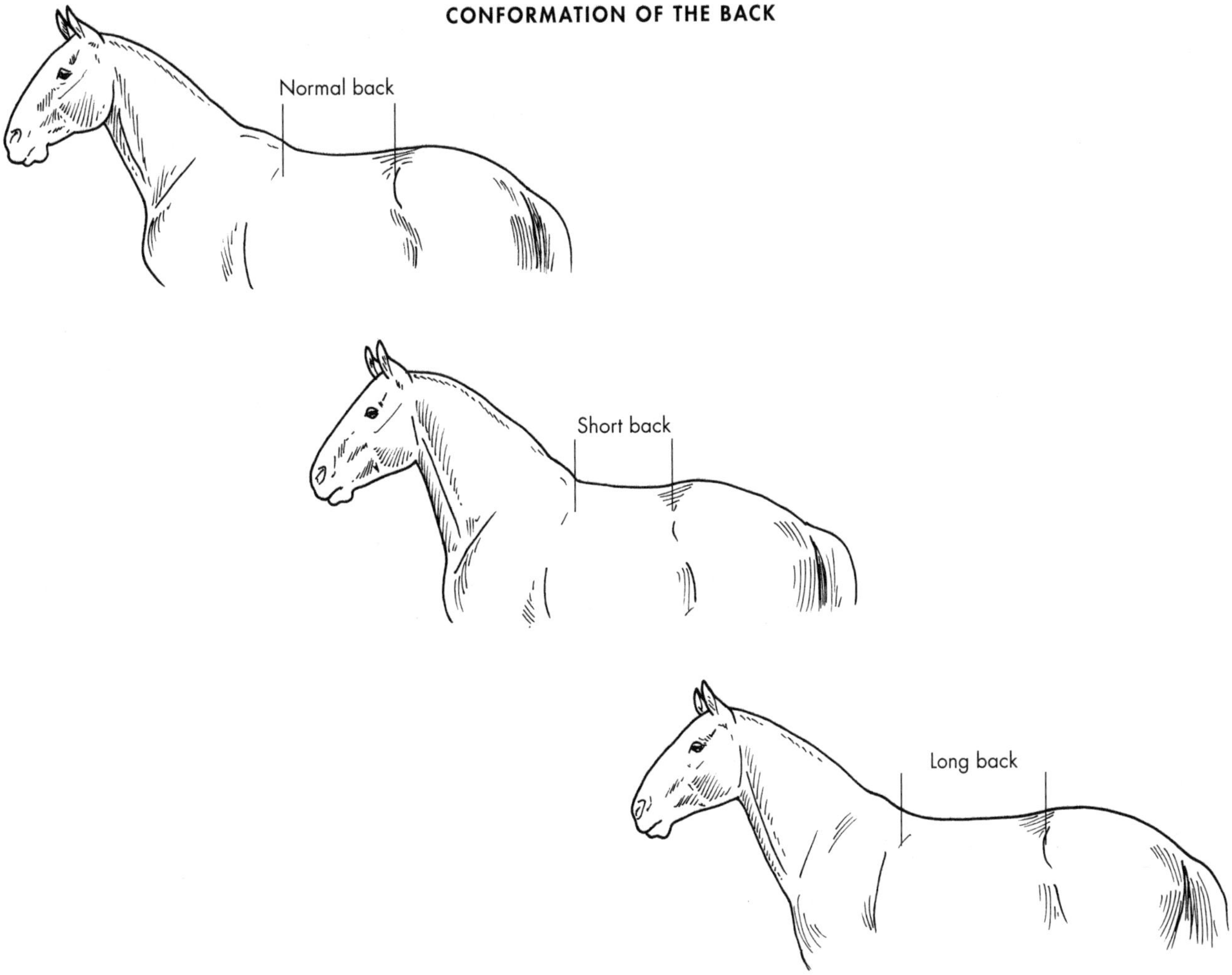

*Normal back*

*Short back*

*Long back*

**The barrel** of the draft horse should be deep and wide, amply accommodating the heart and other organs and giving the large lungs room to expand. The length of the barrel should be several inches more than its depth from the withers to the underline, which in turn should be several inches more than the distance from the belly to the ground. These dimensions create a powerful, stocky horse capable of moving great weight.

**The loin** is a portion of muscle extending from the last rib to the point of the croup, with the point of the croup about as high as the peak of the withers. The distance between the last rib and the point of the hip typically should be no more than three inches. A short, well-developed loin provides strong forward movement.

**The back** of the draft horse should be broad, short, and straight, acting as a bridge for the barrel's ribs and muscles. Since a draft horse should be short through the legs, a back measuring less than one-third the total body length gives the horse an overly stocky appearance. A short back causes a choppy stride and impedes the horse's flexibility from side to side, but may increase his agility in activities requiring power and coordination. A back longer than one-third of the topline and a loin measuring more than a few inches are conformational problems that limit forward movement and therefore are undesirable in pulling horses.

*Swayback*, in which the back dips down excessively from the wither to the loin, is often linked to the long back/loin problem, but may also be caused by old age, nutritional deficiency, or pregnancy. *Roach back* is the opposite of the swayback. Here the back humps up between the withers and loin, decreasing the horse's range of motion.

**FINDING THE IDEAL EQUINE**

## EXAMINE THE HIND LEGS

The draft horse's hind legs are powerhouses of muscle, providing more pulling power than the forelegs. The hind legs drive hard when the animal is pulling forward and are designed to evenly spread concussion, making good conformation of the hind legs important for soundness. The hind leg must be viewed from all sides to determine correct conformation. Imagine a line dropped from the point of the buttocks viewed from the side. As the line travels to the ground, it should touch the point of the hock and the back of the cannon bone.

An animal with excessively straight legs is said to be *straight behind* or *post legged*. In this case, the imaginary line falls through the cannon instead of behind it. Heavy work may cause a post-legged horse to be injured from the intense forward movement. A horse that is *camped under* has a line that falls too far behind the cannon, indicating that the legs are set too far forward under the horse.

The hind legs consist of the hindquarters, stifle, gaskin, hock, cannon, fetlock, and pastern. The cannon, fetlock, and pastern are similar to those of the forelimb. When viewed from the rear, the hind legs should be wide at the hip and close together at the hock and fetlock. When the animal is pulling, his feet are spread farther apart, so if his legs are too straight they will bow out under the strain and be more prone to injury. As with the forelimbs, hind limb conformation problems, when viewed from the rear, include *base narrow* and *base wide*.

**The hindquarters** — composed of the croup, pelvis, and thigh — are muscled and massively strong. Unlike most light breeds, where a square hindquarter is desirable, the draft horse is typically said to be *double rumped*, because a groove running down the center of the hindquarters creates a rounded appearance. Wide-set hindquarters give the draft horse immense pulling power.

**The croup** of the draft horse may be slightly higher than the withers, although preferably the croup is equal in height to the point of the withers. The croup is the area along the topline from the loin to the base of the tail, and should be fairly long.

In pulling horses, a steep croup is desirable. The angle of the steep croup allows for extra muscling as well as exceptional leverage and mechanical advantage making the draft horse an ideal candidate for heavy pulling. A steep croup also allows the draft horse to get his back legs under his body, a feature most successful pulling horses heavily rely on when starting loads from a standstill.

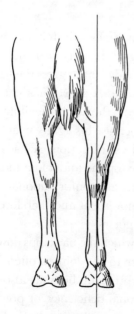

A straight line drawn from the point of the buttock to the ground should bisect the hind leg.

A straight line drawn from the point of the buttock should run through the point of the hock and follow the cannon bone.

**The stifle** should be in a forward position; a stifle that is too far back creates a short thigh and thus less muscling. When viewed from the rear, the stifle should be the widest part of the hindquarters and should be at almost the same height as the elbow in the forelimb. The thigh lies between the hip and the stifle. Ideally, the thigh should be almost perpendicular, creating room for heavy muscling.

**The gaskin,** also called the second thigh, lies between the stifle and the hock, and should be only a little shorter than the thigh. The gaskin should be well muscled on the inside and outside of the leg, for these muscles enable pulling, as well as running and jumping.

**The hock** should be quite large and well defined, not "meaty" looking, and, when viewed from the side, should be as wide and strong as the gaskin. The point of the hock should occur almost halfway between the ground and the stifle. The angle of the hock is correct if an imaginary line running from the point of the buttocks to the ground touches the point of the hock and runs down the back of the cannon. Hocks that are too straight may predispose the animal to inflammation and/or lameness.

Conformational problems in the rear include cow hocks and bowlegs. *Cow hocks* are closer together at the hock than normal and are accompanied by a toed-out stance. *Bowlegs* are the opposite of cow hocks — the fetlocks are closer together than normal, while the hocks are farther apart.

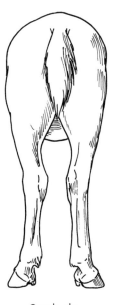
Cow hocks

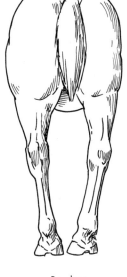
Bowlegs

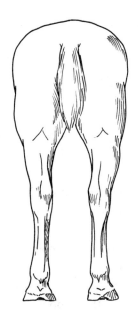
Base wide

Base narrow

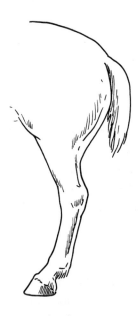
Camped under

Post legged

**POOR HIND LEG CONFORMATION**

## THE FEET ARE CRITICAL

Draft horses are famous for their large feet, which must support a massive weight. Each time a foot touches the ground, it absorbs much of the concussive force that otherwise would reach the leg. One old saying warns, "No foot, no horse." Like the lighter and more athletic sport horse, the draft horse must have correct conformation of the feet for soundness.

Due to their large, shallow soles, many draft horses are predisposed to poor foot conformation and often have feet that crack easily, giving them a splayed and unkempt appearance. However, abnormal hoof conformation is often the result of human influence. Improper shoeing and trimming are especially to blame. Proper care of a draft horse that works regularly on hard surfaces, such as gravel and asphalt, has traditionally called for a regular shoeing program. Recent research into a more natural approach to hoof care suggests that if the hoof walls are properly conditioned and maintained by a skilled farrier, shoeing may not be necessary.

The hoof wall is the horny outer covering of the foot and is divided into three areas: toe, quarter, and heel. It should be healthy and flexible, with enough moisture to allow expansion of the heels that is required for weight bearing. The toe should be round and the heels broad to allow for maximum weight dispersion. Horn growth is about one fourth to one half inch (1 cm) per month, with faster growth occurring in summer.

The foot on a draft horse must be large for maximum support of the animal's bulk. Symmetry of the feet is needed for equal weight dispersion. The front feet should also be symmetrical with each other, and the rear feet should be symmetrical with each other. The front feet of horses and mules are normally one size larger than the rear, and the rear feet should be narrower and more pointed and upright than the fronts.

When viewed from the front, the coronet should run straight across the front of the foot, and then slope symmetrically down toward each heel. The slope and height of the hoof wall should be symmetrical on both sides. Viewed from the back, the bulbs of both heels should be the same height. From

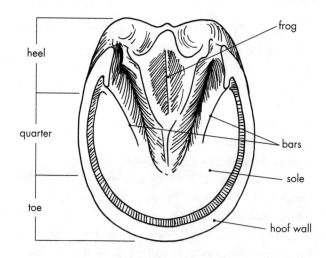

**PARTS OF THE HOOF**

the bottom, a line running straight down the center of the frog should evenly bisect the heels, bars, and frog, with all being proportional and in the same position on both sides of the line.

The angle of the hoof wall should be the same as the pastern: approximately 47 to 54 degrees for the forefeet, and a steeper 49 to 56 degrees for the hind feet. The angle itself, however, is not as important as the pastern and hoof angle being the same.

Feet that are too big are commonly called *mushroom feet*. The spreading of the hoof wall causes a loss of concavity, compromising support. Large, wide feet are prone to sole bruises. Draft horses used for showing often undergo shoeing procedures that lead to lameness. In the show ring, the bigger the foot, the better, so owners commonly let the hooves grow out to enormous sizes — much larger than is healthy for long-term soundness. Show shoes are quite heavy, as well, to accent the animal's foot action. Show shoeing predisposes an animal to navicular syndrome, ringbone, sole bruises, and separation of the wall from the sole.

Feet that are too small are often far more troubling, as they must bear proportionally more weight per square inch. Small or narrow feet do not expand when bearing weight, making them poor shock absorbers. Feet may also be mismatched, often a result of chronic pain that causes a horse to bear more weight on the unaffected foot. The sore foot has reduced heel expansion, the frog shrinks, cir-

A mushroom foot is excessively large and shallow, causing loss of concavity.

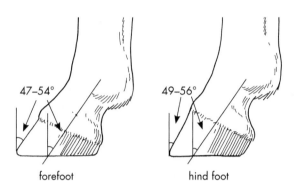

The angle of the hoof should be the same as the angle of the pastern.

culation decreases, and the foot becomes taller and narrower.

Hoof cracks — also called toe, quarter, or heel cracks — can be deceptive in their extent, and should be treated promptly by a veterinarian or farrier. They may be caused by improper shoeing, poor nutrition, abnormal foot conformation, injury to the coronet, or abnormal moisture levels, as well as excessive concussion from, for example, trotting on hard surfaces.

## Ready to Purchase

Now that you know what to look for in your ideal horse or mule, the obvious next questions are: where do I find this animal and how much will he cost? To be sure, you don't just drive over to the nearest dealership, tell them your specs, and drive off with your purchase. Looking for the right animal takes time. Don't be in a hurry while you are looking. You will want to be part detective and part shopper during this process, and the detection will present itself as your largest challenge.

### WHERE TO LOOK

Start your search by contacting draft breed associations and asking whether there are breeders/trainers in your area. If these folks don't have horses or mules that fit your criteria, they may know of someone who does. They may also know of grade animals for sale that would fit your criteria. Always keep your ears to the ground for word-of-mouth sales. If you hear about someone with a promising-sounding animal for sale, call around and get some info about the owner's training skill and the animal's skill level.

Auctions are great places to find rejected animals being sold by anonymous sellers. At an auction you could find a gentle, calm, well-mannered, beautiful animal one night, but wake up to a perfect devil the next morning, as a result of a little sedation used during the auction. Animals bought at auctions may be perfectly sound when going through the sale ring, but come up dead lame the next day, due to a good shot of anti-inflammatory drugs used during the auction.

Obtaining a thorough and accurate history on an animal being sold through an auction is extremely difficult. At most auctions you will not have the opportunity to observe the animal working or to work him yourself. So, although you may pay less for animals purchased at auction, you get what you pay for.

With all that said, you do have some good auction options, providing you take a seasoned teamster with you to thoroughly check things out. The Amish of Ohio, Iowa, and Indiana regularly put on great auctions where each animal is sold with a detailed list of what he can and cannot do, as well as his problems. This system, of course, relies on the seller's honesty, but in general the Amish stand behind their word, and their animals are some of the best working drafts around.

Check the advertisements in *Rural Heritage, Small Farmer's Journal,* and *Draft Horse Journal*. The sole purpose of these publications is to support the draft horse and mule scene, and they often feature great animals for sale. The Horse Team for Sale and Mule Team for Sale directories on the *Rural Heritage* Web site regularly feature well-broke teams that have been trained and used for such jobs as farming and carriage service, and drafts are listed for sale on several other equine-related Web sites.

## WHAT YOU WILL SPEND

You have finally found an animal that fits your must-have list relative to personality, temperament, conformation, age, and ability. Now the big question: What will you spend? If only horses and mules were nicely subdivided into heavy truck, light truck, sport utility, sedan, and sports car. Like motor vehicles, they come with every imaginable price tag, which are based on many factors: location, coloring, breeding, disposition, training, trainer, and more. As with any other good or service, supply and demand drive the market and thus the price of the draft horse or mule.

You can get a general idea of current prices, ability, and concentrations of draft horses and mules at many Web sites listing draft animals for sale. Attending auctions is another way to learn about current prices. If you want to purchase better bloodlines for breeding, expect to pay much more. A well-shown Belgian hitch gelding can easily go for $25,000. Grade horses, particularly geldings, run in the ballpark of $2,000 to $6,000 for an excellent team trained for working.

You will quickly learn that the less common breeds, such as the Shire, American Cream, Suffolk, Clydesdale, Fjord, Haflinger, and American Spotted Draft cost quite a bit more than the average-bred Belgian or Percheron. The bloodlines of these less common horses are less diluted, meaning that they are still dominated by the breeder of excellent pedigrees and thus command a higher price.

Versatility of ability also influences price — you pay for experience. A well-trained, trustworthy team with years of work invested in them are naturally worth a good bit more than a green team.

At one time a good draft mule was far more expensive than a good draft horse (hence much criticism of federal overspending after the United States Army's decision to use only mules during the Civil War), but that does not hold true today. The price of a good, well-trained draft mule team, without any fancy coloring, will be in the same ballpark as a good draft horse team. If you are assessing your animals by the pound, however, you *will* pay more for mules, since they weigh less than their draft horse counterparts. And unusual color or flashy markings definitely boost the price.

> ## Prepurchase Exam
>
> **When a horse or mule** meets the criteria on your personal list, and the price is right, have a prepurchase exam done by your veterinarian. Many people believe the vet should be called out only in times of emergency, a belief that can lead to a less-than-optimal purchase. You do not want to wake up the morning after bringing home your perfect equine, only to discover that he had been tranquilized and is now a maniac that you can't get near. The insurance value of a prepurchase exam by an equine veterinarian is always worth the cost.

Finding a good team of draft equines is a matter of careful research and study.

# DRAFT POWER

A team of draft mules pulls a side delivery rake.

**A.** Many teamsters use their draft horses for carriage hire for weddings, funerals, and other occasions.

**B.** A well-trained team may be handled by even a young driver. This pair is pulling a maple sap wagon.

**A.** What could be more fun than gliding over the snow in a horse-drawn sleigh?

**B.** This handsome Clydesdale is turned out in fancy show harness and put to a Meadowbrook cart.

DRAFT POWER

**A.** A team of Percheron draft mules pull a walking plow at a field day.

**B.** This pair of Brabant Belgians with their foal are pulling a six-row weeder.

**C.** A team of bay Percherons pulls a corn harvester.

**A.** A sturdy team of heavyweight draft ponies puts all their might into moving a load at a pulling competition.

**B.** This driver and his handsome pair of Spotted Drafts are demonstrating the use of a manure spreader.

**A.** Draft horses and draft crosses make excellent riding partners in a variety of disciplines.

**B.** Amazing scenery awaits those with a good pack string.

**A.** A Percheron team with fly nets pulls a sulky plow.

**B.** Two Belgians bring harvested grain to the threshing machine.

**C.** The unbridled power of the draft equine is truly awesome.

**A.** A team of Belgians pulls a disc to break up the soil.

**B.** Handling four horses in a hitch is not a feat for the novice teamster.

**C.** This farmer is cultivating corn with three abreast.

The Percheron is an ancient breed that originated from a small district of France known as Le Perche. Legend has it that the Percheron is the only draft breed with Arab blood, and its grace and liveliness attest to that heritage. The Percheron has size, action, spirit, and a good dose of tractability. It has a compact, muscular body, a fine head with straight profile and large eyes, and clean legs with little feathering on pasterns. It is generally lighter in weight than the Belgian, but is an excellent option for the small-scale farmer who enjoys spirited and beautiful horses. This breed is typically black or gray, although occasionally you'll see a sorrel or a bay.

## Black Percherons

The gene for true nonfading black hair and skin pigment is rare in horses. More common is gray, a dominant gene occurring frequently in the Percheron, particularly the American Percheron. Gray horses are born dark — black in the Percheron's case. As with any horse born genetically gray, the dark foal lightens to gray by the time he is around four years old, and then continues to lighten to nearly white as he ages. A true black Percheron is hard, but not impossible, to find.

## CLYDESDALE

| | |
|---|---|
| Origin | Clyde River Valley of Scotland |
| Influences | The Great Horse of the Middle Ages, Flemish and Dutch stallions, Shires |
| Weight | 1,700–2,200 pounds (770–1,000 kg) |
| Height | 16.2–18.2 hands |
| Color | Bay, brown, roan, or black with white blaze and stockings |
| Temperament | Active, gentle, responsive |
| Uses | Heavy draft; carriage hitches; show |

The Clydesdale, or Clyde, is the third most numerous breed of registered draft horses in the United States. It is a native of Scotland, where it was bred for heavy agricultural and transportation work. The Clyde is probably the most well-known draft breed, made famous by Budweiser advertising.

With its straight, angular body, the Clyde makes an excellent halter showing horse and is often seen as a high-performance show horse. It has a long, well-muscled neck, straight legs with strong, large hooves, a straight profile, and large eyes. Its flashy coloring, long silky feathering, and animated way of moving make it a popular hitch horse.

The imposing Clydesdale is seen more often in large show hitches than pulling farm equipment.

The long, flowing hair on the legs of Clydesdales and Shires, known as feathering, is beautiful when kept clean and white, a feat that's all but impossible in agricultural situations. In addition, heavily feathered horses are susceptible to dermatitis verrucosa, an irritation to the lower leg commonly known as scratches or greasy heel, which is treatable if the underlying problem (often leg mites) is correctly diagnosed.

## SHIRE

| | |
|---|---|
| Origin | England |
| Influences | Friesian, Great Horse, Old English Black |
| Weight | 1,600–2,200 pounds (725–1,000 kg) |
| Height | 16.2–19 hands |
| Color | Black, brown, bay, sometimes gray; often white markings on the legs and face |
| Temperament | Docile, kind, hardworking |
| Uses | Heavy draft; crossbreeding with hunters and Thoroughbreds to produce large, tall jumping/riding horses; driving; show |

As with most draft horses, the Shire's huge size belies its kind nature and willing temperament.

One of the largest horses in the world, the Shire is a massive, English-bred horse that looks much like the Clydesdale. The Shire needed steadiness and strength for working on docks and moving loads through badly paved city streets, the breed's main purpose in the 1800s. Often standing as tall as 19 hands, the Shire has an enormous barrel, a muscular body with long, powerful legs, and heavy feathering. It has a powerful neck and a surprisingly delicate head with long, fine ears. American Shires are most often black with white points, although some are bay; gray and chestnut also conform to the breed standard.

The immense size and copious feathering of the Shire made it a less desirable breed than non-feathered choices for agricultural work in the United States, and thus few were imported. Today the Shire is most often used as a hitch horse or for carriage service, where its long legs and flowing feathers show off its flashy gaits.

## SUFFOLK PUNCH

| | |
|---|---|
| Origin | Suffolk and Norfolk counties, England |
| Influences | Flanders Horse, Flemish, Norfolk Trotter, Cob, Thoroughbred |
| Weight | 1,500–2,000 pounds (680–900 kg) |
| Height | 16.1–17 hands |
| Color | Chestnut, in seven shades from light to nearly black liver; small white markings on the face or lower legs are permitted |
| Temperament | Energetic, docile, willing |
| Uses | Agricultural; draft work |

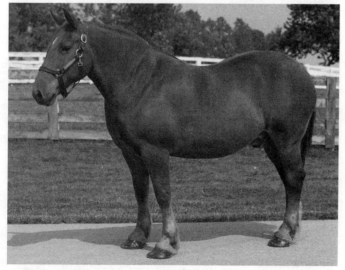

The Suffolk is a British breed that remains relatively rare in North America and throughout the world.

The Suffolk Punch, or just Suffolk, has a tenuous hold in North America, which is home to only 900 to 1,200 of these attractive horses. The exact number is not known because not all Suffolks are registered. Suffolk Punch draft horses are on the American Livestock Breeds Conservancy's critical list. A breed is considered critical if it has fewer than 200 annual registrations and a global population of fewer than 2,000 registered animals.

This breed was developed in England specifically for agricultural work, rather than hauling heavy loads, and is the only draft horse breed able to make this claim. The Suffolk is stout and low to the ground, with a compact, well-rounded, deep body, and a long, chiseled face. There is no feathering on the pasterns. Suffolks are a one-color breed — always sorrel (also known as chestnut), in seven recognized shades: red, gold, copper, yellow, liver, light, and dark. Some roans do exist and may be registered.

The Suffolk horse walks quickly and with purpose, rather than with the action and flash of other draft horse breeds. Bred to work in muddy, swampy conditions, its lack of feathering makes it a clean plow horse and its relatively small feet allow it to step nimbly between rows of crops. With their short coupling (often an attribute of a horse that eats less) Suffolks are excellent keepers.

## AMERICAN CREAM DRAFT

| | |
|---|---|
| Origin | Iowa; single foundation mare |
| Influences | Belgian, Percheron |
| Weight | 1,600–1,800 pounds (725 to 820 kg) |
| Height | 15–16.3 hands |
| Color | Shades of cream; pink skin; amber eyes; white mane and tail; white markings |
| Temperament | Calm, trustworthy, willing |
| Uses | Light agricultural work; pulling |

An original American breed, the American Cream Draft coloring is not albino, but the result of a dilution gene.

The American Cream is the only existing draft horse to boast 100 percent American roots. As a young girl, I once visited an older Amish man in Tennessee who had a team of American Cream horses. The animals intrigued me with their amber eyes, delicate coloring, and glossy, creamy coats. I did not know then how fortunate I was to see those horses firsthand; I haven't seen one since. As a fairly new arrival on the scene, the American Cream is the least numerous of the draft horse breeds, with fewer than 400 horses currently registered. The breed is on the American Livestock Breeds Conservancy's critical list.

According to history, a mare named Old Granny was the grandmother to this breed. She was purchased in 1911 by Iowa stock dealer Harry Lakin and later sold to the nearby Nelson Brothers Farm. By 1946 she had produced enough light-colored foals to be considered the foundation dam of a new breed. Her description has become the standard for the breed registry: rich cream color with white mane and tail, pink skin, and amber-colored eyes. These horses are not albinos. Their color results from the champagne gene, one of four dilution genes found in equine coat genetics. The champagne gene produces pink-pigmented skin. While the color of the coat is the primary requirement of the breed standard, the pink skin, which produces the particularly rich cream coat color, has been designated as the second most important feature.

Muscular and compact with powerful shoulders and hindquarters, the Cream has strong legs; a long, muscular neck; and a medium-sized head. The Cream is known for its docility, uniform physical characteristics, and willingness to work.

The Spotted Draft is a relatively new breed in the United States, but draft horses with spots have been around for centuries.

## NORTH AMERICAN SPOTTED DRAFT HORSE

| | |
|---|---|
| Origin | United States |
| Influences | All draft horses, pinto, or Paint saddle horses |
| Weight | 1,200 pounds–2,000 pounds (545–900 kg) |
| Height | 15–17 hands |
| Color | Any base color, most commonly black, bay, or brown, with white; frequently tobiano, sometimes overo |
| Temperament | Willing, docile, quiet |
| Uses | Heavy to light agricultural work and pulling; parades; commercial vehicles; riding; show |

## The Genetics of Color Dilution

The study of equine coat genetics is fascinating and complicated. Two researchers are mainly responsible for what we know today about equine coat genetics: The late Dr. Ann T. Bowling of the University of California, Davis, and Dr. D. Phillip Sponenberg of Virginia Polytechnic Institute.

Four genetic combinations are responsible for the dilution of coat color. The cream gene's two alleles (individual genes) combine in three different ways to alter red and black pigment. The first combination dilutes the red pigment to a coat gold with cream to white mane and tail (palomino).

The second combination dilutes the black pigment. Depending on the combination of the two separate alleles, the colors produced will be a lighter body with dark points, such as the buckskin, or a smoky black color.

The third combination of the cream gene, a double dilution factor, dilutes the red pigment to a pale cream. The eyes are often amber, blue, or green. This pigment characterizes the American Cream draft horse.

The fourth gene responsible for dilution of coat color is the dun gene. The coat body color evidenced by the dun gene dilution factor is pinkish-red to yellow-red, yellow, or mouse gray. The points are dark, and often a dorsal stripe, shoulder strip, leg barring, or other primitive markings are present. The coloring of the Norwegian Fjord horse is a result of the dun gene.

Heavy horses with splashy pinto coloring have been documented for centuries, but it was only in 1995 that the North American Spotted Draft Horse Association was formed to preserve and promote colored draft horses and establish standards for breeding. In 2002 a second organization, the Pinto Draft Registry, was formed. Spotted Drafts, also known as Spots, are a color breed, meaning that they are registered based primarily on color rather than conformation or bloodline. A flashy pinto horse is often used as one of the parents. The other is usually a Percheron, since Percherons throw colored foals more frequently than do other draft horse breeds.

These horses are classified according to size, percentage of draft blood, and amount of white coloring in the coat. Breeding stock does not all have to show any color, but must have at least one parent that does. These horses may be bred from any heavy breed, as well as certain registered light breeds specified by the individual registries.

### Overo Lethal White Syndrome

**The Spotted Draft** usually has some American Paint Horse or other color breed genetics. Therefore, the appearance of the overo lethal white syndrome gene is always a possibility. This syndrome results from two copies of the overo gene, producing all-white foals with malformed gastrointestinal tracts. Such foals always die. Before purchasing a spotted draft for breeding purposes, ensure that a DNA test has been done to rule out the presence of this gene.

## HAFLINGER

| | |
|---|---|
| Origin | Austrian Tyrol |
| Influences | Arabian |
| Weight | 700–1,200 pounds (325–545 kg) |
| Height | 13–15 hands |
| Color | Any shade of chestnut, from honey blond to dark chocolate; often a deep golden color; may have some dappling; flaxen or white mane and tail |
| Temperament | Willing, hardworking, gentle, docile |
| Uses | Light agricultural work; packing; riding; jumping; show |

The Haflinger is a small, stout horse native to the Austrian Tyrol, a precipitously mountainous region. The breed was used for packing along steep, narrow, winding mountain trails, as well as for plowing, driving, and riding. This all-around horse was developed to show strength, agility, and a great willingness to work. The breed's compact, robust, and clean build, along with a pleasant temperament and desire to please, make this the perfect option for the small farmer with a few acres.

The handsome Haflinger is small but strong for its size and a good choice for a beginner teamster.

Contrary to their appearance, Haflingers are not miniature Belgians. Other than their wonderful personality, willingness to work, and similar coat color, the Haflinger horse shares no heritage with the Belgian horse. Haflingers are not "half" any other breed, either; they are their own breed.

The popularity of this breed is on the rise in the United States, where some 18,000 Haflingers are registered. Many people see this breed as an affordable option to the bulk and appetite of a heavier draft horse. Haflingers also have an avid following as driving, riding, and packing animals. All of the Haflingers I have worked with have been gentle and willing. They also have plenty of pep, and with a little convincing will run barrels and become great jumpers.

The Haflinger comes in only one color — chestnut (sorrel) with flaxen mane and tail. Its height is measured in inches, like a pony, rather than in hands, like a horse, but because its accepted height of 54–60 inches may exceed the 58-inch maximum height for a pony, the Haflinger is referred to as a horse — albeit a small one. Many Haflinger breeders today are shifting their emphasis toward lighter, showier horses, with the result that the breed is losing some of its stout draft qualities. Therefore, like the Belgian and the Percheron, the Haflinger is generally recognized in two types: the old-style, stout, short-legged type and the modern, taller horse with more leg action. Both types are well muscled with a powerful build, sturdy bones, and relatively large hooves. The head is delicate and Arab-looking; the pasterns have light feathering.

## DEDICATED TO DRAFTS

**Tommy Flowers**
**Blackville, South Carolina**

Type of Farm: 28 acres (11 ha) planted in hay, corn, and oats; 1 acre of garden crops

Animals Owned: Brabant Belgians

You do not have to be raised with draft horses to use them and to use them well, as proven by the lifestyle of Tommy Flowers and his family. Although Tommy was raised with Tennessee Walking horses, he is not impressed by their gait, their artificiality, or their use. While he was growing up, he decided that he wanted nothing to do with horses.

When Tommy was 40 years old, his family went to a Christmas tree farm to collect their annual tree. A team of Belgians pulled the tree-harvesting wagon. Tommy became enamored with the animals and proclaimed to his family that he could do something like that. Eventually two well-trained draft horses came to live on the Flowers' small farm. These well-educated horses, Jed and Nell, taught Tommy much of what he knows about driving and working horses. In addition to acquiring these horses, the Flowers family went to numerous plowing days and living history farms, always asking questions. Sometimes Tommy had the privilege of working a seasoned team with a seasoned teamster. In this manner he learned about equipment and learned the lingo of the draft horse teamster. With this combined education from his first team and from experienced teamsters, Tommy has trained all of the horses he has subsequently owned.

One year Tommy saw a pair of Brabant horses on a calendar and decided they looked like a draft horse should; he was going to own them. He did some serious searching and made a 1,000-mile drive to Vermont to purchase a pair of weanlings, filly

The Norwegian Fjord is easily recognized by the traditionally roached mane with a black stripe running down the crest.

## NORWEGIAN FJORD

| Origin | Norway |
|---|---|
| Influences | Ancient wild horses |
| Weight | 900–1,200 pounds (400–545 kg) |
| Height | 12.2–14.2 hands |
| Color | Dun in various shades of cream, silver, red, gray, or (rarely) yellow; dark dorsal stripe; dark mane blending to silver at the neck; dark tail; often dark legs below the knee, frequently with leg stripes |
| Temperament | Agreeable, lively, gentle, willing, easily trained |
| Uses | Light draft; riding; driving; packing; show |

Bulah and colt Rocky. Thus began a fulfilling relationship with Brabant horses that has led Tommy and his family to correspond with other Brabant owners in European countries.

A main topic of interest in their conversations is the Brabant's genetic predisposition to chronic progressive lymphedema (CPL) a condition affecting the legs of nearly all Brabant horses today. Rocky, with 100 percent European bloodlines, had CPL, but his frequent exercise seemed to keep the disease from advancing too rapidly.

Unfortunately, Rocky died at the age of 11 from an undetermined cause. As Tommy and his family dealt with the loss of their pride-and-joy stallion, they considered importing a stallion from northern Europe where the breed is less afflicted with CPL. And so, in late 2007, Tommy spent four weeks touring northern Europe in his second serious search for a Brabant stallion, and in early 2008 a new colt arrived at the Flowers' farm to be paired with Rocky and Bulah's youngest filly.

Tommy and his family have passed on their dream of horse power each year at their Old Time Horse Farmer's Gathering, an event they held from 1996 to 2007. Twenty or so draft horse and mule teams and two yokes of oxen (along with their teamsters) came together to show visitors how to farm "the old way." Tommy planted hay, corn, oats, potatoes, peanuts, sugarcane, peas, and buckwheat in the spring and harvested those crops with horses at the gathering. School children and adults flocked to this event to watch a grain binder, corn binder, manure spreader, mower, plow, hay rake, hay tedder, and many other pieces of equipment in use.

Promoting Brabant horses continues to be a focus for Tommy. He has participated in numerous Horse Progress Days where thousands of folks from around the country are awed by his beautiful, powerful, and gentle breed of draft horse. Tommy's two daughters Annie and Hannah are both involved with the horse-drawn lifestyle. Annie, who graduated from the University of South Carolina with a Bachelor of Science in business administration, has attended numerous plowings with her father. She competes in plowing and driving events, and often wins. Hannah also competes and once experienced the thrill of winning a driving event against her older sister.

The Norwegian Fjord boasts a long and rich heritage. It is one of the oldest and purest breeds, closely resembling ancient horses painted by Ice Age artists 30,000 years ago. This breed most likely originated with Przewalski's horse, the wild horse of Asia, the primitive influence of which is evident in the Fjord's unique coloring. The zebra stripes on the legs, dark stripes across the withers, and dorsal stripe running from the forelock down the neck and back to the tail are all markings passed down from this primitive heritage. This marking is due to the dun gene, a pigment dilution gene. (See box on page 70.) Ninety percent of Fjords are brown dun in color, with the other ten percent being red dun, gray, pale dun, gold, or yellow dun.

Like the Haflinger, the Fjord may be categorized into lighter riding stock and heavier draft stock. The Fjord makes an excellent choice for the small-acreage farmer and is known for its willingness to work, stamina, and vigor. It has a compact, muscular body with deep girth and short, closely coupled back; heavily boned legs and medium-sized black feet; low, strongly muscled withers; crested and well-muscled neck; medium-size head with broad, flat forehead; and a straight or slightly dished face with large eyes and small ears. The thick, wavy mane is typically roached for showing purposes. The Norwegian Fjord Horse Registry estimates a current population of 5,250 registered horses in North America, with approximately the same number of unregistered animals.

## Fjord Mules

Fjord mules are few and far between, because they are contraband in the eyes of the Norwegian Fjord Horse Registry. Article 1, Criteria for Registration, of the association's Rules for Registration states that the registration of any mare or stallion used for crossbreeding with another breed or species will be withdrawn. Members owning Fjords used for crossbreeding will also lose their membership. The breed standards remain strict to retain purity. Few members are willing to risk censure to discover what a Fjord mule is like.

## What's a Grade Horse?

A grade horse does not qualify for breed registration because of mixed breeding, failure to register (some registries have deadlines for registration and will not bend the rules if those deadlines are missed), or failure to meet breed specifications. In short, if the horse does not have papers, even if the owner says he can get them, then it is not registered, cannot be sold as registered, and cannot be bred as registered.

Does failure to belong to a breed registry doom these horses to a life of thankless drudgery and obscurity? Absolutely not. Many teamsters choose grade horses because they are capable of exactly the same amount of work and often cost far less than a purebred horse.

A good team of grade geldings may prove more economically feasible to the small farmer who does not want to raise foals and therefore doesn't want

A draft cross makes a nice driving and riding horse, combining the power of a draft horse with the refined looks and flashier gait of a lighter horse.

to spend the money on mares. The grade gelding was by far the most popular draft animal in the old-order Mennonite community I lived near in Kentucky, providing cheap labor unhampered by fancy pedigree, foaling, or other interests. The horse was there to work.

The draft cross, technically still a grade horse but often considered more refined if the exact parentage is known, is the result of crossing a draft horse and a light saddle horse, such as a Quarter Horse or Thoroughbred. The resulting animal typically has a fairly docile spirit, willingness, strength, and considerable functionality.

Draft crosses are popular with dude ranches, packers, and folks interested in an all-purpose horse. Many hunter/jumper riders are choosing the draft cross for its amazing jumping prowess and strength. The balance, grace, and size of the draft cross also make it a superb dressage horse. This horse is heavier, stronger, and gentler than the typical saddle horse, yet lighter, quicker, and easier to feed than the draft horse. Draft crosses make great carriage horses, since they fit between traditional single-horse shafts.

## Preservation of Breeds

Although it sounds ridiculously intuitive, many people don't realize that without producing and registering foals each year, a breed will eventually die out. Suffolk Punch and American Cream owners, whose respective breeds have dangerously low populations, know this truth and are aggressive and determined to keep their breeds alive. An informed, safe, and responsible breeding program is essential for breed preservation.

The role of a novice in the draft equine world is not breed preservation. However, if you have had previous experience with equines, in a light horse breeding program, for example, and are interested in breeding your draft mare, below are a few suggestions to keep in mind.

### BREEDING WISELY

Like any other routine health care, reproduction should be managed carefully and thoughtfully. Stay abreast of the general equine market, as well as the niche you plan to fill with your foals. The excess of poor quality horses listed in every *Nickel Trader* and

# A KNOWLEDGEABLE JUDGE

**Gene England**
**Winder, Georgia**

Show Experience: 30+ years

Judging Experience: 20+ years

Equine Experience: Has raised Belgian hitch horses and Belgian mules

Gene England knows a thing or two about the way an equine should act and look — he has judged horses and mules for more than 20 years in various shows around the Southeast and can quickly sum up their faults as well as their strong points. Gene grew up in Georgia, farming with mules on his parents' place. When he was young he had an uncle who judged equine events at county fairs and Gene took a liking to the activity.

Eventually Gene began showing his own hitch of high-quality Belgians. The combination of their conformation and his driving skill allowed him to place first in many shows around the region. As his wins became common, folks started asking him to judge. After all, a winner is winning because he has already judged his own animals as capable of being number one. And so Gene's career as a draft horse and mule judge began. Each year he travels to five or six shows in Georgia and surrounding states, handing out ribbons to the most qualified animals.

The horses that Gene sees at shows are usually driven by skilled teamsters. However, he is passionate about introducing youngsters to the world of drafts and has a good idea of the qualities that a novice should look for in a horse. To start, he likes to see a face that is not dished, but without too much of a Roman nose; a prominent Roman nose can indicate a high-spirited, aggressive horse. The eyes should be gentle and set wide apart, the ears clean, long, and rolled up tight (deeply concave).

The neck should be long enough to hold the head high and allow plenty of neck to show despite the collar, while the back should be short to allow for better agility.

Viewed from behind, the hocks and feet should be close together; from the front, look for a wide, powerful chest and large hooves; small feet on a big horse are like small tires on a big truck. From the side the leg should be straight up and down from the hocks to the ankle.

So what breed has all these characteristics, as well as the personality a novice needs? Hands down, the Belgian wins the blue ribbon, says Gene. The Belgian is easy-going (although some of the hitching horses can be quite high-spirited), dependable, forgiving, and willing. Gene suggests that you would do yourself a favor by finding an old, experienced team of Belgians, Suffolks, or Haflingers that have had regular, steady work under good hands. Such a team will get you started in the right direction on your journey to draft horse use.

Based on his experience with showing, judging, and working draft horses, Gene England rates six of the common draft horse breeds as follows:

| | |
|---|---|
| **Belgian** | Docile, forgiving, dependable, and a great horse to train the novice |
| **Suffolk** | Gentle, smart, strong, and willing |
| **Haflinger** | Sweet, dependable, and quite strong for its size |
| **Percheron** | High-spirited, but manageable for the confident novice |
| **Clydesdale** | Too high-spirited for the novice handler unless exceptionally well trained |
| **Shire** | Much like the Clydesdale; not for a novice unless exceptionally well trained |

existing in too many backyards is a testament to the concept of breeding without thought of the future. Buy only quality stock, breed only to quality stock, and produce only quality stock.

Breeding horses is no light matter. You must address many serious considerations and preparations in order to be ready. In addition to ensuring that the mare and stallion are genetically healthy, you must consider breeding options, hauling and boarding, the financial commitment of caring for a pregnant mare and her foal, and the potential complications both in management and foaling.

## CONSIDERING GENETICS

If you do decide to breed your mare, consider genetic problems associated with the breed, including conformational and other defects. You hold the fate of a potential foal in your hands, and must ensure that it will not grow up malformed, sickly, or diseased because of a poor genetic choice in the mare or stallion.

Any animal with known genetic disorders should not be bred. For certain breeds, a DNA test that picks up defective genes is a good idea. In particular, consider DNA tests in a breed predisposed to genetic disorders like JEB and lethal white syndrome. Equine polysaccharide storage myopathy (EPSM) has ravaged the draft horse world, causing muscle wasting and related symptoms. This disease is likely to be hereditary, although a DNA test has not yet been confirmed. However, if a horse is suspected of having EPSM, think carefully about the management involved and the ethics in passing this serious problem to another owner.

## Why Breeds Matter

People become attached to certain breeds for different reasons. Some choose a breed because it is on the brink of extinction and they feel compelled to preserve the genetic heritage. Others raise a certain breed because their parents did. Still others pick a breed because it is flashy and shows well. Choose a breed because it will work well for you. If you are unsure about which breed to choose, consult a well-respected draft horse judge who will tell you what you need to hear.

---

### Aiding the Cause

Although as a novice you should gain equine experience before getting involved in the complex world of breed preservation, you can support the effort to preserve an endangered breed by purchasing your horses from an established breeder.

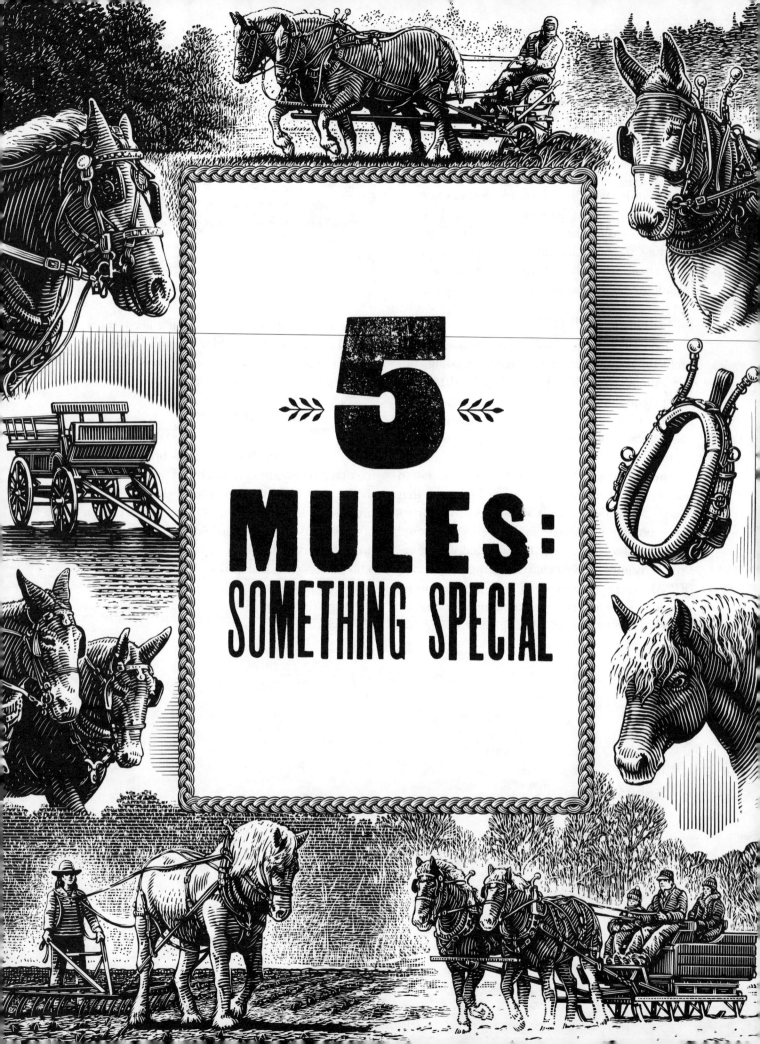

Bobby was a loud, grizzled, beer-belly-sporting native of Kentucky, an enormous mountain of a man who I had the distinct pleasure to know a number of years ago. Bobby was getting up in age and down in health, and as a result no longer used any of his 50-odd horses, mules, or donkeys to work the land. Bobby was the stereotypical horse and mule trader; many people held a lifelong grudge against him for cheating them out of good money for a worthless horse or mule. Although you could never be quite sure if he was telling a big one or the honest-to-gosh truth, the fact remained that Bobby had some of the nicest Percheron, Belgian, and Mammoth Jack stock for miles around.

As I got to know Bobby, I realized he had these top-quality Percheron and Belgian mares for one reason: He loved a beautiful mule. He was plenty fond of his high-stepping mares, but in his words, the mules they produced made the money. Bobby's dad had begun the tradition of owning fine horses and mules. He had been a pioneer in the area, using horses and mules to carve fields out of dense deciduous forests to provide hay, pasture, and food for his horses and his family. He used the horses for the flashier work around the farm and town. He used the mules for times when push came to shove. The mules could stand the heat and humidity much better than the horses, but the horses caught more eyes. For Bobby and his dad, horses and mules coexisted, each with a valued and appreciated role.

Let's not attempt to settle the age-old debate between horse and mule owners. Instead, let's consider some positive and negative qualities of an animal that often gets short shrift from horse owners. No animal is saint or devil; each has its individual personality traits. In learning about the differences and similarities between draft mules and horses, remember that generalizations don't always apply to any one specific individual. Whether you choose horse or mule, both sides have plenty of arguments in their favor.

As with many historically controversial subjects, mule ownership has generated numerous clichés. Mule owners have been dubbed more patient, stubborn, mean, or conniving than horse owners. Without going into the topic of character, it's safe to say that a mule owner faces an extremely intelligent animal that can meet or match any tricks the human can send its way. The horse, on the other hand, is often more willing to go along and avoid making waves.

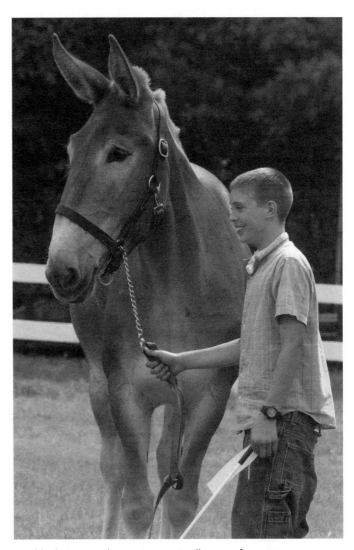

Just like horses, mules participate in all types of competition. Here a young boy handles his draft mule in a halter class.

## Origin of the Mule

A mule is a cross between a horse mare and a male donkey (jack). This breeding usually produces a sterile hybrid because of the difference in the numbers of chromosomes in the cells of the donkey (62 chromosomes, 31 pairs) and the horse (64 chromosomes, 32 pairs). The resulting mule has only 63 chromosomes, an odd number that prevents the mule from forming matched pairs in the early stages of reproduction, thus killing reproductive cells. The rare occurrence of a mare mule producing a foal, reported every now and then, is considered a fluke.

A similar but less common animal is a hinny, the result of mating a female donkey (jennet) and a horse stallion. Some people swear that they can identify a hinny as being slightly more horselike than a mule; others are just as adamant that the difference between a hinny and a mule is too subtle to detect. Some so-called mules, in fact, are actually hinnies.

Just like horses, mules may be registered to acknowledge and account for quality bloodlines. And like horses, they come in a wide range of sizes and colors. A draft mule weighs 1,200 pounds (550 kg) or more. In contrast, draft horses often top out at a ton (900 kg) or more.

### Where Does that Sound Come From?

The mule's unmistakable hee-haw — a cross between the donkey's bray and the horse's whinny — is enough to put a smile on the faces of both knowing and unknowing listeners. The sound is quite difficult to describe, but might best be defined as a long, ragged intake of air accompanied by a series of notes ascending both the musical scale and in volume, followed by a jerky, choky, expulsion of air. Although the hee-haw is by far the most distinctive noise that mules make, they can also snort and squeal.

The mule uses his hee-haw similarly to the horse's nicker, as a greeting to other animals or his owner, when anticipating food, or to let you know that you are off schedule. Donkeys and mules both develop incredible clocks in their heads and will vocalize at nearly the same time each day when they expect food.

While often seen as a companion animal and farm pet, the versatile donkey has many talents.

## THE DONKEY SIRE

Donkeys have a mixed origin and do not come in particular breeds, as do horses; rather, they are characterized by place of origin (such as the Spanish donkey), size (such as Miniature or Mammoth), and coloring (such as spotted). The Mammoth, most often the sire of any draft mule, originates from Spain, France, the Cape Verde Islands, and Portugal. Other influences on the Mammoth worth noting are some of the key sires from Catalonian, Andalusian, Majorcan, Maltese, and Poitou stock. The Mammoth is among the largest donkeys, with adults weighing between 900 and 1,400 pounds (400 and 650 kg) and standing 15 hands.

The Mammoth donkey has a shaggy, bulky, and impressive body which, coupled with his resounding vocal ensemble, has frightened more than one unsuspecting person nearly witless. The jack has a disproportionately large head compared to his body, an enormous convex Roman nose, staggeringly long ears, and long, coarse hair draped over a lean body with a high, narrow, pointed croup. Mammoths may be gray, black, spotted, sorrel, or other colors. True to their native ancestors, they are extremely cautious, deliberate, persistent (also called stubborn), willing, and loyal. Be assured that if this animal does not want to do something, you won't change his mind.

Mammoth donkeys are used for draft work, riding, packing, driving, and, primarily, making draft mules. Mammoth jacks and jennets may also appear in a number of show events.

## THE HORSE DAM

The dam of a draft mule may be any draft breed, and the offspring is therefore referred to by his dam's breed — Belgian mule, Percheron mule, and so forth. The mule retains many of the mare's physical characteristics, including structure, muscling, and color.

One interesting result of breeding a jack and a mare is that the mule may wind up being a few inches taller than his dam, but somewhat leaner through the front muscling. Compared to the dam, the mule has less leg feathering, far longer ears, a more Roman nose, and generally a longer back. The mule's feet are much smaller than the draft horse's.

## BIRTH AND REPRODUCTION

One significant difference between horses and mules is that the horse can reproduce, while the mule (except in extremely rare cases) cannot. The conception rate of a mare with a mule foal is approximately 10 percent lower than that of a mare with a horse foal. The gestation period of a mule foal may be anywhere from 11 to 14 months — a considerable period of uncertainty spent wondering when your mare will foal. Mule foals are smaller than horse foals, making them a great option for the first-time foaling mare.

Even though mules are sterile, they have active sex hormones that affect their behavior, just as do mares and stallions. Molly mules cycle just as mares do and can exhibit the same mood-swinging hormone fluctuations that occur with estrus. John mules, since they have no reproduction potential, should always be castrated; an intact male mule is usually difficult to handle and can be a dangerous liability on a farm.

### The Issue of Anesthesia

**An interesting difference** between a horse and a mule is that a mule can take up to three times the amount of anesthesia needed by a horse. Yet the mule tends to recover more rapidly from anesthesia than the horse, which could create a potentially dangerous scenario if your veterinarian isn't familiar with working on mules.

Don't hesitate to ask your veterinarian about dosage and her experience with mules before your own mule is put under general anesthesia. Friends of mine learned the hard way, when they castrated their mule, that their veterinarian was not accustomed to working with these equines and found herself dealing with a half-awake animal in the middle of losing his studhood.

# Physical Characteristics of the Mule

The mule sports ears that are longer than a horse's, but shorter than a donkey's. Its tail and mane are sparse and coarse, and its coat not as shaggy as a donkey's. The mule inherits its long, sensitive ears and Roman nose from the donkey side of the family. The Roman nose is particularly desired by the mule producer — a mule is not a mule without the distinctly bulging facial profile. The length of the mule's head is determined by the length of both the mare's and the jack's head, and a good mule has a head that looks large for its body.

The mule inherits size and bulk from its dam, though a mule's neck is, on average, narrower than a horse's, requiring the use of a smaller collar than that used for the average draft horse. The mule also has a narrower, steeper croup than the draft horse, because the donkey lacks the fifth lumbar vertebrae. The lack of a fifth vertebrae causes the pelvic bones to be tilted at a steeper angle, and while the mule usually inherits the fifth vertebrae, it also inherits the donkey's steeper croup. A good mule has long, clean legs, sloping shoulders, and light muscling.

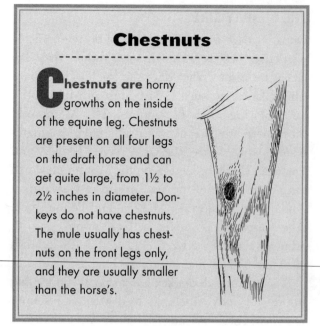

## Chestnuts

**Chestnuts are** horny growths on the inside of the equine leg. Chestnuts are present on all four legs on the draft horse and can get quite large, from 1½ to 2½ inches in diameter. Donkeys do not have chestnuts. The mule usually has chestnuts on the front legs only, and they are usually smaller than the horse's.

The mule, in general, has a far better foot history than the horse, inheriting tough, elastic feet from its donkey father. The mule's foot is narrower and smaller than that of the draft horse, and needs to be shod in only the roughest conditions.

# Healthy as a Mule

The typical mule displays a beautiful trait called *hybrid vigor*, making him incredibly hardy and disease resistant. Mules can withstand dry climates and irregular feeding far better than their colic-prone cousin, the horse. The belief that mules won't colic or founder, however, is a myth. They are just less susceptible, largely because of their inborn self-control when it comes to eating. Healthy mules sometimes leave feed in their boxes; if a horse did the same, you would wonder what was wrong. Although mules are less likely than horses to gorge themselves and get laminitis, they require less protein than horses, so excessive calorie intake leading to excessive weight gain can be a potential hazard for owners who are used to feeding horses.

Mules require the same health maintenance as horses, including vaccinations and parasite control. But thanks to the mule's strong sense of self-preservation, it will generally need to visit the vet less often than the more accident-prone horse. A

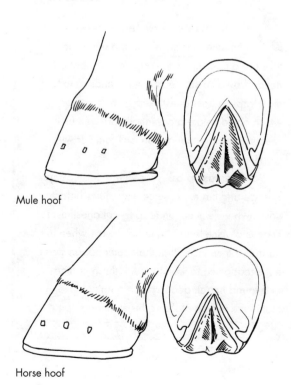

Mule hoof

Horse hoof

horse, treated well, usually lives from 24 to 30 years. A donkey may live from 30 to 50 years. If you cross the two, you will inevitably wind up somewhere in between, which is where the mule lands, living between 30 and 40 years.

My father and I once visited an elderly gentleman in Kentucky, whose middle-aged son was cultivating tobacco with a slow, carefully stepping mule. We learned that the son had been using the mule since he was a ten-year-old boy. I did some quick math and, incredulous, asked the mule's age. "Oh, I don't know, somewhere around 37 years old, I'd guess." And still hard at work.

## That Famous Mule Temperament

Mules are often associated with stubbornness, but muleteers like to think of the animal's ability to think for itself as intelligence. Indeed, the mule is an extremely analytical animal. Being a hybrid, the mule takes half its character from its mother, the mare, and half from its father, the jack. Donkeys are widely known for their careful and methodical approach to life. Donkeys may kick and snort, but their reaction to stimuli is vastly different from that of horses. The horse is an animal of flight; when it is threatened, it runs. The donkey, on the other hand, tends to use its brain to outsmart danger.

The mule retains this incredible sense of self-preservation through intelligence, and its methodical approach to life comes from its well-developed sense of self-preservation — a trait that keeps it from getting into as much trouble as the horse. If, for example, a mule is caught in a fence (which rarely happens), it will usually stand quietly, waiting for help. Most horses, by contrast, will panic, thrash around, try to break free, and often wind up with serious injuries.

Most muleteers will tell you that, while a horse may be bullied into doing something, a mule will not do anything until it accepts the activity being proposed, knows that it is safe, and feels good and ready to proceed.

Some muleteers find that the molly mule has a sweeter disposition than the john mule, but is moodier and more unpredictable from day to day and requires an easier hand during training. John mules are usually solid and dependable, but will test their owners for their entire lives. A john will work like a dream for days on end and then suddenly put an interesting twist on things just to see if you are

A well-trained team of mules can plow a field almost by themselves.

awake. A john can also develop a marked independent streak as it ages, and should be watched more closely. For example, a john that has been trusted to hang around a pack camp unrestrained for many hunting seasons, may, when he reaches the upper teens in age, suddenly decide that the next mountain range needs exploring and never bother to ask your permission.

### POUND-FOR-POUND WORK

Folks who work mules and horses routinely declare that the mule can outpull and outwork any horse of the same weight, and even some weighing far more. Pound for pound, the mule is stronger and more heat- and drought-resistant than the horse, and therefore was chosen as the beast of burden by many farmers in the South who could not afford the bulk and appetite of a draft horse, but still needed to accomplish a lot of work.

While visiting the Appalachian Mountains of North Carolina, I frequently saw older men in tattered, unbuttoned overalls, slowly guiding their mules between rows of tall, dark green tobacco. The plots of land clung to steep hillsides, and trying to navigate those fields with a tractor would have been quite foolish. The versatility of the mule was apparent on these plots of between one and two acres. Many of these mules had been working these same plots of land for 20 or more years. They could, in fact, just about cultivate the plots on their own. When I asked why the farmers chose mules over horses, I was told, in a tone used for young children, "Why, ma'am, this here mule is smarter than any ol' horse. He won't stomp the 'baccy and he'll work all day long without complainin'. The horse — why, he thinks he needs a break."

## Training a Mule

Mules can be so smart that they think they know better than their handler. In reality, they often do. If treated with respect and gentleness, the mule will be a constant and hard-working partner. The mule has a reputation for being a devout and almost doglike friend to the good owner, but vengeful and destructive to the abusive master. They can become quite attached to people who they respect. Consider this trait carefully when looking to purchase a mule; you do not want to inherit a mule that is the product of a bad previous owner, and you need to realize that developing a relationship with a mule will take time.

Mules seem to *want* a leader. They will often form dangerously strong bonds with horses, particularly mares, which can interfere with their respect for the human as a leader. For this reason, many muleteers pasture their mules separately from horses. The bond has its advantages, however. Many packers ride a horse and pack mules because they can tie the horse and let their mules loose at camp and never worry about the mules leaving.

The mule's intelligence allows it to pick up quickly on flaws in a trainer's technique. While a horse may take a while to realize that a trainer is not using proper technique, a mule will figure it out almost immediately and then exploit the weakness. Let's say a mule discovers that you are nervous around his back feet. He will use this weakness to intimidate you at every opportunity. If you usually turn right going out of the barn, after two or three times you will wind up with a mule that will *never* turn left out of the barn. If you drive your mule a few times to a certain point in the road, perhaps because of time constraints, and then turn around, by the third trip you could easily have a mule that will refuse to continue any farther down the road. If you typically stop fieldwork for noon lunch, then on the one day that you're in a hurry to beat an impending storm, your mules will stop at noon anyway, because you taught them to stop at noon.

---

### ◀ CAUTION ▶

#### Mules and Other Animals

Mules do well as pasture mates with horses and other mules, but are sometimes aggressive to smaller animals, including foals, and can be quite fierce with dogs and even children who enter their domain.

Mules require a handler with patience, humility, and intelligence.

## DIFFERENCES IN DEVELOPMENT

Although opinions vary, many muleteers suggest that since mules mature more slowly than horses, they also should be taught more slowly (in shorter increments) than horses.

The mule may take three or four years longer than the horse to be considered fully developed and well trained. A horse may start working at two years of age, but because the physis (area of cartilage near of the ends of the bones in young horses) closes later on mules than on horses, many trainers believe that the mule should be started later to avoid injury. A mule is not fully mature until it is around seven years old, and many mules are started too young.

Training a mule requires great structure, creativity, diversity, and variety to ensure that it does not develop bad habits. Most mule trainers agree that a mule needs a shorter, more succinct lesson than a horse. A distinct key to properly training mules is their inherent need for consistency. Mules quickly establish and rigidly adhere to patterns. Make sure the mule you purchase does not already have, and never develops, patterns that impede your goals.

Another difference is that most horses can stay with tedious repetition much longer than the mule, although the training of a horse should not be dull, either. For example, many horses do quite well with the round-pen method of training; unless the trainer seeks to be exceptionally creative, a mule can quickly become fed up with what it perceives as a senseless routine of going in a circle and can become quite obstinate. A mule needs short, creative lessons. Each training lesson should last no longer than 15 minutes for the first three years of the mule's life. Horses, in contrast, generally have at least a 30-minute attention span by the time they are two years old.

## THE MULE CONNECTION

Since mules outshine horses from health to hee-haw, you might wonder why everyone doesn't own mules. Reproduction is an obvious reason: Someone has to breed the mares that foal the mules. And some people just flat don't care for the smart-ass (pardon the pun) attitude of the mule, or just dislike the way its long ears swivel around. One owner of both mules and horses quipped that he owns mules for their personality and horses for their beauty.

# A MAN WHO LOVES A GOOD MULE

**Hal Novak**
**McArthur, California**

Type of Farm: 30 acres (12 ha); retired hobby farmer
Drafts Owned: Mules, minis, donkeys, and cutting horses

Hal Novak knows the value of a good mule, because he has spent many years studying mules, attempting to understand them, and learning to work with them. One thing Hal makes clear when speaking about mules is that a beginner must think long and hard before acquiring a mule. Additionally, anyone with a big ego had better not even think of owning a mule.

The mule will constantly find ways to upset the applecart. A bully, a completely green beginner, or a person with a big ego may be able to work with a horse but will never get along with a mule. The old saying that a mule will wait a lifetime to get back at an unkind owner isn't quite true, Hal says. "The mule will take the first opportunity to right a perceived wrong." For the beginner, who is bound to make mistakes, the mule's unforgiving nature can prove disastrous.

Hal was raised on a dirt farm in Indiana where his parents farmed with a team of mules and a team of crossbred Percheron horses. Hal's job was to cultivate crops with the mules. At nine years of age Hal thought that he was a pretty accomplished mule teamster, for he never tore out any of the crops. In retrospect, he now attributes this skill to the mules knowing their job well. Later Hal worked with his uncle Jim and Jim's two Missouri mules, Whitey and Louie. These big mules were 17 hands tall and 1,300 pounds (590 kg) apiece. These mules and Uncle Jim knew each other well and depended on each other. As Uncle Jim became older and could no longer climb up onto the hay wagon, the mules learned to pull the wagon by themselves to the hay field to where Hal and Uncle Jim would meet them, load the wagon, and send them back to the barn. The mules walked alone, with the lines tied to the front board, back and forth between the barn and the hay field.

These mules were accustomed to a schedule, and mules don't take a schedule upset lightly. One night Uncle Jim disturbed their evening schedule by coming back from town late, and when he finally walked into the barn to say goodnight, Whitey caught him in the chest with both back feet.

So who should buy a mule? Hal advises that someone who is smart, likes animals, likes to learn, and has a small ego and lots of patience might try a mule. In addition, the mule can't be "pretty pleased" into good manners; he needs a strong and rigidly consistent leader. A novice determined to buy a mule needs to find a team of older mules, in the 17- to 19-year range, that have been worked regularly and treated with dignity and respect, without any abuse.

Over the years Hal has observed that mules learn more slowly than do horses, but retain more. A horse being trained must constantly be reminded about the lesson or he will regress. A mule, on the other hand, will remember a lesson from six months ago and pick up right where the trainer left off. In addition to the good things, however, a mule also remembers the bad lessons, so the mule handler must carefully consider how the mule is trained.

Disciplining a mule is a tricky business, as well. A mule, according to Hal, will accept discipline immediately after the errant deed, but not 30 seconds later; the mule draws a fine line between discipline and abuse, and the trainer will learn to remember this line.

Hal has mules because of their personality and their character. To the beginner who wants to take the plunge into mule ownership, Hal says, "Be humble and ready to learn, because the mule will test and teach you for the rest of your life."

The main reason is that the intelligent, unforgiving, unforgetting, schedule-seeking mule with an extremely high expectation for excellence in its leader is just too much for some folks to handle. People with big egos, who hate to look stupid and who don't like to be tested, are not muleteer quality. Even the best mule will constantly test the best muleteer — not to hurt the handler, but just to see if he is on his toes. A horse is an intelligent creature, too, but is much more accepting than a mule, forgives grievances much faster, and is a far better animal for the novice seeking a positive early draft equine experience.

Most muleteers I have spoken with have had and continue to have both mules and horses as part of their operation. Both animals have their places and their jobs, and both are quite good at what they do. One is just a lot smarter about it than the other.

## The Mule as Army Mascot

**T**he United States Army chose a mule as its mascot in 1899, in answer to the Navy's goat. The choice reflected the mule's usefulness for hauling weapons, ammunition, and supplies during military operations. In 1956 the United States Army deactivated its mule detachments, ending their long and successful contribution to warfare in desert and mountainous areas. Still in residence at the West Point Vet Clinic in West Point, New York, the Army mules remain one of college football's oldest mascots.

Today the Army mule is little more than an ornament at parades and football games. Over the years, sixteen mules have filled the role, only one of which was a molly. Named Buckshot, she served from 1964 to 1986. Currently, three mules serve as mascots. Raider has held the role since 1995 and is the senior mule. Ranger II joined up in 2002, as did General Scott, known as Scotty.

The original Ranger was given by veterinarians to the First Ranger Battalion as a two-year-old, but after dumping the executive officer at a public event, the mule was banished to West Point in 1975. There he wowed the crowds, occasionally entered offices unannounced and uninvited, and retired from active duty at a half-time ceremony in 1995.

Mules take extra time and effort, but once you're hooked, you might not be able to stop with just one.

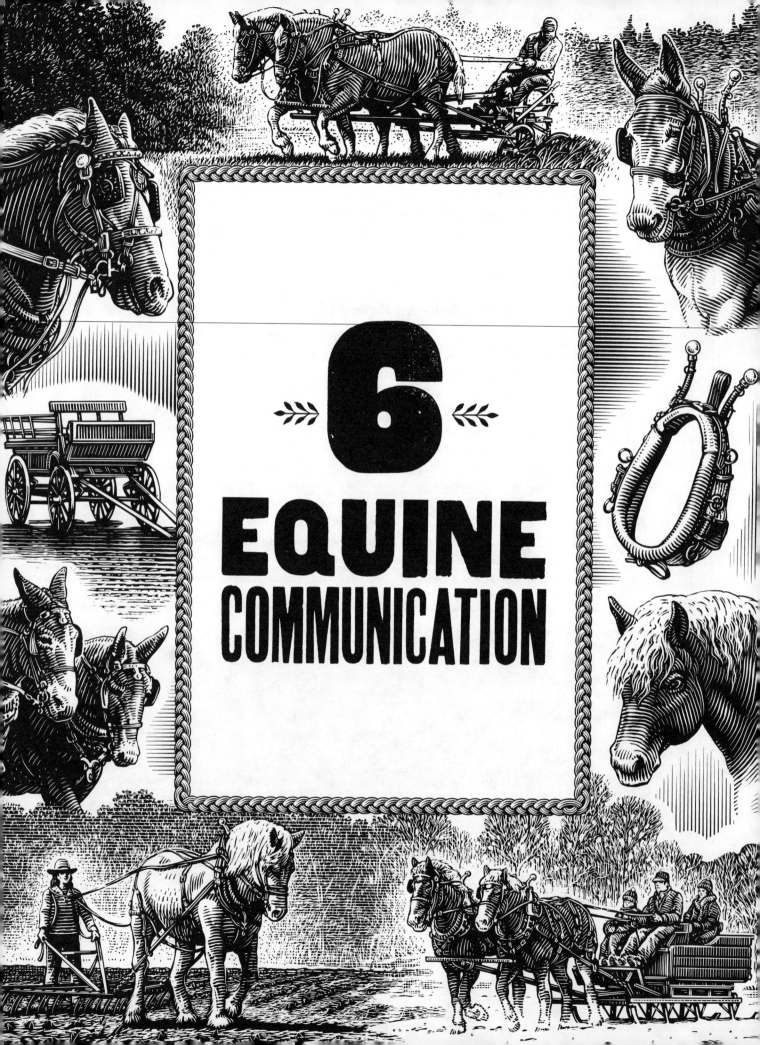

# 6
# EQUINE COMMUNICATION

On a crisp fall morning with frost still clinging to the brittle corn leaves, my sister, my father, and I went out to pick the field corn. Two of our mares were hooked to the wagon that we were throwing the ears into. These mares were well tested and field-proven, yet even they could be a little high-headed and frisky on a cold morning like this one.

Since my row was closest to the wagon, I had the responsibility of keeping the horses and wagon in the right place for the pickers to target. Any teamster knows that jumping on and off a wagon dozens of times is not fun. I solved the problem by hooking the lines to the side of the wagon bed so that I could easily grab them from the ground each time the team needed to move up. I expected my experienced mares to stand quietly until asked to move.

I had used the same method the previous year, and the mares eventually got the hang of moving based on my voice cues alone. I was always ready, however, to make a dash for the lines. Today I noticed that Maude was acting a little squirrelly; perhaps something had spooked her overnight in the barn?

We moved steadily down the first rows, but I was not comfortable enough with Maude's behavior to trust the team to stand or move ahead unless I was within grabbing range of the driving lines. We were working down our sixth row when I saw Maude reach over and nip her teammate. Then I saw her tail rise and realized that she was not only feeling the crisp morning, but was also in season. The next time she nipped her teammate, the other mare nipped back, Maude squealed and kicked, and they both started to walk. I yelled, "Whoa," but they weren't interested in stopping. Fortunately, I was able to grab the driving lines and stop them.

I had picked up on Maude's strange behavior during the harnessing and had continued to watch her after we were working in the field. Had I not noticed the signs, or just assumed she would work as she always did, my caution index may have been low enough to let a bad accident happen. Horses and mules can't speak our language, but they are constantly communicating with us, with their teammates, and with other animals.

## Thinking Like an Equine

Reading human minds is probably not on the list of the achievements of horses and mules, but they do use their brains much more than the minimal level required for eating, sleeping, and thermal control. Teamsters often have the uncanny feeling that a horse or mule is anticipating our commands; we just *know* he is picking up on what we're thinking.

Equines are extremely intelligent creatures, and their brains have enabled them to survive over the millennia. As a domesticated animal, however, the equine has had to adjust to an additional variable, the human, and learn to understand what this two-legged creature expects. The equine's anatomy and social structure dictate how it communicates with the

Equines and humans have to learn each other's languages to communicate effectively and work together.

human. As mammals of a higher order, we humans have a big job on our hands in learning how to read communication signals from the equine. In short, we have to learn how to think like the horse or mule to achieve a balanced working relationship. After all, we expect the horse and mule to read us and our intentions; surely we should extend the same respect to them?

Learning to think like an equine is not something you can master solely from reading a book. Rather, it results from a lifelong commitment to studying methods that have worked for other people, studying your own animals, and studying yourself. Close attention and observation of your animal will help you understand some of the fundamentals of equine communication. Ear position, tail movements, and body language are all key to determining what your horse or mule is planning next. Understanding what your equine is trying to tell you will expand your comfort level with your animal and improve your ability to work smoothly as partners.

## EQUINE COMMUNICATION

Perhaps to their advantage, equines cannot discuss their intentions verbally. Rather, they use their bodies to communicate their brains' intentions. The brain transmits signals of fear, aggression, submission, boredom, and the like to the body, which responds with postures that, interpreted correctly, allow the human to understand what is happening or about to happen, and thus solve a potential problem *before* it balloons into a disaster.

Although the word *draft* implies a cold-blooded and thus calm animal, don't be lulled into thinking that because you are handling draft equines, you will not experience the nervous, excitable behaviors exhibited by the hotter, lighter sport breeds such as the Quarter Horse, Arabian, and Thoroughbred. An equine is an equine. All horses are wired in a similar pattern; some are just less reactive than others. As you work with your draft equines, understand that while they are often less likely to exhibit the sensitivity of lighter horses and mules, they still have the potential for moving with lightning speed when frightened or hurt.

## EQUINE WORLDVIEW

Your horse or mule has a view of the world that is entirely different from yours. Equines are, by nature, jumpy animals. Where you see a puddle of water, the equine sees a bottomless sea. Where you see a bridge and hear rushing water, he sees a monster making a horrible noise. As the handler, you are responsible for trying to understand the way your horse or mule sees his world, and being ready to firmly and patiently remind him that you can be trusted to know what is best because you are the boss.

Why do equines spook so easily? Why does it seem like they see bears behind every tree and alligators in every puddle? As prey animals that have always relied primarily on flight in response to threats, equines have survived the millennia because they anticipated dangers, real or imagined, and fled before being caught. The domesticated horse has learned to trust man, and the well-trained horse should, with cues from his trainer/owner, be able to distinguish real from imagined dangers and learn not to spook and run from every leaf that stirs. Many animals learn to conquer nearly every fear; some can even handle the sight and sound of a semi-truck barreling along a highway past a field.

How can you recognize the need to reassure your animal that all is well? Eventually, you will see the signs subconsciously and correct accordingly, but for starters, let's explore the impulses transmitted by the equine brain as the animal encounters a possible threat. Keep in mind that although the horse brain and the mule brain are similar, they react differently to stimuli. In general, an equine's first reaction to something frightening is to assume the startle posture. The head goes high into the air as the animal tries to get a good look at the object. The legs are braced and, if allowed, the animal stands stock-still. Here is where the mule and horse diverge.

Once a horse has had time to assess the situation (which can occur rapidly), he usually makes one of three choices: he runs away as fast as his legs can go; he stands his ground and threatens or attacks; he walks on by, realizing that the perceived threat is not real. The mule, on the other hand, typically takes much longer to assess the situation. Many muleteers

# A SAGA OF COMMUNICATION

**Vince Mautino**
**Colorado Springs, Colorado**

Type of Farm: Small training facility for pack, riding, and show animals

Animals Owned: Haflinger mules

Vince heads out on a snowy trail in a two-wheel cart converted for winter.

Vince Mautino and his wife Judith have enjoyed having mules as a part of their lives for more than 30 years. Vince did not grow up using draft power. In fact, when he became interested in using horses and mules in the 1960s, he was largely on his own. A generation gap existed between the old cowboys who used horses and mules as working partners and the subsequent increase in interest in the '80s and '90s. Vince was fortunate to work with an 86-year-old cowboy who taught him a great deal about horses and mules.

Much of his experience with training both horses and mules, however, has come from his own observations and from distilling what he has seen and learned from other folks into his own training methods.

Vince shapes his training to the animal he is working with, tailoring it to each animal's unique needs. The observant trainer sees the differences in the behavior of his animals and works with each one individually.

In the 30 years Vince has been working with mules, he has yet to come across two that are alike. Communicating with each one is a unique experience. With their short attention span and later maturity than horses, mules can be a challenge for the most patient of trainers. Vince has learned that communicating in short, concise lessons is imperative for successfully conveying a concept to the mule.

Mules can be one-person animals, a factor that can seriously affect their settling-in time with a trainer or new owner. Vince tells about the new mule owners who refused to take his advice to let their new mule settle in for a couple weeks before working him. Instead, eager to work their animal, they saddled up and headed out and were repeatedly dumped into the dirt by this reputably well-trained mule. The mule did not trust them yet and was happy to tell them so. Most horses, on the other hand, are nervous and upset for the first few days of being relocated, but quickly settle in.

Vince has trained many horses and mules for pack, saddle, and harness work. While each animal is different in its own way, he understands and accounts for the ways in which the mule and the horse evolved, and tailors his training method to the individual, to create an animal that is trustworthy and willing to work. He becomes frustrated with mule owners who feel that they have figured out their one or two animals and are now experts on all things mule. By contrast, Vince finds that the individuality of each animal can provide any equine owner, particularly the mule owner, with a lifetime of learning.

A mule in the startle pose

A mule in the attack pose

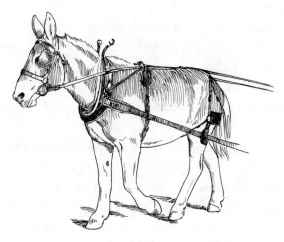

A mule in the walking-on-by pose

have learned that you cannot rush this period of assessment. A mule will refuse to do anything unless he feels that doing so is for his own good. If the situation truly appears dangerous, a mule is more likely to attack than run. A mule will accept the decision of a good and trusted leader and usually obey, but may follow his own mind if he thinks that the situation could result in harm. A mule that does not accept his leader cannot be moved at all.

The well-trained animal should understand that even if he is afraid, you as the human have the last word on the danger of the situation; just remember that the mule takes a little longer to come to this decision than the horse. Whether horse or mule, the animal must learn to walk on by, unless you give the signal to flee or attack.

## THE STRUCTURE OF THE HERD

The equine brain allows the horse not only to sense danger and react, but also to engage in a gregarious herd social structure, react to the world around him, learn, solve problems, and anticipate events. Equines are social creatures, hardwired to communicate with their herd mates. This characteristic is key to our ability to train and work with them. All horses and mules need a defined leader in their social structure. In a herd, this animal is the most dominant individual. In the working environment, the leader should be you.

In the United States, the horses running on the open range on Bureau of Land Management or Forest Service managed land are feral. Many people refer to them as wild, particularly those who wish to "save," "rescue," or "adopt" them. These horses are, however, descendants of domesticated animals that once belonged to Spaniards, Native Americans, and ranchers of the Old West. Their return to an undomesticated state allows us a glimpse into the social structure of equines surviving in an extremely harsh environment without human contact and influence.

Each herd or band generally consists of 2 to 20 animals. The social structure of the feral equine herd is composed of dominant and submissive animals that adhere to a pecking order. The hierarchy is understood by all and enforced by the more dominant animals (usually the stallion and a lead mare).

A dominant horse will challenge the handler's personal space.

A submissive horse approaches with lowered head.

Although adherence to the pecking order is rigid, the order itself can be fluid over time, constantly challenged by other herd members as animals age or relationships among them change. This structure is so well understood by individual horses that many trainers use techniques that incorporate herd dominance behavior and bypass more brutal techniques that rely on force to acquire submission.

For example, the dominant mare in the herd uses a combination of body language and vocalization (nickering in approval, squealing in warning) to relay her message. To push away a subordinate as a means of discipline, she may use her head (ears back and teeth bared) or her body (presenting her hindquarters) to move the animal out of her space (our arms serve to convey the same message of "quit crowding me"). Conversely, she may allow an animal to approach her and then accept him with nickers and a scratch on the withers as a reward. The human can use all the cues of the dominant mare to become the defined leader, and then use this leadership role to proceed with training.

By controlling an animal's movements, the astute equine handler can, in a completely nonforceful and nonfrightening way, convince the animal of the need to accept the leadership of the dominant human. A naturally dominant animal may challenge the fact that the human is boss, but the handler's authority may be asserted with proven techniques.

If your animal ever begins to indicate that he is dominant in your relationship, stop this unwanted behavior in the early stages. As you develop your equine skills, you will learn to recognize and head off behavior that could lead to a role reversal. For example, a horse that rubs his head on you is being pushy — don't allow it. If you need to, seek professional help. An enormous draft animal with respect issues can be dangerous. Some key signs — particularly the position of the ears, lips, nostrils, and mouth — will help you to determine the animal's attitude.

## Physical Methods of Communication

How does an equine communicate his intentions to his herd members and to humans? While vocalizations play a role, body language is more important. Physical signs are the only way you can know what your horse or mule is thinking. These clues are indicated by the eyes, ears, mouth, tail, feet, and legs. Becoming intimately familiar with these key signs could save you from a nasty kick, a runaway team, an injured animal, or worse.

## WATCH THE HEAD AND NECK

Positions of the equine head and neck can tell you volumes about what the animal is thinking. The equine head is typically the first body part to react to stimuli such as fear, threat, or interest. Noticing the position of your animal's head, in addition to the position of the ears, mouth, and eyes, will help you understand and predict what may happen next. Here are some signs to watch:

- Level neck and relaxed head: content, resting, relaxing
- High head, arched neck, in motion: interested, frightened
- Head thrust out, neck level, snaking head from side to side: extreme aggression, high-threat level
- Head reaching forward, neck arched, lips moving: happy and content with touching, grooming, or scratching

## HOW EQUINES SEE

Vision is an important sense for the horse or mule; it is critical for assessing the nature of danger. The placement of the eyes allows the equine to see almost completely around his own body. *Almost* is the operative word — there are blind spots directly behind and directly in front of the animal, which cannot see his own front feet. Never approach a horse or mule in either of these positions without alerting it to your presence and moving cautiously. In addition to these blind areas, horses and mules have difficulty viewing things above their heads and easily become alarmed if something overhead moves or falls.

Equines have monocular vision, meaning that their eyes are set on the sides of the head in such a way that most of the viewing comes from two pictures, one from each eye. Because the animal sees two pictures, his brain may register two different stories, such as a cliff on one side of the trail and a river on the other side. A horse or mule often moves his head from side to side or attempts to look with one eye when trying to get a good view of an object of interest.

Monocular vision does not allow the equine eye to determine depth, which exacerbates the tendency of horses and mules to shy sideways when an object may appear to jump suddenly into the field of vision. The animal cannot determine the real picture without viewing it with both eyes. The ability to move his head to get a good view of the item of interest, and thus be reassured it is not a bear, is a critical variable for equine peace of mind.

The check lines of a harness restrict movement of the neck and head, alarming many young horses the first time they are harnessed. Many teamsters put blinders on their bridles, forcing the animals to look straight ahead, minimizing the two-picture view, and keeping the animals focused on the task at hand. Other teamsters believe that the natural inclination of an equine is to have freedom of head movement without restricted vision, and a well-trained animal will remain focused on the job even with unrestricted vision.

Binocular vision is available only when a horse looks straight ahead. Binocular vision allows the

---

### Monocular versus Binocular Vision

**T**o illustrate **monocular vision** versus **binocular vision**, consider the case of a horse trotting down a road for the first time. To the left he sees an approaching car. As the car nears and moves out of his binocular vision, he can no longer determine how close it is and he swerves to the right. Just as the car goes by, he sees a cow on the right and, not recognizing the bovine, assumes it is a predator and tries to get away by swerving to the left — except that the car is just now going by, which frightens the horse. His choices are to panic and run, to swerve all over the road, or to trot on by. With the handler's reassurance and firmness, this horse can learn that he must trot on by *no matter the circumstance*. The teamster must be the dominant party in this relationship. Just as the horses in a herd know that the dominant mare would not lead them into danger, your horse must accept that you will not take him past a horse-eating monster.

> ### Seeing in Color
>
> **Although equines** have poor color vision, they are not color-blind, as an old wives' tale suggests. The equine eye distinguishes green, red, yellow, and orange well, but has difficulty with blue and violet. The ability of a horse or mule to distinguish and associate the green of grass as a preferred food indicates that he can see green quite well, although smell also plays a large role in feed selection.

Ears pricked forward with interest

animal to have depth perception and judge where he is going. This vision is what animals experience when they have blinders on their bridles.

Here are some signs to look for in the eyes:
- Whites of the eyes (sclera) showing: frightened, nervous
- Cold, hard look: sign of aggression, seeking to intimidate or dominate
- Sunken, dull, unresponsive eyes: signs of overwork, broken spirit, pain, sickness
- Gentle, soft look: sure sign of a happy horse or mule

## HOW EQUINES HEAR

The equine's sense of hearing is far keener than the human's, detecting high- and low-pitched sounds that the human cannot distinguish. Mules, with their larger ears that funnel sound well, may have a better sense of hearing than horses. The position of the ears enables a horse or mule to hear almost completely around his body. Ten sensitive muscles control each ear, allowing it to swivel rapidly to the source of a noise. When pointed directly forward, both ears can focus on the source of the sound. For this reason, a horse or mule often orients himself to face the sound to ascertain its source. The ears frequently move independently, so that the animal can track sounds from different directions or pay attention to the driver with one ear while listening for potential danger ahead.

Low, quiet sounds are calming to the equine. The nicker is a soft sound that reassures all members of

Ears held loosely forward in normal position

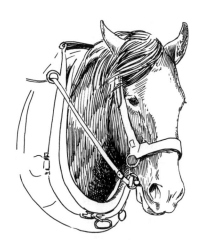

One ear angled back, one angled forward, paying attention to surroundings

Ears pinned to neck in anger

Ears airplaned out to side showing boredom or tiredness

receives grain each time he hears the bin door open and close will likely hee-haw with anticipation. A herd of horses that know a long, shrill whistle results in hay will come pounding up for a snack.

The position of the ear is a critical communication method for the equine. Familiarize yourself with these postures of the equine ear:

- Pricked forward: interested in a sight or sound
- Loosely forward: healthy and alert, but not overly interested
- Angled back: listening carefully to the sounds behind, such as those made by a rider or driver
- Pinned flat against the neck: feeling threatened; ready to move angrily and aggressively
- Drooping to the side: fatigued, bored, relaxed, napping, submissive, or in pain

## THE EXPRESSIVE MOUTH

The equine mouth serves a number of important functions. The mouth and teeth are not only critical for eating, but are also important for communication, for grooming, and as tools for fighting and breeding. Awareness of the signs of your equine's mouth will help you to interpret his behavior. For example, licking and chewing of the lips indicates submissive behavior; soft, loose lips may indicate relaxation, fatigue, or boredom; tight, pinched lips may indicate aggression or fear.

The dominant equine controls the movement of submissive animals in the herd, often with flattened ears and the threat of exposed teeth. A team that typically behaves well together may have a tiff and take a swipe at one another. No matter how well behaved the animals are toward you, you may

the herd. Conversely, the high-pitched scream of a stallion's whinny is meant to be heard far and wide, and to strike terror to the hearts of opponents and awe in the hearts of mares. When you speak to your equine, keep your voice low and calm to encourage a positive response.

Horses and mules typically do not panic when introduced to a new noise. Rather, they take note of the sound and remember the consequences. A horse that has been shouted at before being struck will remember the result of the tone of the voice and flinch in anticipation of the stick. A mule that

> ◀ **CAUTION** ▶
>
> ### No Hand Feeding
>
> Feeding your horse or mule from your hand encourages your animal to become pushy or nippy and can result in an unanticipated bite from an otherwise well-behaved animal. Refrain from hand feeding your animals. Treats can go into feed buckets and still be appreciated.

possibly become the recipient of an ill-timed potshot aimed at another equine. Although I always try to watch carefully for aberrant behavior, I once nearly lost an ear by getting in the way of a swipe from a mare intended for an uncooperative teammate. The biting mare never intended to strike out at a human, yet I happened to be in the wrong place at the wrong time.

Your equine may try to dominate you by threatening to nip or by actually biting you, a sign of a weak leadership position on your part. An equine bite directed at you — whether you are walking by the stall, putting on the harness, or tightening the belly band — or directed at stall mates is a hazard that must be dealt with the instant it happens. A loud shout from you, with an arm held up as you step sharply forward, will indicate to the aggressor that such behavior is unacceptable. If you do not feel comfortable addressing this problem, seek the help of a professional who can resolve the issue without creating fear in your animal or you.

Horses and mules often engage in mock fighting, which includes biting. This behavior usually occurs when animals are in a relaxed environment, the weather is cool, or they are feeling frisky. This behavior is normal between two equines, but *never* let such behavior enter into the relationship you have with your equine, for it is absolutely not safe for you. Many owners make the mistake of letting their foals roughhouse with them; when the foal is a yearling or a two-year-old, he still enjoys the play that is no longer safe for the owner.

Watch for these signs of the lips and teeth:
- Open mouth, teeth exposed: aggression or active fighting
- Licking/chewing: sign of submission, often occurring just as pressure has been released
- Nibbling along the rump or withers of another horse or mule: bonding behavior between friendly animals
- Upper lip flared back over the nose (flehmen response): reaction to an intense or unusual smell, typically viewed in stallions when sensing a mare in season
- Snapping (rapid opening and shutting of jaws by foals accompanied by drawing back the corners of the mouth): sign of submission, usually displayed to older horses

## A KEEN SENSE OF SMELL

Combined with sight and sound, the equine sense of smell allows a horse or mule to size up a situation quickly and to determine if flight is necessary for survival. Equines identify poisonous plants by scent as well. The domestic equine learns that most scents are not associated with harm, but as with sight and sound, if a result worth remembering follows a particular scent, the animal will take note. Say, for example, that you give your mules hay in the morning and hide the grain behind your back. They will likely smell the grain and refuse to eat the hay until you bring forth the rest of their breakfast.

The foal and dam know each other by scent. The stallion recognizes a mare in season by her scent. When introduced, one of the first things horses do is carefully smell each other. The equine's sensitive nostrils will pick up the odor of an agitated or nervous human, which along with the human's physical signs of unease, causes the animal to adjust its behavior accordingly. Some horses and mules behave more carefully with a frightened novice, while a dominant animal may recognize weakness and seek to bully the human. Conversely, a submissive equine may become nervous and distracted by the human's agitated state.

Upper lip curled in flehmen response

Relaxed, loose nostrils

Flared, taut nostrils

Pinched, angry nostrils

Watch for the following signals of the nostrils:
- Relaxed, loose: happily engaging in life
- Flared, taut: when accompanied by pricked ears shows intense interest; could also mean that the animal is winded
- Pinched: angry or in pain

### THE TELLTALE TAIL

The draft horse tail has quite a history. During the boom of the draft horse, most animals had their tails docked at an early age, and many breeders claimed that the foals experienced little or no pain. The reason given for docking then, as now, is that the copious hair on a draft horse's tail is difficult to keep clean and can become tangled in large hitches or around equipment. Today, however, the draft horse scene includes many show animals that will never be worked in an agricultural environment, yet tail docking is practiced as a way to showcase the muscled hindquarters and is considered traditional protocol for many hitch and halter classes.

In some European countries tail docking is outlawed as being painful, unnecessary, and even harmful, because a docked horse cannot effectively swat flies. Although tail docking is legal in some states, many American draft horse breeders opt to leave the tail long and natural, since no one has been able to substantiate the claim that a short tail equals safety.

In stark contrast to the beautiful, flowing draft horse tail, the mule tail is a rather scanty clump of hair attached to the end of the backbone. For showing, mule tails are often trimmed close a few inches down from the top and then left to hang naturally. The hair is coarse and stiff, thanks to the donkey's cowlike tail, but the mule can be grateful to have inherited a few extra hairs from the horse. Draft mule tails are not docked.

The equine tail, history aside, is an important feature of the animal, providing a versatile and effective fly swatter and one more method of communication. Whatever your view, equine tails can tell you a true tale of intent, so keep an eye on the rump. These are a few of the messages conveyed by the tail:

- Slightly elevated: alert or excited
- Low and close to the rump: fearful, exhausted, submissive, or in pain
- Held high over the rump (flagpole): seen in playful foals and adult equines, or equines in a high state of alarm or excitement
- Rapid swishing: irritated; may precede a bite or a kick
- Twirling: irascibility or irritation
- Held to the side: mare in season, possibly ready to be bred
- Swishing lightly: relaxed and comfortable

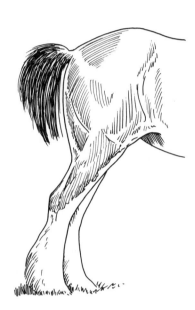

Tail slightly elevated in excitement

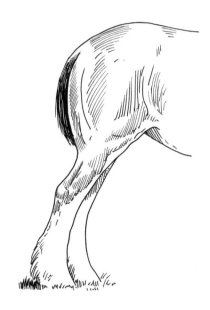

Tail low and close to rump in submission

Flagpole tail held high over the rump, showing alarm or playfulness

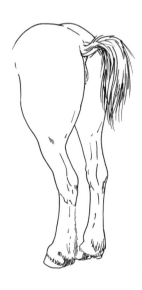

Tail held to side, mare in season

## WATCH THE FEET AND LEGS

The equine's powerful, quick, and agile legs are used for three main purposes: flight, transportation, and fighting. The equine's legs can be deadly for anyone in the way; never underestimate the power and precision of these tools in the equine survival package. In addition to being protective, the feet and legs carry enormous weight and endure massive concussive forces. Signs of injury and pain may be read in the positioning and cadence of the feet and legs. The position of the feet and legs can also signal equine emotions, so always pay attention to what your equine is doing with his feet.

To avoid compromising your own safety, learn the proper way to approach your equine and the appropriate place to stand when he is stationary. When moving toward your animal in the pasture, at a hitching rail, or in a box stall, always approach from the front. If you have to approach from the rear, speak to him to let him know you are there. In the pasture or the box stall, train your animal to face you when you approach.

The safest place to stand in relation to a stationary equine is near the shoulder, facing the animal so that your shoulder is at an angle to his body. From this position you can see his head, and the croup is in your peripheral vision, allowing you to notice movement from both the front and rear. When leading your animal, lead from the left shoulder.

By standing near the animal's shoulder, you avoid his front and rear legs. Avoid standing directly in front, where he can't see you; if the animal were to spook and run, or aggressively strike out, you would be directly in the path. Standing to the animal's rear will also put you in a compromised position should he kick out. Since the equine has a blind spot directly behind, he cannot see what you are doing and may become anxious about your activities.

While both horses and mules kick, folklore and muleteers alike warn us of the precision of the mule's kick. Mules use their feet to defend themselves in many situations where a horse will flee. A mule may try to kill a dog that attempts to chase him, while a horse will just scamper away. Mules, being animals of rigid consistency and pattern, can use their feet to let you know of an upset in schedule, as well.

If you must stand or walk near the rear of your horse or mule, remain as close as possible to the rump. Should the animal kick out, the force will be far less in close range than if you are caught by an extended foot with full momentum behind it. As the animal's handler, obviously you will often need to be in compromised places. When you are at the front or rear of your equine, stay alert to his body language. If you notice signs of aggression, fear, or anger, immediately — but calmly and slowly — move out of the way. Always let your animal know where you are by first speaking and then touching him. Keep your eye out for the following signs:

- Back leg tipped up: relaxed, dozing or sleeping; possibly injured
- Front leg tipped up: injured; equines don't typically rest the front leg
- Pawing: bored, restless, in pain, angry, or inquisitive; may also precede rolling
- Stamping: removing flies, restless, angry, bored
- Striking forward with foreleg: extreme aggression and trying to clear personal space
- Shifting weight from one foot to the other: injury to both legs, possibly bilateral laminitis

Back leg tipped up in relaxation

Front leg tipped up, indicating injury

Striking with foreleg in anger

## Why the Left Side?

**Working an animal** from the left side while saddling, bridling, harnessing, or leading is strictly traditional. This tradition likely originated in Europe with medieval fighters who held their swords in the right hand. A fighter carried his sword on the left hip, which made mounting from the left side the easiest option for throwing the right leg over the saddle. The animal's left side has therefore come to be known as the near or nigh side, while the right side is called the far or off side.

Although no scientific reason has been suggested for working with animals from the left side, animals and humans both appreciate consistency. An animal must be comfortable being handled from both sides, but for the sake of consistency and convenience, trainers the world around work equines from the left side, though this tradition is changing. For example, many natural horsemanship clinicians saddle from the right.

## Equine Vocalization

In addition to their many kinds of body language, horses and mules communicate through vocalization. They use a variety of sounds, each with a specific meaning. Horses and mules diverge dramatically in their vocalizations: The many different sounds produced by the horse can be musical, shrill, frightening, or reassuring, while the mule's hee-haw, also called braying, is one of the stranger sounds made by animals. The mule is capable of varying his vocal sounds to some degree, and often starts out with a convincing bray, only to end up with a sickly whinny. Like the rest of the animal, the mule's vocal confusion is a product of his mixed heritage.

**The blow** of air through the nostrils with the mouth closed is often a sign of contentment. A happy animal may blow softly through the nostrils while eating grain or while being turned out to pasture. Blowing is also done after or during a period of hard, short pulling, such as moving a load up a hill or plowing, but should not be confused with the winded breathing of the overworked animal. A short, soft, blow into another animal's nostrils is a way to get the other's scent as well as a manner of greeting. A contented blow is not rhythmic or repeated regularly; a winded equine blows in short, quick breaths accompanied by heaving flanks. This sound may last for some minutes and can be a sign of respiratory distress.

**The snort** is a guttural sound of alarm and should not be confused with the blow of the contented horse. Snorting in the wild horse herd is a signal for all animals to stop, look, and be ready for immediate flight. The snort accompanied by an elevated tail may be quickly followed by flight.

On cold winter mornings, you may see your herd playfully racing from one end of the pasture to the other, now running full tilt, now standing stock-still, snorting vigorously at some strange object, and then racing away again in mock terror. In addition to the warning snort, a shorter, less violent snort can be an attempt to clear the nostrils or may accompany a respiratory irritation.

**The nicker** is a soft, low-pitched *putt-putt-putt* noise emanating from the nostrils. A stallion sounds a nervous, fast nicker when courting a mare. This sound is often heard as a mare greets her foal, as you enter your barn in the morning to feed, and as herd members greet each other. A nicker lets you know that your animal recognizes you or a companion and is pleased and relaxed.

**The whinny** is a shrill, loud, and intense call used to send important messages such as "Where are you?" and "What are you doing in my territory?" A foal often whinnies shrilly for his dam if she moves out of sight. Companions may whinny for one another when separated. A stallion can challenge an intruder at a great distance with a bone-chilling whinny. Horses in intense pain, fear, or rage will whinny with incredible energy and decibels.

**The groan** can be a sign of intense pain or distress. A painful colic is often accompanied by deep groans. Groaning accompanied by any other signs of pain and discomfort means that it is time to call your veterinarian.

**The grunt** is a normal sound made by horses or mules lying down and experiencing short-wave sleep. A foal sometimes grunts softly when suckling. Grunting, or a cross between a grunt and soft groan, can also accompany relief, such as might occur while an animal is defecating or urinating, scratching, rolling, or lying down after a hard day's work.

**The squeal** is often accompanied by kicking, striking, or other aggressive posturing by both the mule and the horse. A squealing equine is *not* to be messed with. Mares in season squeal when around other horses, and stallions squeal loudly when fighting. Foals may squeal in fear when approached by an unfamiliar animal.

## Listen, Watch, Act

As you observe your equines, you will notice patterns in the way that they learn, communicate with you and other animals, and react to their environment. While you observe, try to behave more like the mule: listen, watch, and act only after making a careful decision. Equines can teach you a lot if you are patient enough to learn what they are trying to tell you. Noticing subtle cues your animals give will help you avoid a potential accident and can greatly increase your training skills.

When you ask your animal to do something, notice and reward behavior such as:

- Turning toward you
- Looking at you (paying attention)
- Licking lips

As you work your animals, listen to and learn from them.

## ◀ CAUTIONARY TALE ▶

### Dealing with a Hypersensitive Equine

You have owned your draft horse for two weeks. You know he needs exercise and reinforcement of his training, so you decide to take him out on the spring wagon. You have done this several times now and you both know the routine. The horse stands quietly to be harnessed, backs into the wagon's shafts with little trouble, and continues to stand while you fasten the traces and hold-back straps. Feeling not quite ready to do everything solo, you ask your buddy to stand at the horse's head while you climb into the wagon. You chirp to the horse, signal to him with the lines, and head up the driveway.

As you proceed down the lane and onto the road, you are congratulating yourself on your smooth transition into horse teamster when your horse veers to the right and the spring wagon almost dumps over in a ditch. As you scramble to gather your wits, you spot a cow outside the fence, poking her head under a bush. Okay, your horse had reason to be a bit nervous. You continue down the road past a herd of cows quietly munching grass on the other side of fence. Your horse slows to a walk and then stops. He has decided that the bovine species is not to be trusted.

Your baloney detector shoots up. You know your horse has seen plenty of cows before, and while he may suddenly decide that he can't pass any cows for the rest of his life, you have to convince him otherwise. You firmly tell him to get up and keep moving past the cows. Being a well-trained horse, he likely will not try this shenanigan again, and you will have reinforced your position as the leader of the team.

Lesson learned: If you had allowed the horse to have his way, thinking, *Oh, we'll just go another way today*, you would have lost a major battle. Such a battle only escalates with time.

- Gently blowing through the nostrils
- Showing alertness and listening with the ears
- Displayed a relaxed tail, head, and lips

When your animal is preparing for a full-blown rebellion, notice the following signs and immediately attempt to correct the stimulus causing the problem:
- Avoiding eye contact
- Showing tense, taut nostrils, lips, tail, and jaw
- Pointing the ears away from you or pinning them back
- Tightening the body muscles
- Moving away from you or avoiding your touch

## How Equines Learn

Key elements for the survival of feral horses are learning and remembering, as they may not have a second chance to escape. Consequently, all equines possess outstanding memories, which makes training both easier and more difficult. The mule, in particular, has an amazing memory. Careful, correct training will never be forgotten, but neither will rough treatment. For the domestic equine, problems with training commonly originate from poor handling, making it critical for the trainer/handler to understand clearly how equines learn. Mules learn in much the same way as horses, but differ in a few crucial areas. Horses are animals of reaction — their action follows a certain stimulus, often quite rapidly. The mule, on the other hand, models its behavior after the donkey in that it usually first evaluates the situation and then acts. Because the mule typically does not spring instantly to action, it is less likely to spook and less likely to respond to force.

The horse may instantly jump ahead when it feels the touch of a whip or just hears it crack, for example, while the mule may stop dead still to evaluate the situation. If pressed, the mule may endure harder and harder strikes without budging a step until it feels the situation is safe. If you want your mule to learn and to accept you as a leader, first teach yourself to think quicker and more reasonably than he does; otherwise, you may be standing in the same spot for a long, long time.

### ACTIONS ASSOCIATED WITH CONSEQUENCES

The equine brain is adept at associating actions with consequences and adapting subsequent behavior. If, for example, a young horse learns that the mares in the pasture will chase him away if he gets too close to their babies, he will wait until the foals move away from their mothers to play with them. Once a horse associates the squeak of a gate opening, he knows that one of two things is going to happen: he will either receive feed or be put to work. Most equines will make this distinction by watching to see whether you have a halter. Whatever the case, they should stand still as you approach.

Horses and mules also learn a lot by watching pasture mates, other animals, and humans. A new mule, unfamiliar with the layout of a pasture, will hang back and let the rest of the herd show him where to find the best grazing areas, where the water is, and when to go to the barn. Horses in a large and unfamiliar range area will follow cattle to water. And equines most certainly watch us humans. By watching us, they learn where the grain is kept, how to open gates, and whether or not to be frightened of that scary plastic bag. By observing our behavior in approaching and touching an object, a skittish horse may often be convinced that the object is harmless. Patience and attention to equine signals allow you, the handler and trainer, to associate the correct idea with the intended result and reinforce desired behaviors.

### THE CONCEPT OF PRESSURE AND RELEASE

Along with the concept of associating ideas comes the theory of pressure and release. Horses and mules respond to pressure, usually by moving away from it, and are rewarded by the release of that pressure, whatever it is. For example, if a foal wanders too close to a strange mare, the mare may threaten (pressure) the foal by baring her teeth and pinning her ears. The point of release occurs when the foal moves away and the mare relaxes her threat; the foal realizes that he has been released from the mare's discipline, because he did what she wanted him to do.

When an animal's movement is controlled by another force, he submits to this force. He will remember this submission and the results. When you say "over" while pushing against your mule's hip and you release the pressure just as he steps over, the mule will begin to associate the release of pressure with the action of stepping over, and realize that if he steps over, no pressure will be forthcoming. Eventually he will learn to step over in response to the word only.

Likewise, if your equine chooses to ignore you and your leadership, you can attempt to regain your position by pushing him away from you consistently until he submits. This technique works because equines are herd animals and therefore have a strong desire to be part of the "herd" for which you are the leader. It is best done in a confined space, such as a round pen, where you can push the animal away, usually by tossing a rope at him, in circles around you until he chooses to turn toward you, licks his lips and, when you approach, stands quietly. The idea is that putting pressure on the animal to continually move away from the handler creates discomfort for the horse, while standing still results in a release of the pressure and becomes the preferred course of action.

## THE IMPORTANCE OF REPETITION AND CONSISTENCY

Don't expect your horse or mule to associate the release of pressure with the reward concept the first time you use it. For example, you will likely have to put pressure on your horse's hip 10 or more times before the idea sinks in that when you say "over," he should step to the side. Successful training requires repetition. Without consistency, however, all the repetition in the world will not help. The animal learns what you want only if you *always* repeat the same action and expect the same result.

As the handler, you set the tone of your animal's behavior with the tone of your voice and your body language. If you have had a bad day and are snappy, short, and pushy during your animal's training lesson — even one as short and simple as teaching the "over" command — you not only violate the consistency rule, but make your animal jumpy, nervous, and possibly rebellious. Expressing frustration or anger, no matter your mood, will never

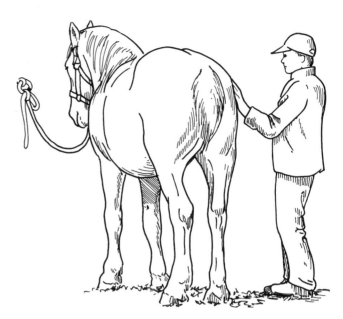

An example of the concept of pressure and release is pushing on the hip and releasing the instant the animal moves away.

improve the situation, but will rapidly damage any success that you have had. Keep your attitude positive, encouraging, and praising, and your animal will make the reciprocal effort to please.

With their exceptional memories, equines can learn complex and sequential processes, such as advanced levels of dressage and the complicated tricks performed by circus horses. They do not, however, learn all those steps at once. Each step is taught separately, with much repetition and consistency. But it must be *good* repetition. Going to the end of the driveway and always turning around at the same spot is not a good form of repetition. Good repetition consists of asking the animal to do the same process in the same manner and expecting the same result whether you are in an arena, on the trail, or in the field.

Make your lessons creative so that your horse or mule does not become bored. As you vary the environment in which you give your lessons, use repetition and consistency in the way you ask your animal to complete a task, the cues you give, and the results you expect. Remember that the mule learns in largely the same way as the horse; it just takes longer. The mule also needs shorter, more creative, and more concise lessons than the horse. He will, however, remember much more — both your victories and your failures — than the horse, and for a far longer time.

Horses and mules learn through many small steps, and your ability to follow through successfully with the small steps will help you refine your animal to whatever level you desire. Since your animals will be well trained when you receive them, the large lessons that they need should be few. Introducing a new watering device or some other barn feature will give you an opportunity to help the animal to learn. Let's say that the waterer is an automatic one with a ball to push to access the water reservoir. The equine will not know to push on the ball unless taught. You can teach him to push the ball by first familiarizing him with the waterer. Then bounce the ball up and down; the animal may startle at first, but will soon recognize the ball as harmless. Push the ball down so the water comes to the surface and splash your hand in it. Remember, you can lead a horse to water, but you can't make him drink. You will probably have to repeat this process more than once.

If you can teach yourself to apply pressure up to the precise point where your animal does what you ask, he will learn to associate the release of pressure with the requested behavior change. This technique takes practice, discipline, and patience on your part but will teach your equine to associate the release of pressure with the improved behavior. Remember always to repeat your lessons consistently, using a reasonable tone of voice and manner, and with a creativity that ensures that repetition does not equal boredom.

With consistency and practice, your team will learn to respond to light pressure when asked to move into position for hitching.

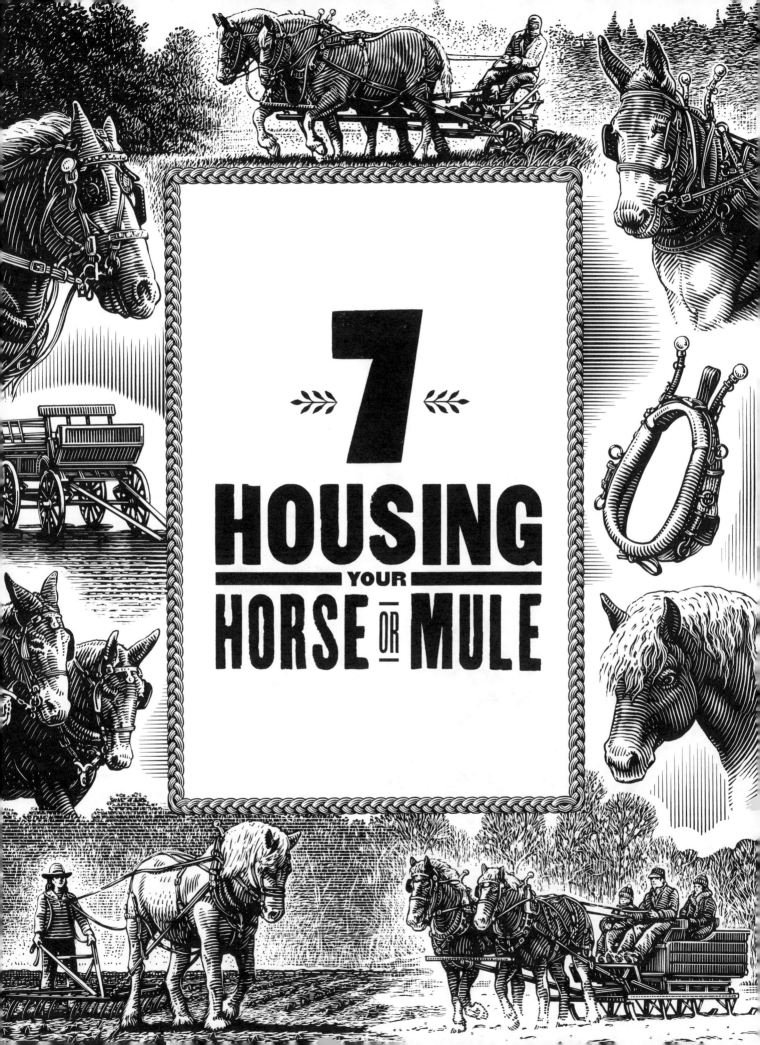

The day finally arrived for my friend Tom to bring home his first draft horse. He had anticipated this moment for a number of years, but the opportunity to purchase the horse had come before he completed his barn. He had electric fencing up around the pasture, but not knowing how the horse would react to unfamiliar conditions, he decided to use an old shed at a neighbor's place to confine the animal for the night. This shed did not have stalls, so the horse would be tied.

During the night, Tom decided to check on his costly investment. As soon as he entered the shed, he knew trouble was brewing. The horse obviously had not approved of standing in a new shed without his usual stall buddies and was down on the ground, with the lead rope wrapped around his neck. He was having trouble breathing and was wet with sweat from thrashing around.

Tom had the good sense to realize the gravity of the situation. He cut the lead rope and gave the horse a slap to refocus him and encourage him into an upright position. The poor animal had rope burns around his fetlocks and neck, and raw patches on his face where the halter had rubbed off the hair. The event shook Tom considerably; he spent the rest of the night holding the horse on a lead rope, letting him graze in the pasture.

What had Tom done wrong? He had not been properly prepared to bring home a horse. The previous owner probably would have been happy to keep the horse until Tom had accommodations ready. The urge to bring home a new animal is strong, but the consequences of being unprepared can be severe.

Even before you start your search for a draft horse or mule, ask yourself:
- Is my land zoned for livestock?
- How will I house this equine?
- Where will I store hay and equipment?
- How will I feed and water the animal?
- Is my pasture safe?
- Are my fences strong enough?

Draft horses and mules are large, strong animals with many needs: shelter, pasture, water, and hay. All these amenities are required and all cost money, space, and time.

## The Legalities of Equine Ownership

Certain legal issues, such as zoning regulations, pertain to all livestock, including equines. Others, such as liability statutes, pertain more specifically to equine activities. Regulations tend to be similar from one area to another but vary in detail with the locality, so it's always a good idea to check with your local authorities to find out the specifics pertinent to your area.

### CHECK OUT ZONING REQUIREMENTS

Although it may seem crazy to those of us who love equines, some people don't want to see or smell

A sturdy fence is an essential component of your equine pasture.

horses, mules, or any other kind of livestock near their homes. Zoning laws restricting livestock are often enacted in areas experiencing rapid suburban encroachment into farmland. Although historically people may have kept animals in your area, new laws may have been passed that restrict livestock allowances. Before purchasing a team, you need to know if you can legally keep equines on your property.

Some of the more common laws regulate the number of animals you can keep, as well as area, shelter, and fencing requirements. Horses and mules need room to move around. Unfortunately, some owners ignore this fact and squash horses together in small pens that are unsanitary and inhumane. This disregard of basic needs has caused lawmakers to set limits for how many animals you can have on a given amount of land. Different cities/counties/states have different zoning laws. As a potential equine owner, you must ensure that you fall within the required land allowance.

## Inherent Risks of Equine Activity

**Anyone injured** as a result of an equine activity and who wishes to sue must prove that the injury was beyond the inherent risks associated with the activity. What, then, is an inherent risk? Most statutes illustrate these risks with situations that describe dangers or conditions that are integral to equine activities. Tennessee, for example, lists the propensity of equines to behave in ways that may result in injury, the unpredictability of an equine's reaction to stimuli, surface and subsurface hazardous conditions, collisions with other equines or objects, and the potential of a participant to act in a negligent manner that may contribute to injury. Each state is different, so to make sure you are within the limits of the law when, for instance, you invite friends to enjoy a wagon ride with your team. Before you launch into equine ownership, obtain the full set of statutes for your state (most are posted on the Internet).

An equine activity sponsor or equine professional who is deemed responsible for an injury resulting from an incident outside of the inherent risks associated with the activity probably lacked caution or awareness of the equipment, tack, or participant. You could be liable if, for example, you provided faulty tack or equipment that resulted in injury, whether or not you knew it was faulty (under the law, you should have known). You could be liable if you failed to safely match the skill of a participant with your equine (such as handing the driving lines of your team to a rank novice). Even if the participant fails to divulge prior experience to you, it is your responsibility to determine whether the person is capable of safely handling your equipment and/or animals. You could be liable under the general term "negligence, or failure to use reasonable care." This broad statement, referring to incidents occurring outside the inherent risks, could get you into trouble pretty fast if the injured party was dead set on proving you were in the wrong.

Know what your state's statutes require and post the mandatory notices where they are supposed to be posted, using the exact language prescribed by the statute. If your state requires a release or waiver, have copies on hand for each participant to sign. Each state has different laws, so check with your own state for current legal requirements.

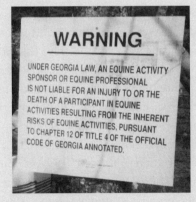

This example is typical of an equine activity liability statute notice to be posted where equine activities take place.

While you might believe you have the right to decide what kind of fencing you use, your neighbors or local lawmakers may not share your opinion. In rural areas, you will generally be allowed to choose your fencing, but if you are in a locale that's being subdivided and parceled off, you will likely run into regulations. So before bringing your animals to their new home, check your local county/city zoning laws to make sure that no one is going to inform you that you can't have electric fencing after you just invested in it.

## SAFETY ISSUES AND LIABILITY

Another important legal issue is the safety of anyone who participates in your equine activities. No matter how careful you are, someone could be hurt by your animals or equipment. Most states have enacted statutes designed to protect people and organizations engaging in equine activities from legal liability as a result of an injury occurring during the activity. Each state has different statutes regarding equine participants and activities, so become knowledgeable about the particulars in your home state.

Most statutes define who and what is covered. As the horse or mule owner, you will fall under the category of an *equine activity sponsor and equine professional*. In most states, this person is defined as anyone, whether or not for profit, who sponsors or provides the facility or animals for an equine activity — including, for example, asking someone over to inspect or evaluate your mule. *Activity participants* are all other people affected by the inherent risks associated with your equine activities. Many statutes define equine activity; the definitions that you need to pay attention to are for *equine training activities, teaching activities*, and, the kicker, "any other equine activities, of any type, however informal or impromptu, that are sponsored by an equine activity sponsor." This clause covers such things as impromptu wagon rides, letting your neighbor's kid sit on your draft horse, or any other seemingly harmless activity.

To confer protection under these statutes, most states require posting of spelled-out notices in specific places. Your state may also require contracts or releases using specific language, to be signed by anyone who participates in equine activities on your property. Determining your responsibility requires research and careful attention to your state's equine activity statutes.

## Sheltering Your Draft Animals

A draft equine needs shelter, and you need a place to store grain and hay. Grain must be kept in a controlled area that cannot be accessed by the horse or other animals, including mice and opossums.

The type of shelter required for your horse or mule varies with your climate and local laws. If you live in a temperate region, your equine will probably be fine with a three-sided shelter to block driving wind and rain. If you live in a colder area, consider something more substantial. Deep snow, icy rain, and driving, bitter winds may create dangerous conditions for animals, even those with long winter coats. Although many horses and mules live comfortably without shelter in northern Montana and other cold places, their owners assume the risk of losing one of them in a blizzard or other serious weather.

---

### Winter Coat

**Horses and mules** look slick and shiny in the summer after shedding the long, protective hairs that make them look shaggy and unkempt in winter. The untidy winter appearance is caused by long outer hairs that grow during cool fall days and nights to create a wonderful natural barrier against the elements. These long hairs allow rain to run off, minimize wind penetration, and reduce the chill of cold air. During particularly cold days, you will see your equine's hair stand almost on end, trapping a layer of air to insulate the animal and keep him warm.

If you blanket your horse or mule during the fall, he may not have the chance to grow this protective hair. Using a blanket creates an unnatural situation, so if you choose to blanket your horse or mule, don't expect him to survive a blowing blizzard without his man-made coat.

# BUILDING FOR DRAFTS

**Pete Cecil**
**Bend, Oregon**

Type of Farm: 5-acre (2 ha) hobby farm
Animals Owned: Retired Belgian gelding

Pete Cecil enjoys making his living building and restoring structures specifically for the equine. Before becoming self-employed in 1995, Pete worked for the United States Forest Service (USFS) as a historical preservation specialist. Many of the buildings that Pete restored were old barns used to house mule pack strings that serviced the fire lookouts. Pete still does historic preservation work. A recent project was the restoration of a beautiful timber frame USFS structure near the Canadian border in Washington state that once housed two 20-mule pack strings as well as an occasional draft horse.

When he's not building or restoring barns, Pete enjoys attending the Dufur, Oregon, threshing bee, as well as local plowing matches, shows, and sales. Pete found the time to build his own dream barn eight years ago. At the time he had a team of draft horses that he and his wife used to cultivate a garden and do work around their place. The barn Pete built for his own animals completely suits his needs and can be expanded to accommodate the new team he is thinking of purchasing in the future. His barn is a free-choice place for his animals: they can come and go in their box stalls as they please. Pete finds this freedom healthy for the horses, which choose to stay outside more than inside, even in inclement weather. He feeds his horses in the box stalls from stout removable feeders. Each feeder has a 2 foot by 3 foot (0.6 m × 0.9 m) hay storage area and a box on either end: one for salt and one for grain.

The barn's floor is compacted dirt, raised a foot above ground level to ensure good drainage. Pete has never had any moisture problems in this barn. He also included adequate hay storage. Pete's barn includes many tie rings, both inside the barn (and in the box stalls) and outside — a feature he has found exceedingly helpful. If he needs to quickly tie a horse, a ring is almost always within easy reach. The barn has much character, too: the wood was horse logged by Dave Mader in northeastern Oregon. The siding is ponderosa pine and the timbers are Douglas fir.

Many of the new barns Pete builds are constructed differently, more expensively, and more airtight than his, although he feels that his design is perfect for his own needs. Most of the barns he builds have box stalls, although some folks follow his suggestions to make portable dividers if they ever want to use tie stalls. Converting a box stall to a tie stall is far easier than the other way around.

When planning a barn, keep in mind that draft horses and mules are large animals and need more space than is usually provided by a barn built to accommodate light horses and mules. Pete suggests a few important considerations for draft horses and mules:

- Doors should be at least 7 feet (210 cm) high and 4 feet (120 cm) wide.
- Windows should be at least 4 feet, 6 inches (135 cm) high and screened on both sides the horses have access to.
- Stalls should be a minimum of 12 feet by 12 feet (4 m × 4 m).
- Keep the barn's flow-path convenient and easy to navigate. Smooth out or round off all sharp edges where wood, metal, or concrete come together.
- The overall stoutness of construction must take into account that an 1,800-pound (810 kg) animal has a lot more heft than a 1,000-pound (450 kg) animal.
- The barn should be convenient and accessible for the working horseman. Feed, hay, and tack should be easy to access.

Age and body condition also dictate the extent of shelter an animal needs. Horses and mules that are less than a year old or more than 15 years old or are in poor condition are more susceptible than others to inclement weather, so plan on barn shelter for animals in these categories.

## BUILDING A BARN

A barn is a loosely defined structure that comes in many shapes and sizes. For some people it is a place to store grain, hay, and the old car. For others it is an elaborate system involving timed lights, heated stalls, and immaculate alleys. Regardless of its specific design, your barn should be practical, with a secure place to store grain, a hay storage area (typically a loft, though many horse people advise against keeping a lot of hay in the barn itself due to the fire hazard), a tack area, and as many stalls as you have or will have animals. If you plan to build a barn, complete your construction before bringing your animal home. Nothing is more dangerous to a horse or mule than a work site strewn with construction materials.

## Proper Ventilation

An essential aspect of any barn is good ventilation. Have you ever noticed mist rising from the small openings of a barn? Have you seen moisture accumulate inside the windows and on other smooth surfaces? Have you noticed a strong ammonia odor when you enter the barn? If you answer yes to all these questions, chances are the barn is too tight and not getting sufficient ventilation. A barn that is too warm or poorly ventilated offers the perfect environment for the development of respiratory illnesses. A cool, well-ventilated barn is far better than an overly warm building with stagnant air, which can damage your equine's lungs.

The opposite extreme of poor ventilation is draftiness. When you're in the barn, do you feel a breeze on your face or other bare skin? If so, the barn is drafty, and this can increase the chances of respiratory illness in confined animals, especially foals, that are immuno-compromised or otherwise at risk. In most old barns, draftiness is unstoppable — you simply can't control wind that seeps through knotholes, around windows, under doors, and through

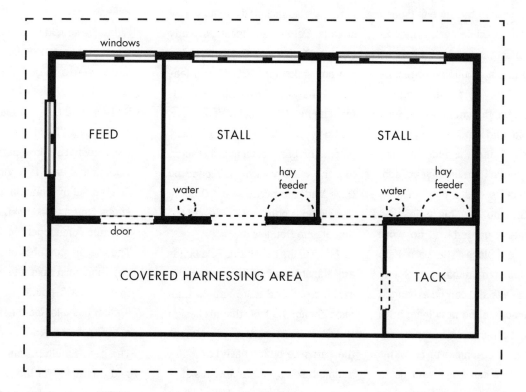

This floor plan includes stalls for two horses, a hay storage area, a tack room, grain storage, and an overhang that allows for harnessing a team in the dry.

> ## The Problem with Opossums
>
> **The North American opossum** carries the protozoal organism *Sarcocystis neurona* in its feces and can cause equine protozoal myeloencephalitis (EPM), a disease that affects the central nervous system of equines and can be fatal. Horses and mules usually ingest the protozoa from hay or grain contaminated with opossum feces. Protecting hay is difficult, but all grain *must* be secured so that animals cannot gain access.
>
> Although a vaccine exists that is said to protect against EPM, no scientific data has been found to prove this claim. Most veterinarians do not recommend it, and some researchers doubt that a vaccine will ever be effective against any protozoal disease.
>
> Fortunately draft horses seem to be less susceptible to EPM than lighter horses. A healthy immune system is key to preventing damage due to exposure. This disease is not contagious. Signs of EPM include depression, incoordination, partial paralysis, and loss of muscle groups, usually on one side of the body. Diagnosis for EPM is typically by the Western blot test, which looks for the presence of antibodies to the protozoa. Unfortunately, many of these tests yield false positives. Treatment is expensive and therefore often unfeasible. EPM can be mistaken for equine polysaccharide storage myopathy, a muscle-wasting disease. Have a competent equine veterinarian examine your horse before deciding on treatment or euthanasia for EPM.

gaps between dried-out planks. In a newer building, however, you should be able to minimize the little breezes that sneak in around openings by putting weather stripping along doors and windows.

## Grain Storage

Grain must be stored where your horses or mules have *no* chance of gaining access. Horses, in particular, will eat grain until they can no longer eat, leading to a potentially fatal case of colic and laminitis. Your grain storage area must also keep out mice, opossums, cats (some cats like to do their business in grain, because it's easy to dig), and other small animals. The grain must be stored in airtight containers that moisture cannot penetrate; moldy grain is unsafe for your equine.

For a small quantity of grain, a new plastic trash container with a tight-fitting lid works nicely. For a larger quantity, a no-longer-functional chest freezer with a lid that still seals is a great option, provided you ensure it cannot be a safety hazard for children. Freezers manufactured before 1970 were made with a latch system that does not allow the door to be opened from the inside. If you use a freezer with this latch system, remove the latch and lock (easily done with a screwdriver and WD-40), and then check to make sure that the lid may be opened by lifting on the front corners from the outside or pushing, even weakly, from the inside. Remember, however, that a wily equine can open trash cans and freezers, so any container *must* be stored in an area from which you can safely restrict your horses or mules.

## Hay Storage

The design of hay storage areas varies with geographic location. Folks in the humid Northeast and Southeast need barns with weather-tight lofts to store hay away from mold-inducing moisture. In semiarid and arid parts of the country, much of the precipitation comes in the form of snow in winter, when dampness is not an issue.

In such semiarid and arid plains and in the Southwest, bales may be stored outside if protected from dew, frost, and soil moisture, and properly covered to shed rain. A good-quality, carefully tied-off tarp provides sufficient top coverage. For small bales, pallets at the bottom prevent soil contact while allowing airflow. Loose hay left outside is traditionally stacked directly on the ground and arranged to shed rain. This system obviously results in some waste, but works well if you avoid feeding equines the moldy lower and upper layers.

If you purchase baled hay, you will soon notice a trend toward the production of large rectangular or round bales versus small rectangular bales. This trend makes hay production easier for the grower, but creates transportation and storage challenges for the small-scale livestock owner. Shoving these enormous bales into a shelter is not an easy operation, so they are typically left outdoors on the ground (resulting in some waste) with a tarp on top.

In the humid East, finding large bales that don't contain weather-damaged hay is a significant challenge. In the drier West, you can often get away with feeding from large bales, provided the tops have been properly covered to shed precipitation. To minimize moisture damage, large bales in many parts of the country are wrapped in sturdy plastic. In humid climates, dry-matter losses range from 35 percent in unwrapped hay to 7 percent in wrapped bales.

A large rectangular or round bale can weigh 1,000 pounds (450 kg) or more, compared to 60 to 120 pounds (27 to 54 kg) for the smaller rectangular bales, and therefore cannot be plunked into your wheelbarrow and rolled to your horses or mules. You will have to store these large bales in a convenient location for feeding. How you store them depends on how you feed. If you have a tractor to move your bales around with, you can stack them. If you feed large rectangles flake-by-flake, or fork out portions from large rounds, the bales must be low enough to be accessible. Regardless of your method of storage, always carefully inspect hay for mold before feeding it to equines.

## Tack Area

Many equine owners keep their grain and tack in the same room. Harnesses, collars, and other tack items are expensive and should be secured against theft and protected from the elements. Leather left outdoors degrades and needs to be kept away from both mold-causing dampness and overly drying drafts. Harnesses left in the company of mice will be chewed, and collar linings plundered for nest material.

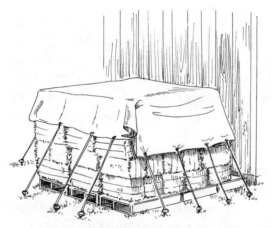

Hay bales stored outside must be covered and should not be laid directly on the ground.

Loose hay can be stacked outdoors with a top layer arranged to shed rain.

Large round bales are typically wrapped in plastic, with the ends left open to prevent mold.

Design your tack area with ample space for storing your harness and other tack, as well as for convenience. To maximize the available space, hang the bridle and collar above its associated harness. My preferred method is to hang the harness on an 18-inch (45 cm) long rack, with the hames sliding on first, followed by the saddle, and the breeching coming last. The collar hangs above on a hook, and the bridle hangs inside the collar on the same hook. To harness a horse, I first put on the bridle and collar. I return to the tack room, slide my right arm through the entire harness, grab the left hame with my left hand and the right hame with my right arm, and away I go. With careful hanging and practice, and a clean, well-organized tack room, you should be able to harness a horse from collar to last snap in less than three minutes.

An alternative way to hang a harness is to have three hooks: one for the breeching, one for the saddle, and one for the hames. Hang the harness on the wall in the horizontal position. Although some teamsters prefer this system, it takes up a lot of space. A single hook with three prongs is a good solution for small storage spaces.

## PLANNING YOUR STALLS

An equine stall may be one of two basic kinds — tie stall or box stall. Both styles are intended to house a single animal, although a large box stall may house a mare and her foal. The geometry and size of your stalls will be dictated by a number of variables:

- How large is your horse or mule? A Haflinger, for instance, does not require the same amount of space as a Shire.
- How much time will your animal spend in the stall each day? A horse or mule should be comfortable and able to lie down in a stall where he will be housed for long periods of time.
- What will your animal do in the stall? If the horse or mule uses the stall only for eating grain and standing when harnessed, a tie stall is sufficient.
- Is your animal a potential broodmare? A broodmare's stall should be large enough for the mare never to feel cramped, even when she and the foal are both lying down.

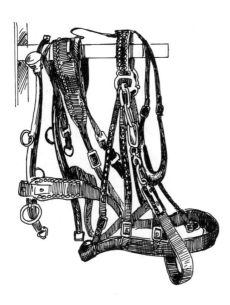

Harness hanging with hames on first (toward wall), then saddle, then breeching last

Harness hanging on a three-prong hook, with hames on one hook, saddle on another, and breeching on the last

## Using Tie Stalls

A tie stall is an open area just large enough for an animal to stand tied by a halter with no more than three feet (90 cm) of lead rope to prevent tangling. For many years tie stalls were used for carriage horses, brewery horses, and other regularly harnessed animals that worked all day in harness and came home to a place where they received their grain and hay, and little else. Because the tie stall is not large enough to allow the horse to lie down flat on his side to rest, it doesn't permit freedom of movement and restricts the animal's ability to lower his head and clear his air passages. Equines that are regularly stalled this way for long periods of time are susceptible to health problems.

Many old-order Mennonite and Amish horses still stand in tie stalls daily after work. But the difference between the Mennonite and Amish horses of today and the draft horses of yesteryear is that the former are usually turned out at night, rather than being left to stand in a cramped stall without getting any real rest. If your horse or mule lives outside and is in his stall only when eating and standing under harness, waiting to work, a tie stall may do the trick for you. To give yourself room to enter the stall to groom and lead an average-size draft horse or mule weighing between 1,500 and 1,800 pounds (680–820 kg), allow a minimum of 6 feet (2 m) in width and 8 feet (2.5 m) in length.

Tie stalls should have a barrier between them so each animal can eat in peace, without any kicking, biting, and general ruckus from animals that don't get along. A hay and grain feeder should be at the front of the stall, anchored solidly so it cannot be tipped over, as the tight space of a tie stall does not allow room for the horse to navigate around a tipped-over feeder. An alley running along the head of the stalls will let you easily put grain and hay into the feeders without entering the stalls. A water receptacle is not commonly part of a tie stall; rather, animals are led to water before and after work.

## Using Box Stalls

A box stall has four walls and a door and provides enough space for the animal to move freely, lie completely stretched out, and live in relative comfort compared to the entirely utilitarian design of a tie stall. Although some box stalls are used simply for holding an animal, most are used for activities such as feeding, harnessing, grooming, and housing the animal overnight.

The stall must be large enough to give your equine room to lie down completely without becoming cast, a potentially fatal situation that occurs when an animal rolls over and hits his feet against a wall, trapping himself on his back so he can't get up.

Many folks feed their horses and mules in box stalls rather than in the pasture. The box stall should be large enough for you to easily and comfortably navigate around the feeders and your animal without putting yourself at risk. You may choose to harness your animal in the stall, as well as do other activities such as grooming and hoof trimming.

A few large rings for tying the animal are a good addition to a box stall. You can position one near the feeder and two across from each other for cross-tying. For an average-size draft horse or mule weighing between 1,500 and 1,800 pounds (680–820 kg), a box stall should be no smaller than 12 feet by 12 feet (4 m × 4 m). A mare with foal should have at least 16 feet by 16 feet (5 m × 5 m).

Each box stall should be equipped with a water receptacle, grain feeder, and hay feeder. Even if you feed your horse or mule mainly in the pasture, you may need to nurse a sick or injured animal in the stall; at these times, the proper equipment makes a difference. The grain feeder may be as simple as a shallow rubber tub, or could be a combination hay/grain feeder. The water receptacle will make a big mess in a stall if tipped over, and must therefore be well secured.

Hay may be fed in a net or rack, but be aware that horses and mules graze with their heads down. It's nature's way of protecting the equine respiratory system by using gravity to keep dust and other irritants out of the lungs. Horses shipped with hay nets placed so high that they must reach up to eat have a high incidence of shipping fever, because dust and irritants reach their lungs. Hay fed in a pasture is typically placed directly on the ground to allow for more natural eating. However, since equines can ingest parasites from direct ground feeding, a

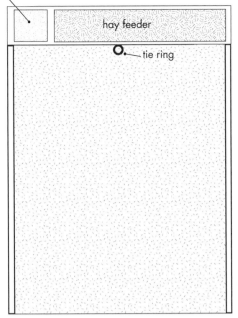

6-foot by 8-foot tie stall

12-foot by 12-foot box stall

pasture feeder — such as a metal feeder available from a livestock supply store — allows animals to eat with their heads down and also keeps the hay off of the ground.

In a stall, hay is usually fed in a container to minimize waste, as it will be trampled and soiled if placed on the floor. To help keep dust out of the lungs, place hay holders and grain feeders low enough that when the animal reaches for a bite, his nose is at least a foot below his chest. When the net is empty it will hang lower, so make sure that it can't hang so low the animal's feet could become entangled in it.

## What Kind of Flooring and Bedding?

The floor of either a tie stall or a box stall may be dirt or concrete. Some older barns have wood floors, which require intense management to keep from rotting away and are not ideal. A dirt floor is difficult to maintain, but concrete is hard on the horse or mule that has to stand on it for long periods of time. A rubber mat designed for equine stalls increases both the functionality of a dirt floor and the comfort of a concrete floor.

Stall bedding is important for maintaining hygienic conditions and should be removed at least once a day in a box stall and twice a day in a tie stall. Bedding that is allowed to remain wet from manure and urine encourages various hoof bacterial or fungal infections, such as canker and thrush, both of which are difficult to eradicate. Clean, dry bedding (in addition to dry corral conditions) is your best insurance against these infections. Frequent manure removal also keeps an animal from having to lie in the manure and removes the threat of manure-associated parasites. Stall bedding may be straw, wood chips, sawdust, or wood shavings.

### ◄ CAUTION ►
### No Black Walnut

Wood products derived from black walnut have been linked to acute and chronic cases of laminitis in equines. Although the specific toxin has not been identified, the cause of this kind of laminitis is attributed to black walnut shavings. Make sure that wood shavings, sawdust, or wood chips used for bedding contain no black walnut.

# Equine Sleep

**Horses and mules** sleep for only short periods at a time. Their natural habitat of wide-open spaces roamed over by predators did not allow them extended periods of off-duty time. The average equine needs 4 hours of sleep per 24 hours and will accumulate sleep in snatches of usually not more than 30 minutes. Researchers recognize three forms of sleep, all of which contribute to the daily sleep requirements.

**Short-wave sleep** (SWS) happens when the brain is inactive but sleep is shallow and the animal may be easily awakened. This form of sleep occurs between periods of work and eating and is similar to dozing, except that the animal is in a slightly deeper sleep.

Equines frequently grab a few moments of SWS sleep in a sternal position, propped up on the chest with the legs tucked underneath, but can also sleep standing up, thanks to the stay apparatus — an arrangement of locking joints and ligaments in their stifles and elbows that allows them to remain upright while asleep. One hind leg may be cocked during upright sleep, but never a foreleg.

**Dozing** is frequently done standing up and while on short breaks from work. These naps are brief and the equine is easily roused.

**Rapid-eye movement** (REM) occurs only when the equine is completely relaxed and lying flat on his side with his legs stretched out. The brain remains quite active, but the body is in a state near languor and the animal may not notice your approach.

During this stage of sleep, the eyelids flutter rapidly and the muscles lose almost all their tone. To experience REM sleep, the animal must feel completely safe, relaxed, and comfortable. An equine is most vulnerable in this position as he may not be easily roused.

An animal that lacks a large enough box stall, or is subjected to other conditions that don't allow him to lie flat out, rarely experiences REM sleep and as a result can become overtired, short-tempered, and neurotic.

> ## Moving to a New Pasture
>
> **W**hen you first put a horse or mule into the pasture, he will smell new scents, see new objects, and want to explore. Checking out the accommodations is natural, but should be done in a controlled manner and with a trusted friend. Snap a lead rope on the animal's halter and lead him to the water receptacle. Splash the water with your hand to make sure he understands the concept. Show him the salt/mineral block. Show him the boundaries of the fence by walking around the entire perimeter. If a stream runs through the pasture, walk him through the running water.
>
> In short, familiarize the animal with his new surroundings so nothing will be too surprising or scary when he is out exploring on his own.

This three-sided shelter for two horses has a framed-in section that serves as storage for tack, grain, and hay.

### PROVIDING PASTURE SHELTER

A three-sided shelter is a great option for healthy horses and mules turned out to pasture. It is not a place to harness your animals, keep grain or hay, or house a sick animal. It simply offers your animals protection from wind, precipitation, and hot sun. It can also protect the salt block.

A three-sided shelter need not be fancy, but must be safe and sturdy. It should not have loose boards or protruding nails, and should not be used to store halters, ropes, or other objects. Horses, mules, and other curious creatures create havoc with whatever is hung within their reach. If you wish to store grain, tack, or hay here, frame four walls to create a completely secured portion of the shelter.

## Planning Your Pasture Needs

Most horse and mule owners do not have enough pasture to maintain draft animals on forage alone. The concept that 2 acres (1 ha) of pasture are sufficient for draft horse management cannot be accepted as gospel because of variability in soils, precipitation, and vegetation. In dry-land situations such as in the Midwest and West, you could be looking at 28 acres (11 ha) or more to provide the proper amount of forage for your horse.

A 1,800-pound (820 kg) draft horse consumes between 25 and 30 pounds (11–13 kg) of forage a day on a maintenance-only program. Although pasture is an important component of any equine's overall management, it provides only some of the calories a hardworking draft equine needs.

### EXERCISE AND ENJOYMENT

If nothing else, pasture gives horses and mules a place to kick up their heels, roll in the dust, and be carefree. During winter, when draft animals usually have less work to do, the ability to kick and snort and run around is important. Horses and mules need to stretch their legs and blow off steam, and if you don't give them the opportunity to do it on their own, they will find a less desirable circumstance in which to do it. Being cooped up in a stuffy stall leads to bad habits, overly excitable animals, and possibly explosive moments for even the best-trained animals. Regular exercise is vital for overall health, especially of the feet and legs, the lungs, and the heart. Equines that don't move around enough are at risk for a variety of ailments.

Horses are built to run, and if they don't get this opportunity in the pasture, you may find them expressing themselves at less opportune moments. A favorite sight of mine is our entire herd racing in a steaming, snorting mass from one end of the pasture

Uninhibited exercise is important for your healthy draft equine.

to the other, running just to hear their own earthquakelike rumble over the ground. Even an animal that is worked hard every day and does not need the exercise still finds the energy to spar playfully with his pasture mates when turned out.

## PASTURE MATES

If you opt for a single workhorse or mule, consider getting him a companion. Equines are herd animals and may develop vices when left alone. Horses generally interact quite well with other livestock. Herds of horses frequently graze amiably with herds of cattle.

Goats and horses are also often pastured together. Mules, on the other hand, have been known to be aggressive toward foals, goats, calves, or other small animals in their pasture. Not all mules are aggressive toward small animals, but if you plan to pasture your mules with a herd of goats, for example, take this factor into consideration.

One horse or mule by himself in a pasture is a sad sight, but an even worse sight is two or more equines that cannot get along. Given enough time, horses develop a pecking order that keeps the most timid one at the end of the line. A small lot does not allow sufficient space for the submissive animal to gradually work himself into the personal space of the dominant animal, so if you don't have enough space for the animals to work out their relationship, keep them separated. Some equines exhibit aggressive behavior toward foals. For this reason, mares with foals should be pastured separately from geldings, mules, or dominant mares without foals.

When two equines are introduced for the first time, they should each be restrained by a capable human in case things turn sour. Another option is to keep them on opposite sides of a well-maintained fence, where they can see and smell each other but not physically interact while they get acquainted. At any rate, never turn horses or mules out together willy-nilly before introducing them, or the result could be a nasty kick wound or damaged fences.

As herd animals, horses and mules are happiest when they are with their buddies.

On our farm, my dad kept a pony stallion named Sampson to tease our mares during breeding season. Because the pony posed no threat of settling any of our draft mares, he remained in the pasture with them throughout the year. The mares never accepted Sampson as one of their own and harassed him continually — until Sally arrived. When we turned Sally out into the pasture the mares, seeing in Sally a threat to their pecking order, strove to make her submissive. Sally would have none of it, and being a massive horse, had no reason to run or fight; she simply put her heels to a couple of the mares, and peace soon settled over the pasture.

Sampson became interested in this new mare and set out to make her acquaintance. Within a few days, Sampson had in Sally a massive wall that stood between him and the other mares. Thereafter, he stuck fiercely by Sally's side. Sally did not seem to notice the pint-sized Sampson, and for years you could see the two grazing contentedly side by side.

## Insect Protection

**During the summer months,** horses and mules spend a lot of time violently swishing their tails and swinging their heads in an attempt to dislodge pesky flies. Flies and other insects are not just pasture pests — they can be catalysts for accidents and vectors for disease. I have experienced more than one horse bucking because an enormous horsefly tried to lunch on his hindquarters. Protecting your equine from insects involves a multifaceted approach to minimize the nuisance level to your animals in the pasture and barn.

### In the Pasture

- Remove anything that holds rainwater — old tires, buckets, and so on. Fill any deep depressions in the ground with soil.
- Keep water receptacles clean and fresh.
- Don't spread fresh manure on pastures during hot weather, when flies are laying eggs and hatching. Instead, spread manure during winter, when flies are not active and when grazing animals eat more hay than grass.
- Deworm your animals regularly to kill parasites before they hit the ground.
- Outfit your animals with fly masks when the flies are bad. Fly masks or nets are also a good idea for sensitive animals that are bothered excessively by insects.
- Make sure that shade is available to provide relief from both insects and the sun.
- Apply insecticide and repellent to your animals. A variety of preparations — chemical or natural — may be found at any equine supply outlet.
- Avoid overcrowding pastures, which causes excessive manure buildup and encourages rapid spreading of vector-born diseases.

### In the Barn

- Install fly tape in the aisles.
- Dust stall manure with fly powder when the animals are turned out to pasture; clean out the manure right before you let the animals back in, allowing time for the dust to work on the flies.
- Remove manure from stalls daily.
- Keep water buckets clean and fresh.

> **◀ CAUTION ▶**
>
> **Insecticide Safety**
>
> Livestock supply stores offer a plethora of chemical insecticides labeled for equine use. When choosing your insecticide, *make sure* the label says it is safe for equine use. Most labels carry a "hazardous to humans and domestic animals" warning. Apply them with caution and always protect your hands and other bare skin. Read the entire label for other warnings. Or opt for one of the many available natural products.

## Building Suitable Fences

Fence height and strand frequency depends on the horses and mules being contained. For a stallion, the fence should be around 6-feet (2 m) high and constructed of a stout material, such as wire mesh on wood posts, that can withstand escape attempts. Foals can squeeze through small spaces and therefore are also best contained by a mesh fence. For the average domesticated mare or gelding, a 4½-foot (1.5 m) fence will suffice, and the options for fencing media are more varied.

Fencing must be highly visible for your equine. Horses have a way of getting injured on fences, no matter how well they are made. If your fence has a weak spot, your horse or mule will find it and exploit it at the first opportunity. Loose wires and boards, rotten posts, and unsecured gates can provide opportunities for your equine to exercise Houdini tendencies. Fallen trees or limbs often create gaps in a fence where equines get hung up trying to cross, or succeed in crossing and go exploring on their own.

Mules are usually more careful than horses and therefore more selective about the fixes they get themselves into. But not much will stop 2,000 pounds (900 kg) of weight at a dead run, and an equine that can't see a fence may go through it, irreparably damaging both himself and the fence.

## DETERMINING WHAT MATERIALS TO USE

Fencing options for equine owners are many and varied. They include electric fencing, flexible rail, vinyl, wood, vinyl-coated wood, wire mesh, and pipe. Conventional wood or pipe fencing may be reinforced with a strand of electric wire to discourage animals from leaning on it.

Barbed wire, an invention of great importance in the nineteenth century, should be avoided. Horses and barbed wire make a nasty combination; most veterinarians have seen plenty of lacerations, severed tendons, and ruined ligaments resulting from collisions with barbed wire. Barbed wire is often strung on T-stakes, handy implements for impaling an unlucky equine. Although it may be the most economical option in the short run, the cost of managing injuries could easily surpass the initial savings. Why take that risk if you can afford to replace an existing barbed fence?

Regardless of the material you prefer, fencing will be a significant investment. However, installing a safe, secure, and effective fence is critical to your success as an equine owner and will help to minimize your trips to the vet clinic. Fencing options vary in price, installation difficulty, and containment efficacy, requiring careful consideration of your land, budget, skill level (if you build your own), and animal safety.

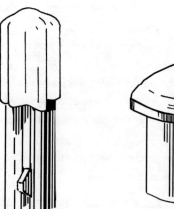
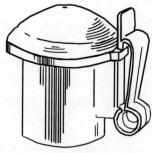

Putting caps on all T-posts reduces the risk of injury to your equines.

In addition to the fencing options, you also have post options for attaching the fencing. The most common is the metal T-post, which has been the cause of many puncture wounds in equines. If you use T-posts, always put a protective plastic cap over the top to reduce the risk of injury. Other options are solid vinyl posts and standard wood posts. Wood posts are expensive, but provide a safe and secure method for attaching your fencing.

## Electric Fencing

Electric fencing for equines comes in a number of styles including tape, braid, rope, coated wire, and mesh. Standard smooth-wire electric fencing is not suitable for equines because they can't see it well, although it can be made to work if at least the top strand is tape, rope, braid, or vinyl coated wire. Electric fencing offers a psychological as well as a physical barrier and is therefore highly effective, although in some areas it either cannot be legally used or must have signs posted identifying the fence as electric.

Most electric options designed for equines will stretch if pressure is applied, thus reducing the chance of injury should your animal run into it. Electric mesh fencing is an excellent option for foals, provided it is *always* live; otherwise, a small, curious animal can easily get tangled in it.

Electric fencing may be installed on solid vinyl posts, wood posts with insulators, or metal T-posts with insulators and protective caps. You will need a minimum of three strands of electric fencing to contain equines. The bottom strand should be no lower than 24 inches (60 cm) and the top strand no lower than 48 inches (120 cm).

### Introducing Equines to Electric Fencing

A **horse or mule** that is not familiar with electric fencing needs to be introduced to it before the fence becomes an effective barrier. The animal should discover the fencing himself (off lead) and in a nonthreatening manner. This discovery often occurs out of curiosity, as the animal leans to get a bite of grass or brushes against the fence. The animal should never be chased into the fence or brought up to the fence and forced to touch his nose to the current. One or two nonthreatening contacts are usually enough for an animal to learn to respect the fence.

Mesh electric fencing

Wire electric fencing

## Flexible Rail Fencing

Flexible rail fencing is made of high-tensile wire encased in polyethylene to form rails of varied widths. This style of fencing is tensioned, so if a tree falls on it or a horse steps on it, the fence will give rather than break. Flexible fencing may be installed on wood posts, solid vinyl posts, or metal T-posts with protective caps. The height and number of rails are the same as for electric fencing.

Flexible rail fencing

## Vinyl Fencing

Vinyl fencing consists of rails and posts made entirely of vinyl. The main benefit of this fence is its low maintenance — it never has to be painted. Posts should be installed on 8-foot (2.5 m) centers, since rails are typically 16 feet (5 m) long, and should be set in concrete. The rails should be fastened to the inside of the posts, to provide added strength in case a horse leans on the railings. Rails fastened to the outside of the posts are more likely to give way. You'll need a minimum of three rails for proper containment. Since vinyl can break on impact, adding a strand of electric wire along the top and middle rail will help to keep your animals away from the fence.

## Wood Fencing

Wood fencing, such as you see around many horse parks and racing stables, is beautiful and sturdy, but also high maintenance and attractive to the wood-chewing and cribbing horse or mule. In some areas, however, it's the only fence option allowed by local regulation. It should be made of a wood that will not quickly decay, such as red cedar or redwood. Set posts in concrete and fasten the rails (a minimum of three) with screws rather than nails. As with all railings, fasten wood rails on the insides of the posts. An electric wire along the top rail will reduce the equines' urge to chew or crib on the rails.

Vinyl-coated wood fencing combines the low-maintenance and anti-chewing efficacy of vinyl with the sturdiness of wood. Install it the same way that you would a wooden fence.

Wood fencing

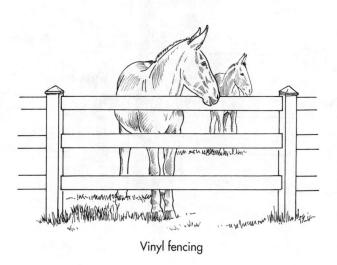

Vinyl fencing

## Wire Mesh Fencing

Wire mesh fencing is constructed of galvanized wire that is highly visible, low-maintenance, and rust-resistant. The mesh openings are 2 × 4 inches (5 × 10 cm), which is small enough that hooves cannot get stuck in them. Since the fencing is installed

Wire mesh fencing

under tension, it springs back into place if an animal steps on it, reducing both the chance of injury and the need for fence repair. The wire mesh comes in 48-inch (120 cm) and 60-inch (150 cm) heights, and may be installed by fastening on the inside of wood posts, metal T-posts with cap protectors, or most other post styles. This fence is excellent for confining stallions and foals.

## Metal Pipe Fencing

Metal pipe fencing is a solid and nearly indestructible but expensive option. However, a horse or mule that attempts to escape could seriously injure himself due to the unforgiving and unbending nature of steel. Mule owners are particularly partial to pipe fencing (a mule is usually not stupid enough to hurt himself trying to get out), which is typically constructed of high-quality tubular metal pipe at least 2½ inches (6 cm) in diameter. Install at least four rails — more for an equine escape artist — and set the pipe posts in concrete.

Metal pipe fencing

## Dealing with Stall and Pasture Vices

A bored horse or mule is a dangerous pasture or barn ornament. The equine body was made to wander freely while grazing 24 hours a day, and an animal that is confined in a stall or a dry lot may become bored enough to resort to abnormal behavior to fulfill these instincts. Such abnormal behavior is called a vice, and can become an almost incurable habit that the animal resorts to to keep from going stir-crazy. Although some vices are merely irritating or annoying, others can cause physical harm to your animal, as well as damage to your property.

**Box walking.** The animal endlessly walks in circles, damaging ligaments and joints and causing fatigue. This vice is almost impossible to cure, but may be minimized by giving the animal as much turnout as possible.

**Weaving.** The animal sways his head and neck from side to side while shifting his weight from one leg to the other, causing unnatural wear on shoes, joints, and ligaments. This vice is easily learned from other horses, and curing it is almost impossible. It may be minimized by stabling the animal as little as possible and providing distractions, such as plenty of forage or toys.

**Kicking.** This particularly annoying and dangerous habit can destroy stall walls and cause leg injuries. The animal frequently learns that this destructive habit receives quick attention, and therefore is encouraged to continue the behavior. This animal will likely never be cured of the vice, but as much turnout as possible helps.

**Digging and pawing.** If your stable floor is dirt, this habit can lead to enormous holes; if it is concrete, your horse or mule can rapidly wear down his shoes or hooves, as well as suffer damaged joints and ligaments. The habit of pawing or digging, typically done in anticipation of turnout or food, can quickly spread to other animals in the barn. Rewarding the behavior with the expected result only reinforces it, so make sure your animal is standing quietly before you feed him or release him.

**Wood chewing.** This habit, too, is easily passed from animal to animal in a barn, and is a sure sign of abject boredom. Wood chewing may lead to impactions, tooth wear, colic, and serious stall damage. Putting a nasty-tasting paste of cayenne pepper or horseradish on the wood stops the behavior, but usually only temporarily. Curtail a serious problem by installing metal guards along the edges of stalls and fences.

**Cribbing and windsucking.** An equine that bites an object (cribs), flexes his neck, pulls back, and gulps air (windsucks) experiences a rush of endorphins that can suppress pain and provide a feeling of wellbeing. Cribbing and wind sucking are debilitating vices, and are virtually impossible to cure. Cribbing and windsucking can occur as individual vices, but typically occur in combination. They may lead to excessive weight loss, gas colic, and excessive tooth wear. For temporary relief, a cribbing collar may be used in conjunction with nasty-tasting pastes applied to surfaces.

Many of the above vices stem from the boredom of confinement and are far less prevalent in horses or mules that are worked regularly or enjoy regular turnout. However, equines that don't have enough grazing or hay to occupy themselves in a pasture may resort to similar vices even when outside. Pasture vices may include the ones listed above as well as some new ones.

**Mane and tail chewing.** This vice is often seen as a normal grooming routine, but becomes a vice when engaged in obsessively. It not only results in an unsightly appearance, but can cause intestinal obstructions. Animals that are underfed can also exhibit this behavior.

**Manure and dirt eating.** Manure eating (coprophagy) is typical in foals two to six weeks of age and helps them establish normal gut bacteria, but in older animals may be a sign of boredom or lack of roughage, protein, or minerals. It may lead to colic, ulcers, and other digestive disorders. Once this vice is established, it is extremely difficult to stop. Keep stalls and pasture areas picked clean of manure and provide plenty of roughage.

Although heavy horses are less high-spirited and therefore less likely to exhibit nervous behaviors than lighter horses, the bored draft horse, too, can fall prey to any of the unpleasant habits described above. Once established, these vices may be nearly impossible to break, and frequently spread to other bored animals in the barn or herd. To keep an animal busy when he is not at work, consider pasture mates (equine or otherwise), music on a radio, horse toys, slow-release feeders, or an increased quantity of a lower protein hay.

# 8
# FEEDING YOUR HORSE OR MULE

The draft horses on our family farm all worked regularly. In winter, they worked at least twice a week; in spring, summer, and fall, they did daily strenuous work such as plowing, mowing hay, or cultivating. During the winter, hay made up the bulk of their diet, since the pasture grasses weren't growing. In the spring, summer, and fall, when the pasture was green and producing large amounts of forage, the horses ate some hay each morning between the time they were put in the barn (after being turned out all night) and the time they went out to work.

We fed grain in proportion to the work they were doing. Our grain of choice was straight oats, but often we mixed our own feed, using oats, cracked corn, molasses, and a nutrient blend. The horses received their grain in the morning and in the evening; when the teamsters came in for lunch, the horses ate more hay. This diet served our horses well and they always performed without trouble. The horses were neither too fat nor too thin; they were well muscled, athletic, and in excellent condition.

Knowing what I do now about the energy requirements of the equine, how the equine digestive system works, and how sensitive this digestive system is to colic and laminitis given the right circumstances and elements, I have changed some of my draft equine management practices. In particular, a diet containing more fat and fewer starches and sugars maximizes equine energy efficiency. Such a diet means feeding less grain while still providing the energy needed for high performance. So where does this energy come from?

Working draft horses and mules must be appropriately fed for the high amount of energy they use.

# THESE HORSES MAKE THEIR OWN FOOD

**Sam and Susan Arbogast
Hillsboro, West Virginia**

Type of Farm: 280 acres (113 ha); beef cattle, hay, small grains, corn

Animals Owned: Three Suffolks, one Belgian, four riding horses

Sam and Susan Arbogast typify the kind of folks who have draft horses not only because they love them, but also because they recognize their simplicity, power, and self-sufficiency. The four draft horses living on the Arbogast farm know what it means to work for their own oats — as well as the oats, corn, and hay of other animals on the farm.

The Arbogast farm is a longtime family farm, where Sam was born and raised. After 20 years in the Navy, he returned to his roots in West Virginia. He wanted to farm, but was decidedly against "all the big tractors." Since Sam was raised with draft horses, they were a natural choice of power for him and his wife. Many draft horse and mule owners dream of utilizing their animals to grow their own feed. Sam and Susan are realizing that dream. Their farm provides not only all the grain and hay for their eight horses and 40 head of cattle, but also some to sell. They mow and rake hay on more than 100 acres (40 ha). The baling (an average of 4,000 small bales and 300 large round bales) is done by tractor. The Arbogasts use about 300 round bales and 2,000 small bales of hay a year. They sell the other 2,000 square bales. They also plant and harvest 5 to 6 acres of oats and 5 to 6 acres (2–2.5 ha) of corn per year, netting them around 340 bushels of oats and 400 bushels of corn. What they don't use for their own animals, they sell.

The grass-finished Angus beef cattle Sam and Susan raise provide a nice diversification for their farm income. They raise around 40 head each year for 10 to 15 customers who regularly buy their beef. In addition to the beef cattle, the Arbogasts have sold shares of two Guinea Jersey dairy cows. Selling shares allows their customers to obtain raw milk and cream, while the Arbogasts remain within the boundaries of the law regarding the sale of raw dairy products.

In talking with Sam, you would never guess all this activity comes from a man 65 years old — an age when many folks hit the TV and golf clubs. Sam and his wife are active in promoting the education of youngsters. Susan teaches second grade and Sam has been involved with encouraging the local high school to recognize the need for horse-drawn equipment. Using an idea he saw at Horse Progress Days, he had the high school welding class build him a horse-drawn round bale mover.

Sam has made a point of obtaining equipment in excellent condition. He owns a corn binder, grain drill, and corn planter bought from the original owner who still knew the dates of purchase. Sam owns three number 9 McCormick Deering horse-drawn mowers that he keeps humming to cut the many acres of hay. Sam's motto of "It's not about what you earn, but what you can save" is evident in every aspect of his farm, and in his dedication to using horse power to produce feed for the livestock on his farm.

# The Equine Digestive System

Horses are primarily grazers by nature, obtaining most of their calories from eating grass and other pasture plants. The equine digestive system is technically monogastric, meaning it consists of a single-chambered stomach, in contrast to a ruminant (such as a cow or goat) and its four-chambered stomach. But the equine digestive system functions like a cross between a ruminant's and a nonruminant's.

Digestion occurs as a result of food being broken down by enzymes, bacteria, and other microbes. Enzymatic digestion of soluble carbohydrates, such as the starches and sugars in grain, occurs in front of the cecum. Bacterial and other microbial digestion of structural carbohydrates, such as the fiber in hay and pasture grasses, occurs in the cecum and large intestine and is sometimes referred to as hindgut digestion.

Digestible energy (DE) refers to the number of calories available to the animal from a food source. Digestible energy is what's left after the energy lost as manure is subtracted from the total energy available in the feed. Calories of DE are commonly reported in calories (also called kilocalories) or in megacalories (a megacalorie is 1,000 calories).

A hardworking draft animal requires an enormous number of calories. These calories come mainly in the form of fats and carbohydrates. Most draft horses and mules should obtain the majority of their carbohydrates from pasture and hay forage. Forage contains cellulose, which provides the structural carbohydrates needed for fatty acid formation. Fatty acids are important components of fat and provide energy for the bacteria responsible for digestion and vitamin production. Equines on total forage diets receive up to 70 percent of their energy from fatty acids.

Many draft horse and mule owners lack sufficient acreage to provide the necessary forage from pasture alone, and as a result much of this forage comes from hay. Concentrated feeds such as grain and pellets provide more fuel (nonstructural carbohydrates such as sugar) for animals that use extra energy, such as broodmares, senior equines, foals, and the work/performance horse or mule. Feed concentrates sparingly, and only if your draft animal needs the extra calories.

## ESTIMATING FEED REQUIREMENTS

We know that light horses in the 1,000-pound (450 kg) range need approximately 15 calories per pound (0.033 calories per kg), or 15,000 calories per day, but no one has definitively studied the feed requirements of draft horses. We do know, however, that a draft horse with twice the weight of a light horse does not require twice the calories. On average, draft mules require fewer calories per pound per day than draft horses. You can estimate the feed requirement in calories of DE for a draft by using the same formula for calculating the feed requirement for a light horse and then multiplying by 0.75 for a draft horse or 0.65 for a draft mule.

A quick rule of thumb is that a draft horse should get approximately 1½ to 2 percent of his body weight in forage (hay, pasture, or a combination of the two),

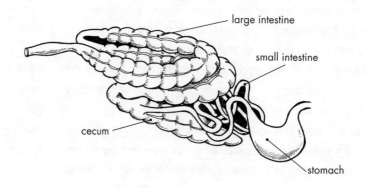

**THE EQUINE DIGESTIVE SYSTEM**

with concentrated feeds (grains, pellets, or the like) given only to animals expending extra energy. A 2,000-pound (900 kg) draft horse, therefore, should eat a minimum of 30 to 40 pounds (14–18 kg) of forage each day to obtain the maintenance amount of 22,500 calories per day. You can calculate the exact weight of forage needed if you know how many calories the forage has per pound.

A draft mule eats at the lower end of the range, or about 1½ percent of his body weight in forage. A 1,400-pound (636 kg) draft mule would need about 21 pounds (9.5 kg) of forage each day to obtain the necessary calories for a maintenance diet.

## Calculating Calories

**Calculating the number of calories** your draft equine needs per day is an important step in determining the required maintenance diet. Here are some examples of how to figure out your animal's needs:

To calculate how many calories a 2,000-pound draft horse needs:

$2{,}000 \times 15 \times 0.75 = 22{,}500$ calories per day

To calculate how many calories a 1,500-pound mule needs:

$1{,}500 \times 15 \times 0.65 = 14{,}625$ calories per day

You can calculate the approximate pounds of forage your draft equine needs by taking 2 percent of a draft horse's body weight, or 1.5 percent of a draft mule's weight. Examples:

To calculate the pounds of forage per day needed by a 2,000-pound draft horse:

$2{,}000 \times 0.02 = 40$ pounds of forage per day

To calculate the pounds of forage per day needed by a 1,500-pound draft mule:

$1{,}500 \times 0.015 = 23$ pounds of forage per day

Because the quality of forage varies dramatically, ensure that your animal receives adequate nutrition by cross-checking the above results. Start with the number of calories your animal needs per day and divide by the number of calories your forage contains, to determine exactly how many pounds per day of that forage your equine needs for maintenance. For example, to calculate how many pounds of grass hay (containing 500 calories per pound) are needed by a 2,000-pound draft horse to obtain 22,500 calories per day: $22{,}500/500 = 45$ pounds of grass hay

| Feed type | Typical calories/lb | Mcal/kg |
|---|---|---|
| Alfalfa hay, early bloom | 800–1,100 | 1.8–2.4 |
| Grass, early bloom | 600–800 | 1.3–1.8 |
| Grass hay, early bloom | 500–800 | 1.1–1.8 |

Because calories required for a draft equine per day is such a large number, the number is often converted to megacalories (Mcal). One Mcal is 1,000 calories.

Metric Calculations:
**Draft Horse**

weight $\times\ 0.033 \times 0.75$ = total Mcal needed per day for maintenance

weight $\times\ 0.02$ = kg forage needed per day

**Draft Mule**

weight $\times\ 0.033 \times 0.65$ = total Mcal needed per day for maintenance

weight $\times\ 0.015$ = kg forage needed per day

## BODY CONDITION SCORE

How do you know whether you are feeding your animal enough calories? How do you know if he is at a weight that maximizes his productivity and longevity? D. R. Henneke and others at Texas A&M University have developed a system of body condition scores that provides a good management tool. Based on observations of various points of the body, the system allows you to determine the optimum amount of body fat for different horses and mules, regardless of the breed. The amount of feed you give can be roughly estimated by formula, but only the body condition score of each equine will tell you if the formula fits that animal.

Once you have a score for your animal, as determined by the accompanying table, the logical question is: What is this animal's optimal score? For most equines, the optimal score is 5, which defines a horse or mule with some fat but not excess weight. An animal scoring below 5 lacks sufficient fat to remain healthy. Below a score of 3, the animal has virtually no fat reserves; protein begins to break

### Body Condition Score

| CONDITION | NECK | WITHERS | LOIN |
|---|---|---|---|
| 1 POOR | Bone structure easily noticeable, animal extremely emaciated, no fatty tissue may be felt. | Bone structure easily noticeable. | Bones protrude prominently from vertebrae. |
| 2 VERY THIN | Faintly discernible, animal emaciated. | Faintly discernible. | Slight fat covering base of bones protruding from vertebrae; remain prominent. Bony protrusions at sides of vertebrae feel rounded. |
| 3 THIN | Accentuated. | Accentuated. | Fat buildup halfway up on bones protruding from vertebrae, but easily discernible. Bony protrusions at sides of vertebrae cannot be felt. |
| 4 MODERATELY THIN | Not obviously thin. | Not obviously thin. | Negative crease along back. |
| 5 MODERATE | Blends smoothly into body. | Rounded over vertebrae. | Back level. |
| 6 MODERATELY FLESHY | Fat beginning to be deposited. | Fat beginning to be deposited. | May have slight positive crease down back. |
| 7 FLESHY | Fat deposited along neck. | Noticeable fat deposited over and along withers. | May have positive crease down back. |
| 8 FAT | Noticeable thickening. | Area along withers filled with fat. | Positive crease down back. |
| 9 EXTREMELY FAT | Bulging fat. | Bulging fat. | Obvious positive crease down back. |

Adapted from: Henneke, et al. *Equine Veterinary Journal*, 1983, vol. 5, no. 4, pages 371–2.

down from the muscles when the equine needs extra energy. Conversely, an animal with a body score of 7 or higher is more susceptible to health conditions such as laminitis and colic, as well as lameness issues. Some animals need a body score higher than 5 due to circumstances such as lactation, exposure to extreme cold, or exposure to severe stress.

In addition to the animal's basic energy needs, his body type is an important factor in the body score. Some horses and mules, typically those with short backs and necks, are easy keepers and can maintain a body score of 6 or 7 even on a reduced diet. Other horses and mules, typically those with long backs and necks, are hard keepers, and hay and concentrated feeds seem to pour right through them without adding any fat. Such an animal may remain at a 4, or even a 3, for most of his adult life.

Some animals are difficult to score and need to be carefully observed to determine their true score. During winter, when long guard hairs obscure the equine body, you will need to feel the body through the hair to get an accurate assessment.

| (see illustrations on next page) | | |
|---|---|---|
| **TAILHEAD** | **RIBS** | **SHOULDER** |
| Bones protrude prominently from vertebrae. | Easily discernible. | Bone structure easily noticeable. |
| Prominent. | Slight fat cover, yet easily discernible. | Accentuated. |
| Prominent, but individual vertebra cannot be visually identified. Hook bones appear rounded, but still easily discernible. Pin bones not distinguishable. | Slight fat cover, yet easily discernible. | Accentuated. |
| Prominence depends on conformation; fat may be felt. Hook bones not discernible. | Faint outline discernible. | Not obviously thin. |
| Fat beginning to feel spongy. | Cannot be visually distinguished, but may be easily felt. | Blends smoothly into body. |
| Fat feels soft. | Individual ribs may be felt, some fat beginning to be deposited between them. | Fat beginning to be deposited. |
| Fat is soft and beginning to be deposited along inner buttocks. | Individual ribs may be felt, but noticeable filling of fat between them. | Fat deposited behind shoulder. |
| Fat very soft, and noticeable along inner buttocks. | Difficult to feel. | Area behind shoulder filled in flush with body. |
| Fat building up. Inner buttocks may rub together. | Patchy fat appearing over ribs. Flank filled in flush with body. | Bulging fat. |

## BODY CONDITION SCORE ILLUSTRATED

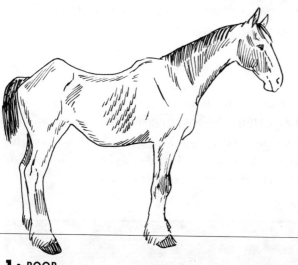

**1 • POOR**

**4 • MODERATELY THIN**

**2 • VERY THIN**

**5 • MODERATE**

**3 • THIN**

**6 • MODERATELY FLESHY**

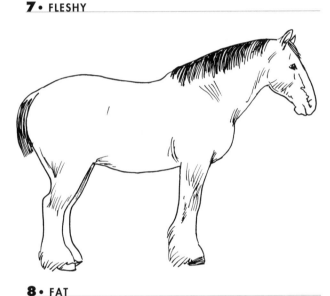

**7 • FLESHY**

**8 • FAT**

**9 • EXTREMELY FAT**

## Figuring Out Hay Needs

The major component of any draft horse's or mule's diet should be grass, or hay if sufficient pasture is unavailable. Animals, like people, have different metabolisms. Monitoring the body condition of your animals will let you know if the suggested 1½ to 2 percent of body weight of feed is too much or not enough to maintain optimal condition.

Feeding plenty of hay is particularly important during cold weather or other times when pasture forage is unavailable. During those times when your animal is on a maintenance-only diet, you will find you can feed less hay, or hay of a lesser quality, than when the same animal is working hard, lactating, or wintering in a cold climate.

Hay quality varies widely, so don't assume that one bale of hay is as good as the next. Good hay typically contains a mixture of grasses, such as timothy and orchard grass, and possibly some alfalfa. Mature grass hay has between 5 and 8 percent protein; alfalfa ranges in protein from 18 to 24 percent. Horses need only 10 to 12 percent protein, and mules need even less. So if you feed straight alfalfa hay, you must feed less of it.

Grass hay generally costs less than alfalfa and also keeps your animals busier eating, since they need to eat more to obtain the necessary nutrition. Because the equine stomach is designed to process frequent small amounts of forage over a period of hours, it constantly produces gastric acid, meaning ulcers may form in a stomach that is empty for much of the time, as tends to occur when an animal is fed smaller amounts of high-protein hay. Also, the more time your animals spend eating, the less time they have to get bored. To reduce the risk of stomach ulcers and boredom, feed more of a lower-quality hay, rather than straight alfalfa.

The ideal equine diet consists of a mixed hay containing both grasses and alfalfa; try to use hay that is no more than 50 to 60 percent alfalfa. An 1,800-pound (820 kg) draft horse getting only minimal forage from pasture should eat at least 25 to 30 pounds (11–14 kg) of hay each day.

## WHAT IS HORSE HAY?

You'll hear folks refer to horse hay and cow hay. Horses tend to have pickier palates than cows. Their digestive system is also more sensitive than that of cows. Good hay is the result of a combination of factors, all of which are important to your equine's health. Here are some factors to consider when buying hay:

**Color.** Green hay is better than brown hay. Green means the nutritive value (particularly vitamin A) is still high, while brown indicates leaching has occurred, usually as a result of rain.

**Mixture.** The percentage of grasses to alfalfa in your hay is not set or critical, although 50 to 60 percent alfalfa is just about ideal for a hard-working draft equine. If your animal is not exercising vigorously every day, is an exceptionally easy keeper, or is prone to founder, the percent alfalfa should be 20 percent or less.

**Quality.** Fine stems indicate the hay was not overly mature when cut. Leaves are more nutritive than stems, thus you want a high percentage of leaves.

**Cleanliness.** Dust and mold can contribute to respiratory problems and colic in your animal.

**Weeds.** Unless the hay is certified weed free (as required by many Forest Service lands and wilderness areas), it probably contains a few weeds. Weeds have little nutritive value and some are toxic. Look carefully at the hay to ensure there are as few weeds as possible.

**Beetles.** Blister beetles, (*Epicauta pennsylvanica, E. malculata, E. immaculata,* and *E. lemniscata*) can cause serious gastrointestinal and urinary tract irritation in equines, leading to severe colic. This problem is typically associated with semiarid and arid regions. The only way to be sure hay does not contain the beetle is to walk the field before it is harvested and check for beetles.

**Price.** Everyone wants a good deal. Unfortunately, you usually get what you pay for. Where you live, as well as bale size, dictates the price of hay.

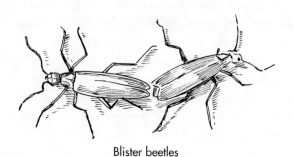

Blister beetles

> ### ◀ CAUTION ▶
> ### Deadly Toxin
>
> *Clostridium botulinum* is one of the most toxic substances known. It is closely linked to the tetanus form, *C. tetani,* but is quite a bit more potent. *C. botulinum* toxicity is associated with hay containing dead animals (which may have hidden in the hayfield and been hit by the mower, then baled along with the hay). Thoroughly inspect your hay as you feed it out, to ensure its quality and purity. Baling small animals and rodents is fairly common, so if you find a dead snake, mouse, or other animal in the hay, discard the entire bale. A vaccination for *C. botulinum* does exist.

## FEEDING HAY IN THE PASTURE

If you feed hay as a supplement to pasture forage, watch to see that your horse or mule is eating the proper amount of forage to maintain good body weight. If an animal picks up weight on the current diet, back off on the supplemental feeding of hay.

Animals that are given supplemental hay in the pasture require a safe and clean feeding area. Many horse or mule owners feed hay on the ground rather than in racks, particularly when feeding a lot of animals at once. Horses and mules do not feed as harmoniously from a single location as do cows. Feeding on the ground lets you spread the hay over a wider area to reduce dinnertime squabbles. When feeding more than one animal, spread the hay evenly so that all animals, including submissive ones, have a chance to eat. Feeding from the ground is also healthier for horses, as it allows them to graze naturally with their heads down.

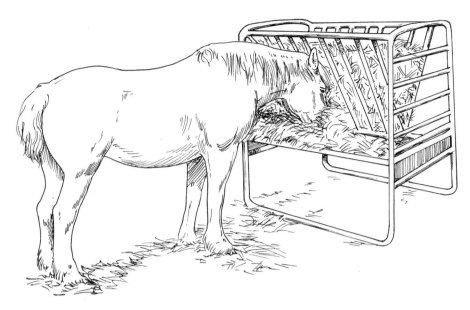

Feeding hay from a rack minimizes waste and spoilage, but feeding from the ground is more natural for equines.

## Figuring Out Concentrate Needs

In addition to the forage needed for maintenance, your draft horse or mule may need supplemental concentrates. Plan on supplementing the pasture and hay diet with a concentrated feed to provide extra energy if the animal you are feeding:

- Is pregnant or nursing
- Is young and growing
- Is a senior equine
- Has poor dental health
- Is in regular hard work
- Needs to gain weight

Concentrated feeds include such products as straight grain (oats, corn, barley, and the like), sweet-feed mixes, beet pulp, fortified-feed mixes (containing grain, alfalfa pellets, minerals, and vitamins), and feeds for the senior equine. Fats are also an excellent source of digestible energy.

Grain can be an important part of a balanced diet, but is unnatural feedstuff for the equine, and its use is often misunderstood. A horse or mule that is not doing heavy work can remain quite healthy on hay and forage alone. After all, feral horses and burros subsist entirely on forages.

A common misconception is that grain is the concentrated feed of choice. Quite the opposite — grain is a poor choice in terms of the energy it provides. Grain has only 35 percent of the digestible energy provided by fat. Grain can cause your draft animal to become hyperactive (and therefore dangerous) because of a sugar high from the large amount of starch, and can also cause a proliferation of bacteria, resulting in an endotoxic reaction that can lead to colic and laminitis.

### ◆ CAUTION ▶

#### Feeding on Sandy Soil

If you have excessively sandy soil, do *not* feed hay or grain directly on the ground. Horses or mules fed on sandy soil may inadvertently consume excessive amounts of sand, leading to sand colic or contributing to the formation of an enterolith — a serious intestinal obstruction. When feeding horses on sandy soil, put the hay in a low manger anchored with end posts set in concrete to prevent tipping.

Avoid feeding grain if you have other concentrated feed options. Instead, supplement your high-energy horse or mule with a fortified feed mix, alfalfa pellets, and/or other concentrated feeds. Here are some tips for feeding concentrates:

- Avoid sweet feed; it contains molasses, which tastes appealing but can result in excessive energy.
- Choose feeds with at least 12 percent crude fiber.
- Choose a fortified feed, rather than a grain mix.
- Look for pelleted feeds, which are more nutritious and often less expensive than grains or sweet feeds.
- Don't feed more than four pounds of concentrate per meal; if necessary, feed more often so you can feed less at a time.

## FEED ACCORDING TO NEED

For the horse, suggested concentrated feed quantities vary depending on his age and weight and the amount of work being accomplished. A horse doing light to moderate work, one to four hours a day, should consume no more than 5 percent of his total calories as concentrated feeds, while a horse working hard for four or more hours per day should consume between 20 and 40 percent of his total calories as a concentrate. The number of hours worked per day must be consistent; do not feed your horse 30 percent of his calories in concentrated feeds if you have just worked the horse for the first time this season. Any animal starting on concentrated feed, such as a performance or work animal coming off winter rest, must gradually work up to the concentrated amount needed for the work level at which the animal regularly performs for the season.

Feeding the senior (15 years or older) draft horse or mule is a matter of determining by trial and error the amount of concentrated feed needed to maintain condition. Some senior horses and mules don't need any, while others continue losing weight even when fed 50 percent of their dietary calories in concentrated feeds. In addition to all the hay they want to eat, many senior equines need concentrated feeds especially in the winter months, when they expend energy just to stay warm.

The mule generally requires less protein than the horse and stays in good flesh with less feed. A mule therefore is overfed when given the same number of calories as a horse of similar size. A mule does not lactate or bear young, ruling out two of the most common reasons for feeding concentrates. An animal on maintenance or light work alone should be fed *no* concentrated products.

A mule being used heavily every day and requiring extra energy from a concentrate should be fed the least amount you would feed a horse of the same weight doing the same work. Even then, monitor the mule's weight and body condition and adjust the amount of concentrate as needed.

## AVOID GRAIN OVERLOAD

Grain overload is a significant factor in laminitis and colic, due to its potential to disrupt hindgut digestion and/or bring about an endotoxic reaction. As the owner of a draft horse or mule, you *must* ensure that your animal receives *only* the exact amount of grain he needs. In no case should grain be fed in quantities exceeding four (three is better) pounds at one feeding.

---

### Oil It Up!

**Fat is an essential part** of the equine energy program. Your hardworking draft horse or mule can safely obtain essential energy through fat, which is 85 percent digestible and does not carry the same hazard for colic or laminitis as grain. One cup of vegetable oil may be substituted for 1½ cups of sweet feed, but introduce it gradually so the animal becomes accustomed to the taste. Top-dressing your equine's concentrated feeds with vegetable, flax, or wheat germ oil will also improve coat condition. In warm weather or if exposed to light, oil can turn rancid, so store it in a cool location.

## Calculating Concentrate

Calculating the amount of concentrate that your animal needs when performing a particular activity for a certain period of time will help you to keep his body condition score at the ideal number 5. Use the table below to determine the approximate number of calories your animal needs for a particular activity, then determine how many pounds of concentrated feed the animal needs to make up the difference in calories not covered by forage.

**EXAMPLE:** To calculate the amount of feed needed by an 1,800-pound draft horse plowing for more than four hours per day: 40,000 × 0.3 = 12,000 calories, or 20–40 percent of the diet, in concentrated feeds.

### Calories Needed for 1,800-pound (820 kg) Draft Horse or 1,500-pound (680 kg) Draft Mule*

| STATE OF EQUINE | CALORIES NEEDED/DAY | % CALORIES OF TOTAL DIET FED AS CONCENTRATE |
|---|---|---|
| Horse working 0–4 hours/day (maintenance amount; all activities) | 20,000–21,000 | 0 |
| Early lactating mare | 35,000–36,000 | 10–30 |
| Broodmare (last 3 months of gestation) | 24,000–25,000 | 10–30 |
| Horse working 1–4 hours/day (light activity) | 24,000–25,000 | 0–5 |
| Horse working 1–4 hours/day (heavy activity) | 27,000–28,000 | 5–10 |
| Horse working 4 or more hours/day (light activity) | 30,000–31,000 | 10–20 |
| Horse working 4 or more hours/day (heavy activity) | 40,000–41,000 | 20–40 |
| Mule working 0–4 hours/day (maintenance amount; all activities) | 14,000–15,000 | 0 |
| Mule working 1–4 hours/day (light activity) | 16,000–17,000 | 0–3 |
| Mule working 1–4 hours/day (heavy activity) | 18,000–19,000 | 3–7 |
| Mule working 4 or more hours/day (light activity) | 20,000–21,000 | 7–20 |
| Mule working 4 or more hours/day (heavy activity) | 23,000–29,000 | 15–30 |

The activities listed in the first column are assumed to be consistent over an extended period of time, rather than erratic hours here and there. Light activities include pleasure driving, pleasure riding, and the like. Heavy activities include plowing, mowing, harrowing, competitive driving, intensive training, packing, and the like.

*Nutritional data for mules is based on experience and extrapolation in the absence of hard research.

**EXAMPLE:** Consult the table below to calculate the amount of 2,000-calorie fortified feed mix needed to provide your plow horse with 12,000 additional calories: 12,000/2,000 = 6 pounds of fortified feed mix.

| FEED TYPE | CALORIES/LB (MCAL/KG) | POUNDS (KG) OF FEED |
|---|---|---|
| beet pulp | 1,300 (2.8) | 9.2 (4.1) |
| fortified feed mix | 2,000 (2.2–4.4) | 6.0 (2.7) |
| sweet feed | 1,500 (2.6–3.3) | 8.0 (3.6) |
| vegetable oil | 4,100 (9.02) | 2.9 (1.3) |

Just as some humans cannot metabolize certain foods, some equines cannot metabolize the soluble carbohydrates in grain. The inability of these animals to assimilate carbohydrates may lead to a severe muscle wasting condition called equine polysaccharide storage myopathy. For such horses, oil makes a suitable source of supplemental energy.

Unless you have a single animal in the pasture, don't feed loose grain. When a herd of horses or mules is fed loose grain, one animal (the boss) is likely to overeat while the others get little or none. Feed grain only where each animal gets a measured amount and will not be rushed while eating.

## ◀ CAUTION ▶
### Unprocessed Corn

The starch in unprocessed dried whole or cracked corn disrupts hindgut bacteria and subsequent digestion, which can lead to colic and laminitis. Avoid feeding corn to horses and mules unless it is in a digestible form — pelleted, extruded, or ground.

## Cold Weather Requirements

**Cold temperatures,** driving rain, vicious snowstorms — any inclement weather — increase the energy your horse or mule requires to maintain his body temperature. Although it was once thought that increasing the concentrate level (such as grain) was important for cold-weather maintenance, research has shown that the energy gained from hindgut fermentation of roughage such as hay actually provides more warmth for your equine.

The coat condition of your equine is particularly important when considering cold weather feeding needs. If your animal has just come off the show circuit and is slicked off, he will need far more digestible energy than a pasture-kept animal that has grown out a shaggy coat for winter.

A general rule for adjusting rations for cold weather is to increase the digestible energy by 1 percent of your original amount for every 1°F decrease in temperature below 45°F (7°C). Precipitation, coat condition, and wind speed are, of course, variables in this equation, but 45°F (7°C) is considered the general break-over point for an increase in digestible energy.

Keep your equines warm during cold weather by making sure they get sufficient calories.

## Does Pasture Provide Enough Forage?

Grass is a small part of the daily diet of some equines, simply because insufficient land is available for pasture forage. For other equines, pasture forage is all they need until winter comes and their diet is supplemented with hay. Whatever your case, you must remain aware of the condition of your pastures and know, at least generally, what your animals are eating.

When a significant part of the equine diet is pasture, healthy pasture forage is essential for a healthy animal. A healthy pasture has a minimum of weeds, is grown on soil that is well fertilized and contains the correct minerals necessary for plant growth, and has a variety of grasses and other forage plants suited to the geographic area.

### ESTABLISHING GOOD FORAGE

Does your pasture look like a manicured lawn? Probably not. Most likely it looks a bit ragged at times, is spotty, and includes a variety of plant species. These aspects are common to all pastures. The trick is to ensure that the variety of species present constitute sufficient nourishment for your animal. The specific species you select to grow will largely depend on what grows best in your geographic area.

To determine if your pasture forage is of good quality — containing the proper (not too much, but enough) amount of minerals and other nutrients necessary for your equine — you have to start with the soil, for that is where plants receive their food for life. Have your pasture soil tested at least every other year; once a year would be better. For help in learning about proper sampling protocol, as well as in testing your soil for critical minerals, consult your local USDA Service Center or state university USDA Extension Service. If you grow your own hay, test the soil on your hay field, as well. If you don't grow your own hay, ask the producer if the soil is tested. Your state Extension Service or USDA Service Center, as well as your veterinarian, can tell you if you live in an area that is deficient or toxic in a certain nutritional element.

The Extension Service or USDA Service Center can also help you select appropriate forage species for your area, including those that winter well in your climate, are drought or moisture resistant, and provide the optimal nutritive value for your animals.

Regular mowing to reduce weed seed maturation and spread is important for pasture health. Also make it a point to keep manure concentration to a minimum. Many equines like to defecate in the same spot in the pasture, a situation that can lead to sterile soil after years of concentration. Keep the manure spread out on the pasture with a harrow or other spreading device.

### ROOTING OUT TOXIC PLANTS

Too many equine owners overlook toxic plants until one day an animal turns up dead in the pasture. Horses and mules generally steer clear of poisonous plants, yet these plants still pose hazards. Some equines develop an appetite for certain toxic plants found in large quantities in their pasture. And all equines are tempted to nibble at potentially poisonous plants when a grazed-down pasture offers little else to snack on. Walk your pasture frequently to see what's growing there. Thoroughly familiarize yourself with the plants known to be fatal to equines.

Some plant species can cause allergic reactions that are not generally fatal. Other plants can be deadly. An all-too-common cause of equine poisoning is pitching ornamental brush clippings over the pasture fence. Many ornamentals are poisonous to a horse or mule. Ornamental shrubs, trees, and plants are the most common culprits in toxicity cases seen by veterinary practitioners. Don't throw clippings over the fence into a pasture unless you know with absolutely certainty they will not harm your animals.

## Common Poisonous Plants

| PLANT | RANGE AND HABITAT | TOXIN AND SYMPTOMS | TREATMENT AND PREVENTION |
|---|---|---|---|
| **Amsinckia species** | The amsinckia species contains several members linked to cirrhosis of the liver. Fiddleneck, fireweed, buckthorn, tarweed, and yellow burr weed are all plants found in semi-arid regions of Idaho, Washington, Oregon, and California. | Poisoning occurs when mature amsinckia with seeds are baled into hay or contaminate grain. In removing the plant's toxins (primarily pyrrolizidine alkaloids) from the equine's body, the overloaded liver becomes scarred and can no longer filter out the toxins. The result is disturbances of the nervous systems, producing signs such as head pressing, recklessness, aimless wandering, ulcers of the mouth, listlessness, yellow membranes, and eventually anemia and death. | Cirrhosis of the liver cannot be cured. |
| **Black locust** *Robinia* spp. | Native deciduous tree of eastern North America often forming pure stands, naturalized in much of West. Tolerates wide range of environmental conditions from wet to dry areas. | Robin, a glycoprotein similar to ricin (see castor bean), found in all parts of the plant. Highest concentrations in new growth, seeds, and bark. Ingesting the leaves, twigs, or bark causes colic, diarrhea, dilated pupils, weak and irregular heartbeat, depression, and death. | No known antidote; provide supportive therapy. Remove *Robinia* species from areas frequented by horses. |
| **Black walnut** *Juglans nigra* | Large deciduous tree native from New England and southern Canada west to Great Lakes and south to Gulf and Atlantic Coasts. Naturalized in parts of the West. | Toxin unidentified, possibly juglone. Nibbling on bark and branches or bedding containing black walnut shavings leads to colic and laminitis. | Treat for laminitis. Remove black walnuts from pastures and other areas around horses; avoid using black walnut shavings for bedding. |
| **Brackenfern (pasture brake)** *Pteridium aquilinum* | Tall fern native to much of the United States and southern Canada except the Southwest. Tolerates many environments including waste ground, wetlands, forest, and burned areas. | Thiaminase, an enzyme that can induce thiamin (vitamin $B_1$) deficiency. Symptoms of thiamin deficiency include blindness, weakness, and depression. Ingestion can also depress bone marrow, dangerously lower blood clotting factors, and produce massive hemorrhaging and possibly cancer. Horse may stand with back arched and legs apart. | Poisoning most often occurs from ingestion over extended period of time. Remove plant from hay, hayfields, and pastures; provide prompt injections of thiamin hydrochlorate when symptoms of thiamin deficiency appear. |
| **Castor bean** *Ricinus communis* | Shrubby tender perennial naturalized in southern regions of North America. Commonly grown as an annual as far north as southern Canada. | Ricin, an extremely toxic protein found throughout the plant but most concentrated in the seeds. A few ounces of ingested seeds can be lethal to horses. Symptoms appear a few days following exposure and include colic, diarrhea, sweating, greatly increased heart rate, and death. | There is no known antidote for ricin poisoning. Administer activated charcoal and intravenous fluids that include vitamin C. Avoid planting castor bean near areas frequented by horses. |

## Common Poisonous Plants *(continued)*

| PLANT | RANGE AND HABITAT | TOXIN AND SYMPTOMS | TREATMENT AND PREVENTION |
|---|---|---|---|
| **Cherry** *Prunus* spp. | Deciduous shrubs and trees native to southern Canada from Atlantic provinces west to Rocky Mountains and south to Texas and Florida. | Cyanogenic glycosides are present in all plant parts except the flesh of ripe berries. Damaged or wilted leaves more toxic than fresh. Ingestion results in release of cyanide that inhibits blood hemoglobin from releasing oxygen to tissues, resulting in death from anoxia (lack of oxygen). Symptoms include difficulty breathing, cardiac arrest, and sudden death. Ruminant animals are more susceptible than horses. | Animal should be treated with intravenous therapy using sodium thiosulfate and sodium nitrite. Remove plants from areas frequented by horses. |
| **Cocklebur** *Xanthium strumarium* | An annual found in most areas of North America. Tolerates a wide range of environments: tilled and fallow fields; waste ground; pastures; and near lakes, rivers, and roadways. | Carboxyactractgloside, a toxin that inhibits cellular metabolism, is present in seedlings and burrs. Symptoms include convulsions; vomiting; liver damage; rapid, weak pulse; labored breathing; leg and neck spasms. | Supportive therapy is only treatment. Cultivate areas where seedlings may appear in spring. Remove plants from pastures and other areas used by horses. |
| **Crotalaria** (rattlebox) *Crotalaria* spp. | A large group of annuals and perennials native to many warm climates worldwide. Most poisonings result from species grown in southeastern United States from Atlantic Coast to Texas. Listed as noxious weed in some Southern states. | Seeds and leaves contain toxic alkaloids that lower blood pressure and heart rate. | Provide supportive therapy. Avoid planting *Crotalaria* species as cover crops near pastures and other areas frequented by horses. Do not use hay from fields containing *Crotalaria* plants. |
| **Death camas** *Zigadenus* spp. | Perennials native to the southeastern United States west to Great Plains and Rocky Mountains and north to southern Canada. Some species also occur in Far West. | Plants contain toxic alkaloids that lower blood pressure. Bulbs and mature leaves contain highest alkaloid content. Symptoms include difficulty breathing, excessive salivation, weakness, and coma. | No known antidote. Less than 10 pounds will kill a horse. Administration of atropine sulphate may be helpful. |
| **Equisetum** (horsetail, scouring rush, jointfir) *Equisetum* spp. | A primitive perennial with an erect leafless stem native to most areas of North America. Tolerates a wide range of environments including waste areas, wetlands, pastures, and roadsides. Listed as noxious weed in parts of Pacific Northwest. | Thiaminase, an enzyme that can induce thiamin (vitamin $B_1$) deficiency, is the suspected toxin. Most often ingested when consuming baled hay. Cumulative effects include weakness, trembling, staggering, excitement, diarrhea, and loss of condition. | May be present in hay. Remove source of equisetum exposure and administer thiamin injections to restore proper levels of vitamin $B_1$. Avoid using hay from fields where equisetum grows. |
| **Fiddleneck** (fireweed, tarweed, yellow burr weed) *Amsinckia* spp. | Annuals native to western North America from British Columbia south into Mexico. Occasionally naturalized east of the Rocky Mountains. Common weeds in fields of wheat and grain. | All parts of the plant contain pyrrolizidine alkaloids that inhibit proper cell division and act as a liver poison. Symptoms include photosensitization, depression, weakness, odd behavior, diarrhea, and weight loss. | May be present in hay. Even small amounts can be lethal if ingested. Administer supportive therapy. Avoid using hay from fields where fiddleneck grows. Remove the plant from pastures and other areas frequented by horses. |

## Common Poisonous Plants (continued)

| PLANT | RANGE AND HABITAT | TOXIN AND SYMPTOMS | TREATMENT AND PREVENTION |
|---|---|---|---|
| **Lantana** *Lantana* spp. | Mostly shrubby perennials naturalized from southeastern United States west to California. Prefer sandy, dry environments but tolerate a wide range of conditions including coastal areas and alkaline soils. | All parts of the plant contain triterpene acids that produce symptoms including photosensitization, reddening of the eyes, difficulty breathing, increased heart rate, and bloody diarrhea. | Horses will eat lantana if grazing is poor; fatal dose is 20 to 30 pounds. Provide supportive therapy, especially fluids. Remove the plant from all pastures, hay fields, and areas frequented by horses. |
| **Larkspur** | Considered the number two cause of death (after locoweed) in cattle, sheep, and other livestock in the western United States, larkspur is fortunately not palatable to horses and mules, and thus not usually consumed in large enough quantities to be lethal. Larkspur is a beautiful plant that makes for a colorful but deadly pasture. | A horse or mule that eats sufficient quantities will experience profuse (often bloody) diarrhea, elevated and exaggerated gaits, leg twitching, prostration, paralysis, and coma. | Immediate veterinary intervention and laxatives to purge the remaining toxin are necessary to save the horse. |
| **Locoweed (crazyweed, milk vetch, poisonvetch)** *Astragalus* and *Ozytropis* spp. | Most common cause of livestock poisoning in western North America especially Great Plains where many species are native. Listed as noxious weed in Hawaii. | Different species of locoweed contain a variety of toxic substances including the alkaloid swainsonine, the glycoside miserotoxin, and selenium in concentrations high enough to produce poisoning. Swainsonine produces heart failure, reproductive disorders, locoism, edema, and birth defects. Miserotoxin produces locoism, difficulty breathing, and incoordination of rear legs. Selenium can produce both a syndrome called the blind staggers where animals walk in circles and have impaired vision and poor appetite, and alkali disease, indicative of chronic poisoning with symptoms including malformed hooves and hair loss. | No effective treatment known. Avoid pastures, range, and hay that contain locoweed. Animals not killed by locoweed poisoning are often permanently impaired. Some animals become habituated to eating locoweed. |
| **Nightshade** *Solanum* spp. | Large group of mostly Eurasian plants frequently naturalized over wide areas of North America. Close relatives include potatoes, tomatoes, peppers, and jimsonweed, all of which are similarly toxic. Common in waste areas, along roadsides and stream banks, and in cultivated fields. | Plants contain alkaloids including solanine that mimic atropine and impair normal nervous system function by inhibiting the enzyme acetylcholinestrase, causing muscle tremors and weakness. Some species also contain saponins that can produce digestive problems including excessive salivation and diarrhea. | Provide supportive therapy as indicated including the administration of intravenous fluids. Remove plants from pasture and range areas. |

## Common Poisonous Plants (continued)

| PLANT | RANGE AND HABITAT | TOXIN AND SYMPTOMS | TREATMENT AND PREVENTION |
|---|---|---|---|
| **Oleander**<br>*Nerium oleander* | Large shrub naturalized and used as a landscape plant in warmer regions of North America. Tolerates a wide range of conditions including drought. | All parts of the plant contain the extremely toxic cardiac glycosides oleandrin and neriine. Plants with dark-colored flowers may be most toxic. The glycosides inhibit normal cellular electrical conductivity resulting in erratic heart rate, rapid breathing, dilated pupils, diarrhea, and heart failure. Less than a quarter of a pound of this common ornamental will kill a horse. | Administer activated charcoal as soon as ingestion is suspected. Treat heart irregularities as they manifest with antiarrhythmic drugs. Provide supportive therapy as needed. Avoid giving the animal calcium or potassium as these elements exacerbate the effects of the toxins. Remove plants from areas frequented by horses. |
| **Poison hemlock**<br>*Conium maculatum* | Large biennial herb naturalized across much of North America. Found in pastures, along roadsides, and in wet areas. Listed as noxious weed in parts of Midwest and West. Not to be confused with water hemlock, which is actually more poisonous but less frequently eaten. | All parts of the plant contain toxins including the potent alkaloid gammaconiceine. Symptoms include slow heart rate, drooling, staggering, uncoordinated movement, and dark urine. | Administer activated charcoal when ingestion is suspected. Treat presenting symptoms and provide supportive therapy. Remove plants from areas frequented by horses. |
| **Rayless goldenrod (jimmyweed)**<br>*Isocoma* spp. | Woody perennials common to arid regions of Western North America. | Plant contains tremetol, which produces muscle tremors, weakness, stiffness, and depression when ingested over several weeks. The toxin is present in the milk of lactating animals and can poison nursing foals. | Remove plant from pastures where possible. Place animal on healthy diet. |
| **Red maple (swamp maple)**<br>*Acer rubrum* | Large tree native from Maritime Provinces to Minnesota and south to Texas and Florida. Common in a wide range of environments from forests to fields, pastures, and a variety of wetlands. | Leaves contain unknown toxin that damages hemoglobin and is most concentrated in wilted leaves, autumn foliage, and bark. Symptoms usually appear about 3 days after ingesting leaves and include rapid breathing and heart rate, weakness, depression, brown urine, and brownish blood. Small amounts can be quite harmful. | Supply supportive therapy including administration of intravenous fluids as needed. Blood transfusions may be necessary. Remove red maple trees from areas frequented by horses. Dried, wilted, or damaged leaves are more toxic than fresh foliage. |
| **Russian knapweed (Russian thistle)**<br>*Acroptilon repens* syn. *Centaurea repens* | A close relative of yellow star thistle and more toxic. Naturalized in waste areas, pastures, and along roadsides through most of southern Canada and Midwest United States south to Mexico and west to Pacific Coast. Reported less frequently in eastern North America. Listed as noxious weed in most of the United States. | Unknown toxin produces "chewing disease" with symptoms including uncoordination, restlessness, and repetitive chewing motions but inability to eat. Poison accumulates over many days of ingesting the plants. More toxic than close relative yellow star thistle. Disease is most common in western North America in early summer and midfall. | Once symptoms appear condition is untreatable. Supportive therapy is sometimes administered in mild cases. Euthanasia often indicated to prevent animal from slowly starving to death. Avoid pastures, range, and hay contaminated with Russian knapweed or yellow star thistle. |
| **St.-John's-wort (Klamath weed, goatweed)**<br>*Hypericum perforatum* | A naturalized perennial of meadows, roadsides, waste places, and prairies throughout much of the United States. Listed as noxious weed in Hawaii and parts of the West. | Toxic agent is hypericin that produces photosensitivity, and in severe cases, convulsions and coma. Unpigmented areas become sore, swollen, and itchy, with peeling skin. | Remove animal from direct sunlight. Mild cases can be treated with anti-inflammatory medication. Fresh leaves are much more toxic than dried foliage. Remove plants from pastures. |

## Common Poisonous Plants *(continued)*

| PLANT | RANGE AND HABITAT | TOXIN AND SYMPTOMS | TREATMENT AND PREVENTION |
|---|---|---|---|
| **Senecio** (includes groundsel stinking willie, ragwort, and tansy ragwort) *Senecio* spp. | At least 70 species that thrive in a wide variety of habitats throughout much of North America. Listed as noxious weed in Hawaii and parts of the West. | All parts of the plant contain pyrrolizidine alkaloids including jacobine and seneciphylline, which when ingested produce a syndrome called "Pictou disease." Symptoms include cirrhosis of the liver, jaundice, abdominal swelling, lung congestion, and agitation. | Liver disease initiated by pyrrolizidine poisoning has no effective treatment. Keep animal away from direct sunlight. Eradicate the plant from all pasture and range land. |
| **Sorghum grass** (includes Sudan grass and Johnson grass) *Sorghum* spp. | Group of drought-tolerant perennial and annual grasses most common in pastures throughout the South and Southwest. Listed as noxious weed in many states across the country. | All parts of the plant contain dhurrin, a cyanogenic glycoside. Ingestion results in release of cyanide that inhibits blood hemoglobin from releasing oxygen to tissues, resulting in death from anoxia (lack of oxygen). Symptoms include difficulty breathing, cardiac arrest, and sudden death. Sorghum can also accumulate toxic quantities of nitrates. | Keep animal calm and administer sodium thiosulfate and sodium nitrite intravenously. Removal of plants from pastures can be difficult. Use of herbicides followed by reseeding with select forage is indicated in severe infestations. |
| **Tall fescue grass** *Festuca elatior* syn. *Festuca arundinacea* | A common grass that is not harmful in itself but is often infested with a toxic endophytic fungus (*Acremonium coenophialum*). Listed as noxious weed in many Mid-Atlantic states. | A problem only for pregnant mares; may cause abortion, retained placenta, and failure to produce milk. | Test field for endophytic content of grasses if poisoning is suspected. In pastures and fields containing high levels of endophytes use herbicides to remove infected grass, and reseed with select forage. |
| **White snakeroot** *Ageratina altissima* syn. *Eupatorium rugosum* | Found in sandy areas in the Midwest. Often stays green in early fall, when horses might be tempted to eat it if other plants have dried up. Listed as noxious weed in Hawaii. | Contains the alcohol tremetol, which damages the heart muscle and liver; symptoms of poisoning include weakness, lethargy, loss of coordination, and stiffness. A foal can be poisoned through nursing if the mare eats the plant. | Remove plant from pastures where possible. Place animal on healthy diet. |
| **Yellow star thistle** *Centaurea solstitialis* | An annual closely related to Russian knapweed and similarly toxic. Naturalized in pastures, fields, roadsides, and disturbed ground throughout much of North America west of the Great Plains, less common in eastern areas. Listed as noxious weed in Hawaii and parts of the West. | Unknown toxin produces "chewing disease" with symptoms including uncoordination, restlessness, and repetitive chewing motions but inability to eat. Poison accumulates over many days of ingesting the plants. Disease is most common in western North America in early summer and midfall. | Once symptoms appear condition is untreatable. Euthanasia often indicated to prevent animal from slowly starving to death. Supportive therapy is indicated in mild cases. Avoid pastures, range, and hay contaminated with Russian knapweed and yellow star thistle. |
| **Yew** *Taxus* spp. | Native and naturalized needled evergreens; some species commonly planted as ornamentals. Native species prefer moist forests along Pacific Coast and throughout eastern North America. Introduced species most commonly planted from eastern North America west to Midwest and south to Gulf and Atlantic Coasts. | Needles, twigs, and seeds contain the alkaloid complex taxine, which depresses the action of the heart resulting in heart failure. The toxin is most concentrated in the needles during the coldest months of the year. | No effective treatment is known. Supportive therapy should be provided as symptoms indicate. Atropine sulfate can sometimes increase slowed heart rate. Remove ornamental plantings of yew in areas where horses have access. Native species should be removed from pastures. |

## Are Supplements Necessary?

Good quality grain, hay, and forage should meet most of a hardworking horse's or mule's nutritional needs, but may not meet them completely because of inherent deficiencies or toxicity level of certain substances in your local soil, or as a result of extreme energy output by the animal. Consult your veterinarian to learn more about supplements that might be called for in your geographic area and for the kind of work your animals do.

### PROVIDING SUFFICIENT VITAMINS

Many vitamins are required for your equine's health. These vitamins are typically present in sufficient quantity in pasture with a good forage base or in good quality hay. If your animal does not have access to good pasture or hay, or has just come from a home lacking these essentials, he may be vitamin deficient. In a rare case, such as when an animal has been driven to over-grazing particular plants because of poor-quality pasture, he may have a toxic level of certain vitamins or other nutrients in his body.

Vitamins are either water-soluble or fat-soluble. Water-soluble vitamins are processed in the equine intestine and include vitamin C and the many forms of vitamin B. Fat-soluble vitamins are stored in fatty tissues and other cells and include vitamins A, D, and E. Because they can be stored, they have the potential for building up to toxic levels. High levels of vitamin A and E are not known to cause any problems and may actually be beneficial. An excess of vitamin D, however, can produce lameness and weight loss if it builds up to toxic levels. Vitamin D toxicity is commonly associated with poor feed mixing or with eating plants that have high levels of vitamin D, such as wild jasmine *(Cestrum diurnum)*.

More typical than toxicity levels are deficiencies. Vitamin A deficiency, usually resulting from poor-quality, leached hay, is often manifested as susceptibility to lice infestation. Vitamin E deficiency is commonly associated with a lack of access to good-quality pasture or to hay containing alfalfa, clover, or other legumes. Vitamin E deficiency causes a horse to age faster than he should and can also appear in a young, growing horse as spinal disease or problems, or muscle dysfunction. A severe vitamin E deficiency can be evidenced by brown urine, a condition caused when the muscle-cell protein myoglobin is released during muscle breakdown.

Many commercially mixed, concentrated feeds provide vitamins. If your horse or mule shows signs of acute deficiency, your veterinarian may suggest supplementing his diet more heavily.

### PROVIDING SUFFICIENT MINERALS

Minerals are needed to keep an animal healthy, and are usually present in sufficient quantities in good-quality hay and grain. The soils in your local area, however, may contain too many or too few of these critical minerals, leading to toxicity and/or deficiencies in your animals. If you determine that your geographic area is low in a certain mineral, such as copper, iron, or selenium, no matter the quality of your pasture or hay, you must then supplement your equines' diet with a mineral supplement or commercially mixed feed ration.

**Iron** is a critical component of the oxygen-carrying protein hemoglobin in red blood cells. Iron deficiencies are typically associated with anemia, but anemia in equines usually stems from a deeper-rooted cause than low iron, and therefore rarely responds to additions of the mineral in the diet. A more likely problem is too much iron. Toxic levels of iron often result from a high iron content in the water, or from oral or injectable supplemental iron. High levels of iron in the blood can interfere with other trace minerals, as well as cause hepatitis and liver damage.

**Copper** is necessary to prevent a developmental bone disease known as osteochondrosis dissecans (OCD), which affects foals. To prevent OCD, both the pregnant mare's and the foal's diets must include sufficient copper. Equines are not susceptible to copper toxicity as far as anyone has been able to determine, but certainly may be copper deficient. Copper deficiencies appear to affect foals more than adult animals.

**Calcium and phosphorus** are required for building a strong skeletal structure. More important than their presence, however, is the ratio in which these two minerals occur in the diet. A ratio of two parts calcium to one part phosphorus is considered ideal for the healthy equine body. When the calcium level is lower than the phosphorus level, the body attempts to correct the ratio of necessary calcium in the blood by removing calcium from the body (primarily from the bones), causing the bones to become thickened and soft.

To ensure the correct calcium-to-phosphorus ratio in your equine's diet, you must provide the correct balance between good-quality hay/forage and grain. A diet containing high levels of grains and brans (which are high in phosphorus and low in calcium) in relation to hay and forage may result in a calcium-to-phosphorus ratio imbalance. To correct an imbalance, feed a commercially mixed ration containing the correct ratio.

While the equine diet must include enough calcium to balance the calcium-to-phosphorus ratio, avoiding too much calcium is just as important. Magnesium, which closely mimics the structure of calcium, can compound the problem of excess calcium. When equines consume alfalfa hay grown in soil inherently high in magnesium, this mineral, mixed with the high amounts of calcium already present in the alfalfa, can cause the formation of enteroliths, creating intestinal problems and leading to severe colic that requires surgery. The high amount of calcium or magnesium in alfalfa hay must be balanced with a source of phosphorus, such as a commercially mixed feed.

## The Importance of Selenium

Selenium is a critical element for the equine. It aids in the proper function of many cells, particularly muscle cells, and is a powerful antioxidant. Selenium deficiency and toxicity are two of the more common mineral problems, and are usually associated with specific geographic areas. Selenium toxicity occurs when animals ingest vegetation grown on alkaline soils, and is common in the western United States in arid states such as Utah and Nevada.

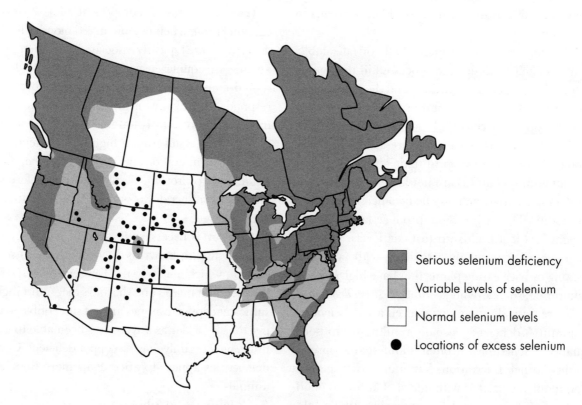

**SELENIUM DISTRIBUTION ACROSS NORTH AMERICA**

Plants in the locoweed and milk vetch families grow well in alkaline soils that are high in selenium, making these plants good indicators of a dangerous level of selenium in the soil. If either locoweed or milk vetch grows in your pasture, other pasture plants are likely taking up selenium at high levels, making the pasture an unhealthy forage option. Animals fed alfalfa hay containing large amounts of selenium can develop signs of selenium toxicity, as well.

Selenium naturally occurs in soil, with the normal concentration estimated at 0.2 parts per million (ppm). Chronic toxicity can occur when forage containing from 5 to 40 ppm is ingested over a sustained period. It may also occur from ingesting vegetation or grain grown in a high-selenium area. Birth defects, wandering or staggering (blind staggers), paralysis, and blindness are all associated with chronic selenium toxicity.

Acute toxicity can occur where ingested vegetation contains 10,000 ppm, or after an animal consumes selenium in the amount of just 0.00005 ounces per pound of body weight (3.3 mg/kg). Signs of acute poisoning include abnormal movement; dark, watery diarrhea; elevated temperature; weak, rapid pulse; labored respiration; bloating and abdominal pain; blue or pale mucous membranes; and dilated pupils. No known treatment will reverse the effects of poisoning, and often the animal dies before a diagnosis can be made.

Acidic soils have the ability to bind selenium in chemical forms that are unavailable for plant uptake. These soils, therefore, do not have vegetation associated with selenium toxicity. In the eastern United States, where acidic soils are common, selenium deficiency may occur. In these regions, supplementation of selenium is necessary for good health. Signs of acute selenium deficiency mainly include muscular dystrophy in foals, causing tongue paralysis that inhibits suckling; weak, unsteady gaits; rapid heart rate with arrhythmia; and labored breathing. Chronic selenium deficiency in older equines is indicated by weight loss, emaciation, general muscle weakness, and diarrhea. Selenium deficiency may be indicated by brown urine, a condition caused when the muscle-cell protein myoglobin is released after muscle breakdown. Salt/nutrient blocks in selenium-deficient areas typically contain supplemental selenium. For equines, the nutritional requirement for selenium is only 0.000002 ounces per pound of body weight (0.10 mg/kg).

## The Need for Electrolytes

Electrolytes are salts and other minerals needed for healthy equine cells, particularly the cells involved in strenuous energy output by the heart, other muscles, and nervous system. A hardworking draft horse or mule sweats a lot during a day of labor under harness; sweat contains salts that need to be replaced. Equines also flush electrolytes from their bodies when they urinate.

A trace-mineral salt block, available at any feed store, is an excellent source of salt for hardworking equine. It contains not only salt, but also trace

### Going Overboard with Supplements

In today's equine world you can find supplementation for just about everything under the sun — a shiny coat, strong hooves, healthy joints, and a multitude of other concerns. Go to any feed store or equine catalog and you will find a plethora of products your animal supposedly must have for peak performance. You could spend a fortune buying supplements your equine may not need.

If your animal has an obvious need for a supplement — if, for example, his hooves are badly cracked or he can't seem to keep shoes on, his coat is obviously dry and unhealthy-looking, or he is a senior with stiff joints — then perhaps you should look into some of these supplements. Most supplements are harmless, although over-supplementing does have the potential for harming your animal. Consult your veterinarian on the potential risks associated with a supplement you wish to use. Above all, avoid the urge to purchase every supplement being trumpeted as a silver bullet.

minerals including iodine, which is necessary for the proper function of the thyroid.

A trace-mineral salt block should be available free choice. The exception is for a horse or mule that has been deprived of electrolytes and therefore craves salt so badly he licks constantly and may consume an entire block in a few days. No equine should spend more than 15 to 20 total minutes a day licking the mineral/salt block. Regulate the salt intake of an animal that craves salt by taking the block away for a day and adding some iodine and potassium (available in regular and light table salt) to his concentrate. Equines that excessively lick a mineral/salt block need large volumes of water, so make sure a steady water supply is readily available.

Just because your trace-mineral block contains minerals *does not* mean it contains the proper amounts of all the minerals your animal needs, particularly if you are in a mineral-deficient area. A trace-mineral block contains mostly salt, plus some minerals. If you live in an area inherently low in one or more minerals, provide these minerals to your horse or mule as an additional supplement.

## Water Is Essential

Water is a critical element of every draft animal's diet and must always be available. A draft horse can drink between 8 and 15 gallons of water per day. The exact amount depends on the humidity, workload, and forage; horses and mules drink more water when eating dry hay rather than fresh pasture forage.

A horse or mule that is allowed to be choosy with his water will be. Animals that have never been in a pasture with a creek, for example, will not want to touch running water. Neither do equines like to step in mud at the edge of a pond. Many overly cautious owners haul water when traveling, because their animals won't drink strange water. Although a person can go overboard in accommodating their equines' whims, water consumption *is* important. Exercise patience with your horse, teaching him that the stream is all right, and that the water therein is wet and good. Do not expect your animal to drink out of a stagnant, algae-ridden water container. Fresh, clean water, whether from a bucket, pond, or running stream, is important for your animal's health.

Veterinary practitioners treat an increased number of equine colic cases during fall and early winter. When water sources become scarce or frozen over, insufficient drinking leads to impactions and severe colic. In winter, you may need to install water tank heaters or routinely break ice. Horses and mules will learn to drink from heated waterers containing a ball they first have to push aside, although you may have to help them to get the hang of it the first few times. Solar-heated models are also available.

My responsibility as a young girl was to take our horses to the creek when dry weather caused the water in the tank at the barn to run low. Watering the horses at the creek on a chilly fall morning, surrounded by the glory of the blazing autumn colors, is a memory I cherish — although riding one horse while leading two more down to the creek could be quite a chore. Even if you must go out of your way to provide a fresh water supply for your equines, take heart in knowing that it is essential to their good health.

All equines need a ready supply of clean, fresh water.

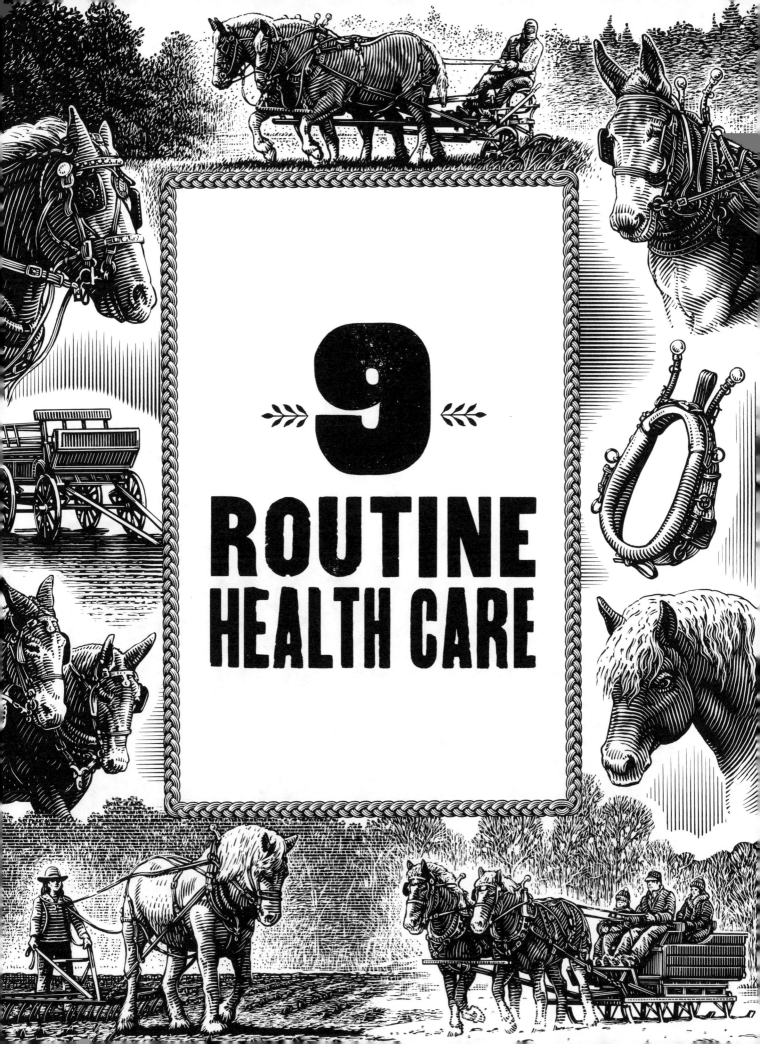

# 9
# ROUTINE HEALTH CARE

My husband came home frustrated about a horse he had examined that day at his veterinary practice. The horse was supposedly 14 years old but was severely malnourished and looked as though his legs would give way at any time. He had normal eating habits, decent quality feed, and plenty of pasture available. The blood work was normal, but a brief look at the horse's mouth told the entire story. A naturally malformed jaw had caused this horse's teeth to erode more quickly than was typical for his age. The upper incisors, in fact, were nearly gone. The normal biting and chewing of forage was not possible for this horse, thus his poor condition.

Yes, nature gave this horse a handicap, but had the owners recognized it 12 years earlier, preventive dentistry could have given the horse a reasonably healthy life. Unfortunately, they were now faced with the choice between keeping the high-maintenance animal or euthanizing him.

Preventive health management is of benefit not only to your equine, but also to you. It will save you time, frustration, and money. During the past few decades, the number of years a typical draft horse or mule can be expected to live has steadily increased. This increase results from many factors, the most important of which is proactive management of problems that previously resulted in premature aging.

The external maintenance of equines was once considered superfluous to the working farm; animals were there to provide labor, not to live forever. Today's teamsters want their horses and mules to live long, active, contented lives, and therefore take such issues as dental health, parasite control, hoof maintenance, and grooming more seriously. Reg-

Keeping your draft animal in good health is more than a matter of providing proper food and shelter; it also includes regular exercise.

ulations governing the transportation of equines have helped to limit the spread of disease across the country. A growing interest in diagnosing, addressing, and treating lameness has also led to improved longevity. The result of increased efforts to keep equines comfortable and content is that draft horses and mules remain in excellent condition and live well into their twenties and even thirties.

## The Basics of Health Care

Just like a car or tractor, your equine requires regular maintenance. Part of this maintenance includes knowing what is normal for each animal in terms of body and hoof temperature, pulse rate, respiration, mucous membranes, urination, and defecation — all indicators of general health. Besides knowing what is normal in your healthy animal, you should vaccinate against bacterial and/or viral diseases, as well as take measures to control internal and external parasites and provide regular dental and hoof care.

### DETERMINING THE NORMAL PARAMETERS

Because your horse or mule cannot tell you in words when he is sick, the only way you can determine when he is not feeling well is by being familiar with his daily patterns. What is normal for your horse or mule? Does he usually have floppy ears, or is ear flopping a sign of distress? Is he typically restless in his stall, or is restlessness a sign of discomfort? Does he often leave some hay in his manger, or is reduced appetite a sign of internal trouble? Become intimately familiar with your animal's normal behavior.

Once you know what is normal for each animal, you have a baseline with which to compare any behavior that deviates from the norm. Recognizing abnormal behavior helps you to determine when to call your veterinarian and allows you to give the vet critical information about the animal's condition. Additional important information includes abnormalities in hydration, temperature, pulse, respiration, mucous membranes, urination, and defecation. Noting any deviations will help you and your vet to assess the seriousness of the situation.

### Hydration Status

Hydration status indicates the water content of the body's tissues. Equines can be fussy about cold water and may have a tendency to become a bit dehydrated during the winter, but dehydration can also be the result of fever, excess sweating, severe pain, trauma, or blood loss. To perform a quick test of an animal's hydration status, pinch or fold a flap of skin at the front of the shoulder and then release it. In the well-hydrated equine, the skin will snap almost immediately back into place. If the skin does not quickly spring back into place, the animal may be dehydrated. Since each animal has different skin and muscle tone, determining a baseline of how quickly an individual animal's skin springs back into position is important.

Provide easy access to clean water to give your animal the option to rehydrate. If the animal refuses to drink, your veterinarian may choose to hydrate with fluids applied intravenously or through a nasogastric tube (inserted into the nose to reach the stomach).

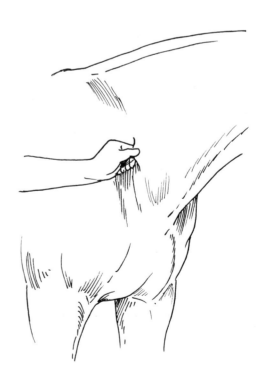

Pinch the shoulder or neck skin to determine hydration status — if the skin does not immediately snap back to normal, the equine may be dehydrated.

Take the maxillary pulse under the jaw.　　　　Take the digital pulse over the sesamoid bones.

## Body Temperature

A normal rectal temperature for the equine is between 99.5°F and 101.5°F (37.5°C–38.6°C). Contact your veterinarian immediately if the temperature exceeds 102.5°F (39°C). A temperature above 103°F (39.5°C) indicates a serious condition.

The equine's body temperature is taken rectally. A digital rectal thermometer from the drugstore will give a readout in a minute or less. Stand to one side of your horse's hind legs, pull the tail away from the anus, and insert the thermometer. Hold it until it beeps and then remove it and read the temperature. Most digital thermometers register within a minute or so. Keep your fingers on the thermometer *at all times*, so it won't get lost in the rectum. One option is to tie a string to the thermometer, a trick more typically done with the old mercury thermometers, since they take longer to register, but still good insurance.

## Hoof Temperature

You won't be able to obtain the precise temperature of your equine's hooves, but as you handle your animal every day, notice if the temperature of the hooves is consistent. Increased hoof temperature is associated with laminitis; recognizing the onset of laminitis as soon as possible allows you to begin early treatment. In rare cases, increased hoof temperature could also indicate an abscess or a general inflammation from, say, an infected nail hole.

## Ear Temperature

As with hooves, you can't easily define a precise value for normal ear temperature, but you should know in general what your animal's ears feel like under normal circumstances. If your animal has been injured or appears to have internal problems, cold ears and legs may indicate shock. This information will help you assess the seriousness of the animal's injury or illness.

## Pulse Rate

The normal equine pulse is 30 to 40 beats per minute. To take the pulse of a horse or mule, stand to the left side of his head and reach under the jaw, feeling for the large maxillary artery on the inside of the lower jaw. An elevated pulse may indicate recent exercise, trauma, fever, pain, or other stress or illness. Excessive exhaustion and severe cold can lower the pulse. Older animals may have a lower pulse rate, which is normal for them.

An elevated digital pulse, which is taken at the midpastern or over the sesamoid bones at the level of the fetlock, may indicate laminitis. This pulse is typically hard to detect. A visible and bounding pulse is cause for concern.

## Respiration Rate

Normal respiration for the resting horse or mule is between 12 and 20 breaths per minute. Assess your equine's normal respiration rate when he is at rest. Observing flank movements or nostril movements are two of the most common ways to determine respiration rate. Count the number of breaths per minute. Counting breaths accurately takes some practice, so make a point of doing it periodically to determine what is normal for your animal. You can also use a stethoscope placed against the underside of the horse's neck, about four inches from the throatlatch.

Elevated respiration rates may result from recent exercise, excessive fat, or stress of any sort, such as pain, illness, or alarm. The respiratory rate greatly increases with exercise, but in a fit equine should return to normal within 30 minutes after exercising. A slower-than-usual respiratory rate may be a result of excessive fatigue or severe cold.

## Mucous Membranes

The mucous membranes allow us to visually assess the function of the circulatory system. Mucous membranes are evaluated in the gums of the mouth. Normal mucous membrane color is pink. An abnormal color of the membranes, such as dark red, white, blue, or yellow, is cause for immediate veterinary attention.

> ### Determining Critical Normals
>
> **W**hen talking to a veterinarian or veterinary technician about an equine examination, you will hear the acronyms TPR and CRT. TPR stands for *temperature, pulse,* and *respiration.* CRT is *capillary refill time.* Normal values are:
>
> **T** 99.5°F–101.5°F (37.5°C–38.6°C)
> **P** 30–40 beats per minute
> **R** 12–20 breaths per minute
> **CRT** 2 seconds

You can assess capillary refill time by pressing against the gums with your thumb and then releasing. The white color that results from the pressure should be replaced with the normal pink color in approximately two seconds. If the refill time is noticeably faster or slower, consult your veterinarian; an abnormality reflects problems with the circulatory system.

## Urination Status

Horses and mules urinate every four to six hours. Some animals are choosy about where they urinate, often preferring deep bedding or grass to reduce the amount of splashing. A gelding usually urinates with his legs stretched out and his back somewhat flattened. A mare stands with her head low, back arched, and hind legs extended and spread apart. Normal urine is pale to dark yellow in color. Abnormal urine color is cause for concern.

**Red urine** may be caused by bleeding or may result from a normal phenomenon caused by oxidizing compounds (pyrocatechines) passed in equine urine. Urine that is red during passage may indicate internal bleeding and should receive prompt veterinary attention. Urine that discolors because of pyrocatechines is often noticed after it pools in snow or discolors bedding exposed to sunlight.

**Brown urine** is most often associated with muscle damage. The abnormal urine color is due to the

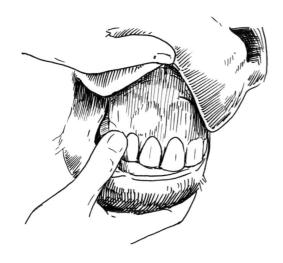

Determine capillary refill time by pressing the gums and noting how quickly the normal color returns.

destruction of enzymes and subsequent leakage of myoglobin into the bloodstream, a condition known as rhabdomyolosis. Because selenium and vitamin E are also essential to muscle function, deficiencies in either of these nutrients can also lead to rhabdomyolosis and thus brown urine.

**Black urine** typically results from a melanoma somewhere in the urinary system and should receive veterinary attention to determine if treatment is possible.

### Defecation Status

Horses and mules defecate every two to three hours, although the rate increases with stress, which may induce more frequent bowel movements. Each movement contains from 5 to 20 fecal balls. Mules typically produce smaller balls than horses. Feces are a tremendous source of parasites, with each fecal ball containing as many as 30,000. Parasite reinfestation is unavoidable where animals are forced to eat in a contaminated area.

Dropping "horse apples" is, for some horses and mules — particularly stallions — a way to mark their territory. Many equines will gravitate toward a particular place in the pasture or stall to defecate.

### Gut Sounds

A normal, healthy horse or mule has quite a symphony of ongoing gut sounds. These sounds, which are best heard with a stethoscope placed over the middle abdomen area, result from normal activities of the gastrointestinal (GI) tract. Become familiar with what is normal for your animal. An absence of these sounds may indicate a serious impaction or other GI obstruction. Hyperactive or greatly increased sounds can indicate a bout of diarrhea.

## ANNUAL VACCINATIONS

People have successfully owned draft horses and mules for years without administering vaccinations, but the chances they take in doing so are far from cost effective. The cost of vaccines is low compared to the loss of a valuable animal.

How terrible would you feel to have your equine come down with a serious illness that could have been prevented by a $12 vaccine? My friend Chuck felt awful the day his veterinarian diagnosed his three-year-old Percheron stallion with tetanus. Chuck regularly vaccinated his mares when they foaled, but he was a little lax with his stallion and geldings. When his stallion started acting stiff and

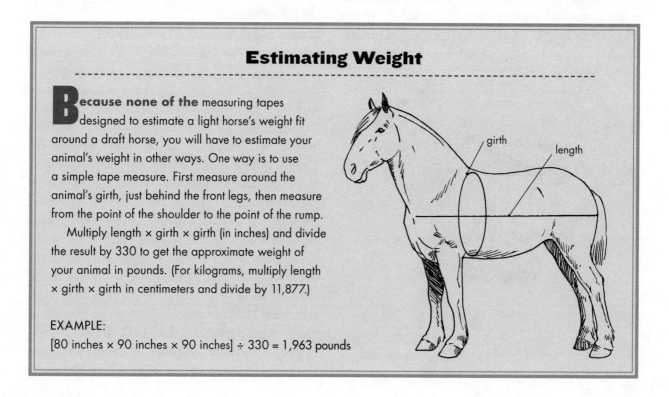

### Estimating Weight

**B**ecause none of the measuring tapes designed to estimate a light horse's weight fit around a draft horse, you will have to estimate your animal's weight in other ways. One way is to use a simple tape measure. First measure around the animal's girth, just behind the front legs, then measure from the point of the shoulder to the point of the rump.

Multiply length × girth × girth (in inches) and divide the result by 330 to get the approximate weight of your animal in pounds. (For kilograms, multiply length × girth × girth in centimeters and divide by 11,877.)

EXAMPLE:
[80 inches × 90 inches × 90 inches] ÷ 330 = 1,963 pounds

unwilling to move, Chuck thought the horse had laminitis and treated him with an anti-inflammatory. As the day wore on, he decided to call his veterinarian. By the time the vet arrived, the horse could barely flex his legs. When the vet attempted to check the stallion's mucous membranes, the horse could not relax his jaws to open his mouth. Tetanus treatment was initiated, but the prognosis was grim, and the valuable stallion was euthanized two days later.

While vaccines offer protection against many fatal diseases, any foreign substance introduced into the body carries some risk for the animal. Vaccine reactions are common in the equine, but are typically mild and temporary, usually involving soreness and slight swelling at the injection site. A more severe reaction includes a large swelling and fever. In the most violent reaction, collapse and death may occur. Severe and violent reactions are rare, but are always a potential risk. Because of the more common potential for soreness and/or swelling, avoid injecting into the area of the neck that comes into contact with the collar. Instead, inject into the pectoral (chest) muscles.

The production of vaccines is an evolving science and the spectrum of diseases covered and the kind of vaccines available changes constantly. Many different vaccines are currently made to protect against a host of bacterial and viral diseases, including western and eastern equine encephalitis, West Nile virus, equine herpes virus, botulism, influenza, rotavirus A, rabies, anthrax, strangles, Potomac horse fever, and others.

Your horse or mule, of course, does not need every vaccine available, but should be given essential vaccinations. Your veterinarian will recommend the vaccines for best protection in your particular area. Regardless of your geographic area or the condition of your animal, he should at least receive a tetanus vaccination. Based on your location, your veterinarian may also suggest vaccinating your equine against rabies.

Vaccines come in two forms, killed (inactivated) and modified live. Some vaccines come in both forms, while others come only in one, depending on the manufacturer and the current technology. A killed vaccine is prepared from inactivated toxins or killed infectious organisms and is generally less likely to cause a reaction. It may also be less effective than a live vaccine. Live vaccines contain, as the name implies, live organisms modified to give immunity without causing the disease. Live vaccines are typically more effective, but are also more dangerous to the animal. Your veterinarian will help you determine which form of vaccine is best for your equine.

---

## Handling Vaccines Yourself

**Vaccines may be purchased** through many different suppliers, including feed and tack stores and mail-order catalogs. These vaccines are usually effective, although problems arise when a vaccine is handled incorrectly, such as being exposed to heat. Since you cannot know the handling history of these vaccines, you take a risk when you assume they are good. Work with your veterinarian to ensure that your vaccine has been properly handled. If you purchase vaccines from your vet to administer yourself, ask about the best way to transport and store them. You may need to take along a cooler to transport the vaccine safely home. Tossing the vaccine onto the dash of your pickup while you go for a burger won't do.

Administration of a vaccine should be done by a knowledgeable person. Many horse and mule owners vaccinate their own animals. If you are interested in learning how, first observe your veterinarian and ask for guidance. Swelling and other complications can arise from incorrect administration. Since an equine's immunological response to a vaccine may be altered by stress, avoid work or long-distance hauling of your animal before or directly after vaccination. Although no scientific data suggest that giving multiple shots can be detrimental, some practitioners split up a series of shots to avoid introducing excessive antibodies all at once.

## Rabies

Although vaccinating against rabies is mandated by law in some species, such as dogs, vaccinating the equine against rabies is not. States with a high incidence of rabies may strongly suggest that all equine owners vaccinate against this fatal disease. Horses are at high risk of exposure to rabies from wildlife with the disease. An equine that develops an unexplained nervous disease may be immediately euthanized if rabies is suspected and the animal's vaccination is not current. Rabies vaccinations must be administered by a licensed veterinarian.

Signs of rabies are highly variable and difficult to definitively diagnose, thus any animal suddenly showing neurological signs should be suspect. Treatment is generally not advisable; when rabies is suspected, immediate euthanasia often follows. Handle any animal suspected of rabies with extreme caution.

## Tetanus (Lockjaw)

Tetanus is caused by *Clostridium tetani*, a bacterium found throughout the environment. Equines are exceptionally sensitive to the disease, which is terribly destructive but easily avoided. All equines, without exception, should be vaccinated against tetanus. Pregnant mares should receive a booster four to six weeks before foaling. All equines require a booster every year and any time a deep puncture wound occurs. Foals should be vaccinated before they reach four months of age and again at five months. Humans should have regular tetanus boosters as well, particularly people who handle horses.

*C. tetani* bacteria can infect wounds, particularly puncture wounds. The toxin affects nerve activity, causing muscles to severely spasm and eventually leading to death from spasms in the diaphragm and other muscles involved in respiration. Treatment for tetanus depends on the severity of the case. If the animal is lying down, the prognosis is grim. Treatment includes sedation, hydration, and nourishment therapy (an animal with a locked jaw cannot eat or drink), antibiotics, and tetanus antitoxin.

## Equine Encephalitis

Eastern equine encephalitis (EEE), western equine encephalitis (WEE), and Venezuelan equine encephalitis (VEE) are all carried by mosquitoes and cause inflammation of the brain in horses and humans. A vaccine is available to provide effective protection against all three diseases. Vaccinate foals at four months and again at five months. Adult horses should receive an annual booster in the spring, prior to mosquito season. This vaccine does *not* protect against West Nile virus.

Signs of encephalitis include acute behavior changes, colic, and (not always) fever. Neurological signs include head pressing, loss of control of body movements, blindness, circling, seizures, paralysis, and dementia. As many as 50 percent of equines in an area may be affected. Inflammation is typically managed with nonsteroidal anti-inflammatory drugs (NSAIDS) and steroids, but complete recovery is unusual.

## West Nile Virus

West Nile virus (WNV), once restricted to the northeastern United States, is now found in nearly every state. It is carried by mosquitoes, affects both humans and animals, and causes encephalitis and death for approximately 40 percent of infected equines. If you are in a high-risk area, vaccinate foals at three to four months of age, again one month later, and a third time at six months.

A horse or mule with an unknown vaccination history requires two vaccinations, three to six weeks apart. After the initial series of vaccinations, adult horses and mules should receive an annual booster in the spring before the onset of mosquito season. Depending on the risk level of your geographic area, adult horses and mules should additionally be vaccinated twice a year, more if the area is considered high risk. Your veterinarian can help you to determine the desirable frequency of vaccination.

Although no WNV vaccine is currently approved for pregnant broodmares, many veterinarians vaccinate them because of the damaging nature of WNV. The potential negative effects of this usually safe vaccination are worth risking against this dramatically damaging disease. Vaccinate broodmares annually, ideally before conception, and administer a booster four to six weeks prior to foaling. The booster protects foals for three to four months after foaling.

Signs of WNV include an inability to control the rear legs that progresses over several days to an inability to stand up, sometimes fever, involuntary muscle contractions of the face and neck, hyperexcitability, and stupor. With proper veterinary treatment, which includes managing the inflammation with NSAIDS, 60 percent of affected animals recover.

### Rhinopneumonitis (Equine Herpesvirus)

Rhinopneumonitis is caused by equine herpesvirus (EHV). Two forms, known as EHV-1 and EHV-4, are associated with equines, and both can cause a mild respiratory disease in adults. Sometimes EHV-1 also causes neurological signs, including weakness and muscle wasting. Vaccines offer short-term protection against EHV, but none protects against the neurological component.

Because EHV has been implicated in abortion, stillbirth, and the death of newborn foals, broodmares should be vaccinated at five, seven, and nine months into the pregnancy. Vaccinating the mare four to six weeks prior to foaling bolsters the maternal antibody found in colostrum (the first milk produced by the mother). Vaccinate a foal at two to three months, and repeat at two- to four-month intervals until the foal reaches one year of age.

Equines that travel regularly or frequently and come into contact with other equines — such as through shows, plowing matches, and other events — should be vaccinated twice a year. A good management practice for the control of EHV is to isolate all new equine arrivals on your farm.

Signs of EHV include an abrupt and symmetric loss of muscle control that may rapidly progress to an inability to get up from lying down, constipation, and nerve damage of limbs, bladder, and brain. Treatment includes management of urinary and fecal functions, antibiotics, steroidal treatment, antivirals, and intense supportive nursing care. The recovery of affected animals is highly variable.

### Annual Coggins Test

**Coggins testing is required** for hauling equines across state lines and to events, but should be done once a year whether or not you intend to ship your horse or mule. Most states and shows require proof of a negative Coggins test within the past year; others require a negative test within the past six months.

The Coggins test was developed by Dr. Leroy Coggins to test for the presence of equine infectious anemia (EIA), a viral disease that may be carried by horses without any symptoms. The signs can appear suddenly and without warning, and include fever, weakness, weight loss, and depression. The virus is spread from animal to animal through biting insects and the reuse of hypodermic needles.

No treatment or vaccine is available for EIA. By law, an equine with EIA must be isolated for the rest of his life or destroyed. Aggressive testing and elimination of carriers have reduced this once common bane of horses and mules.

## The Importance of Controlling Parasites

An astonishing array of parasites constantly assault your horse or mule. Intestinal parasites represent a significant danger to draft equines due to the damage they can wreak on the intestinal wall and blood vessels supplying the intestine, potentially leading to severe colic. A regular parasite-control program is imperative for equine health.

Not all horses and mules are affected by all internal parasites. To determine an appropriate deworming protocol, have your veterinarian examine a sample of your equine's manure every three to six months. Besides saving you money on the purchase of unnecessary dewormers, knowing specifically what internal parasites to target helps to minimize parasite resistance to a particular product.

# SHE KNOWS DRAFT HORSES FROM THE INSIDE OUT

**Beth Valentine, DVM, PhD**
**Corvalis, Oregon**

Professional Experience: Associate professor at Oregon State University and head of the Neuromuscular Disease Lab; researches the treatment and prevention of muscle disease in draft horses

Other Experience: Coauthor, *Draft Horses, an Owner's Manual*; virtual vet at www.ruralheritage.com

Dr. Beth Valentine came to her research focus on draft horses through a circuitous and unconventional manner. She did not grow up with draft horses, nor did she practice as an equine veterinarian. And yet, the path her career has taken led her to become one of today's few research veterinarians studying the draft horse.

Dr. Beth (as she is affectionately known) grew up in a Long Island suburb, a place that does not evoke images of peaceful pastures and grazing draft horses. Still, she dreamed of riding horses. Not far from her home she found a retired dairy where a few horses still lived. One of these horses was an old carthorse named Blondie. She eventually pestered the owner into letting her ride Blondie, and thus the bond between draft horse and future veterinarian was formed.

Her academic path did not steer her toward horses. After graduating from the College of Veterinary Medicine at Cornell University, she practiced for one year in a small/exotic animal clinic, where she decided that private practice was not for her. So she embarked on a career in veterinary pathology, first at Johns Hopkins University and then back at Cornell, where she studied diseases of the muscles. Eventually her research led her to the shockingly common muscle disease of draft horses, once known as Monday morning sickness, shivers, and tying up, and now called equine polysaccharide storage myopathy (EPSM).

Dr. Beth found little to no research in the draft horse world and set out to learn all she could about this muscle disorder. She met with breeders, owners, draft horse associations, and clubs. As she talked with draft owners, she realized how isolated they felt. Her research has led many draft owners to have a paradigm shift in the way they feed their animals.

In her current role on the faculty at Oregon State University, Dr. Beth continues to study muscle diseases. A common sentiment expressed to her as she meets draft owners is that vets don't want to work on their animals. So she helped to organize the OSU Veterinary College draft horse club, hoping to encourage veterinary students to handle draft horses and realize that, in general, they truly are gentle giants.

Although Dr. Beth stoutly denies being a "real" (practicing or clinical) veterinarian, her knowledge of the draft horse, literally from the inside out, has provided a vast resource for the information-hungry draft horse world. Her recommendation for the necessities to ensure optimal health include:

- A diet high in fat and low in starch
- Exercise
- Proper hoof care
- Appropriate deworming and vaccinations
- A good relationship with your veterinarian

Dr. Beth hopes draft horse owners will come to appreciate veterinarians more. In turn, she hopes veterinarians will recognize the growing interest in draft horses and respond with a commitment to their care.

Strongyles show up in a fecal examination, but pinworms and tapeworms do not. Because of the difficulty of ascertaining the presence of pinworms and tapeworms, treat your animals against these worms at least once a year.

## DEALING WITH INTERNAL PARASITES

The following intestinal parasites have particularly serious consequences if left to multiply uncontrolled.

**Ascarids** are roundworms that apparently do not bother a mature equine, but the immune system of an animal less than one year in age has not yet developed sufficiently to eliminate these parasites. Ascarids do not typically cause damage to blood vessels or tissues of the intestine, although when present in large numbers, they can migrate and damage tissue. Large numbers can also alter intestinal activity and cause blockages in the young horse or mule, particularly when killed off through purge deworming, which eliminates large numbers of parasites at once.

**Bot flies** lay eggs on the legs of horses and mules. A bot egg coming into contact with moisture hatches and becomes a threat to your horse. An animal licks the affected area and ingests larvae that eventually reach the stomach, where they attach to the lining — sometimes so thickly that the lining is not visible during scoping procedures. Some equines do not experience complications from bots, while others develop colic, oral ulcers, and perforation of the stomach wall. The end result can be peritonitis, a surgical emergency. Bots may be controlled by the daily removal of eggs from the horse's hair and by a deworming program that targets bots.

**Pinworms** sometimes cause the annoying habit of rubbing the rear end on stall walls, pasture fences, or any other available object. These threadlike worms lay their eggs around the anus, causing irritation and itching. If present in sufficient number, they can be seen around or inside the anus.

**Large and small strongyles** are intestinal worms with potentially serious implications to equine health. Large strongyles (*S. edentatus* and *S. vulgaris*) are more dangerous to a draft horse or mule than small strongyles because they damage intestinal blood vessels, which may result in decreased or blocked blood flow. These parasites pass their eggs in manure, where the larvae hatch, are swallowed by grazing animals, and then burrow into the intestinal wall to begin a journey through a number of tissues, until they return to the intestine as adults. Large strongyles can cause weight loss, poor growth in young horses, colic, and anemia.

Small strongyles neither penetrate the intestinal wall nor migrate through tissues, as do large strongyles, but burrow into the intestinal wall and remain dormant for several months while completing their life cycle. During this time they are resistant to most dewormers. Large numbers of small strongyles can cause colic and diarrhea.

**Tapeworms** are almost always present in a horse or mule. Although these worms are large — ranging from 1 to 3 inches (2.5–5 cm) long — they are difficult to detect except through internal examination after the animal has died. Complications arise when tapeworms attach in large numbers at the juncture of the small intestine, cecum, and large intestine, causing an obstruction.

In the past, tapeworms were not linked to colic. Recent discovery of antibodies produced by tapeworms in the equine body, however, has implicated tapeworms in cases of mild to severe colic caused by the telescoping of the intestines as a result of a parasite load. This condition requires surgical intervention. Tapeworms can also interfere with the natural movement of the intestines, resulting in gas colic. Although tapeworms are not considered a significant threat to digestion or the absorption of nutrients, they can be associated with a general failure to thrive.

## PROPER USE OF DEWORMERS

Deworming frequency should be determined by your veterinarian, who will factor in your climate, the season, the number of animals you have, and the parasites that affect them. Most equine owners deworm using the purge program, which involves giving high doses of dewormers three to four times a year. This system removes all parasites at one time, but reinfestation occurs rapidly.

Animals at particularly high risk for parasite infestation are those pastured with animals that are not on a good deworming program (leading to rapid reinfestation) and animals pastured where fresh manure has been spread. A purge-worming protocol may not do enough for these individuals. Daily wormers are available that help control parasites in animals at risk for high levels of parasite infestation.

Using a dewormer containing the same drug compound repeatedly over a long period of time can allow parasites to become resistant to that product. On the other hand, rotating drug products allows the parasites to become resistant to multiple drugs. Instead of blindly rotating dewormers, ask your veterinarian to help you develop an effective schedule for your geographic area, the time of year, and the overall risk level of your animal. By examining a sample of your equine's manure, your veterinarian can suggest a deworming protocol to maximize the efficacy of intestinal-parasite control for your draft horse or mule.

## ADMINISTERING DEWORMERS

Dewormers typically come as a paste in a tube. Your challenge is to push all the paste into your horse's

Administering a paste dewormer

or mule's mouth and down his throat. If an animal has been there, done that, and decides he doesn't like the procedure, you could be in for a full-fledged rodeo. You may find it useful to have a helper hold the horse's head.

Before beginning this exercise, make sure the dial on the syringe is set to the correct weight for your animal. Open his mouth by inserting your thumb

| Common Dewormers | | | | | | | |
|---|---|---|---|---|---|---|---|
| **DEWORMER** | **WORMS AFFECTED** | | | | | | |
| | Bots | Ascarids | Migrating (larval) strongyles | Adult large strongyles | Adult small strongyles | Pinworms | Tapeworms |
| Fenbendazole | no | yes | yes | yes | yes | yes | no |
| Ivermectin | yes | yes | yes | yes | yes | yes | no |
| Moxidectin | yes | yes | yes | yes | yes | yes | no |
| Praziquantel | no | no | no | no | no | no | yes |
| Pyrantel pamoate | no | yes | yes | yes | yes | yes | yes, double dose |
| Pyrantel tartrate | no | yes | yes | yes | yes | yes | no |

These drugs are sold under many trade names, some of which contain a combination of drugs. Consult your veterinarian to determine the combination that will best work in your area and to develop an appropriate rotation schedule.

into the side. The ideal way to administer the paste is to hold the horse's head high enough that gravity won't allow the paste to fall back out. If you are short like me, try to immobilize the tongue by holding it while you administer the dewormer. Some equines tolerate this procedure and won't give you a lick of trouble. A carrot or other treat offered immediately after the medication can facilitate swallowing and make the process more palatable.

> ### Control the Worms
>
> **Your horse or mule regularly sheds** copious amounts of parasites. One horse apple contains up to 30,000 parasite eggs. Controlling parasites involves a multifaceted approach. The American Association of Equine Practitioners suggests the following excellent management procedures to reduce parasite loads in your draft horse or mule:
> - Pick up and dispose of manure droppings in the pasture at least twice weekly.
> - Mow and harrow pastures regularly to break up manure piles and expose parasite eggs and larvae to the elements, where they have less chance for survival.
> - Rotate pastures by allowing other livestock, such as sheep or cattle, to graze, thereby interrupting the life cycles of parasites.
> - Group horses by age to reduce exposure to certain parasites and maximize the deworming program geared toward that group.
> - Keep the number of horses per acre to a minimum to prevent overgrazing and reduce the fecal contamination per acre.
> - Use a feeder for hay and grain, rather than feeding on the ground.
> - Remove bot eggs quickly and regularly from the hair coat to prevent ingestion.
> - Rotate deworming agents, not just brand names, to prevent chemical resistance.
> - Consult your veterinarian to set up an effective, regular deworming schedule.

## DEALING WITH EXTERNAL PARASITES

External parasites such as mosquitoes, deer flies, and ticks not only make life unpleasant for your horse or mule, but can also transmit diseases that in the worst cases may cause fatal complications. Take time to notice and treat external parasites. Preventing lice, mites, and ticks on your horse or mule both reduces the spread of disease and makes your animal feel better.

**Lice** seem to particularly afflict animals that are short on vitamin A. Vitamin A deficiency is often caused by eating hay that is low in the nutrient, typically in winter. In the northern states, many horses are infested with lice in late winter and early spring, when they are crowded together in close quarters. Lice spread rapidly among horses and mules. Two kinds of lice affect equines: biting lice that feed on bits of hair and skin, and sucking lice that pierce the skin to feed on blood, creating a new irritation each time they feed.

Lice are most commonly found at the base of the tail, on the sides of the neck, on the flanks, and under the jaw, although in a severe case they may infest the entire body. Lice may be seen with the naked eye crawling and jumping, but are quite small so require careful observation. An infestation of lice can make a horse or mule irritable and distracted. The animal will scratch himself against fences, trees, and other objects.

Our draft mares picked up lice every year during wet winter weather, when they hung out together in the barnyard. We provided relief by treating them with an insecticidal powder, which had to be repeated numerous times throughout the season. The lice were always worse under the fetlocks, and the poor mares rubbed their legs on anything handy.

Animals may be treated against lice by dusting them with louse powder, or if they will tolerate it and the weather is warm enough, spraying. Livestock supply stores carry a number of different insecticides, in both powder and liquid form, to treat equines against lice. A treatment of ivermectin or moxidectin paste wormer every two weeks for three doses also controls lice. Harnesses, blankets, stalls, and corrals must also be treated with an equine-safe insecticide to effectively eliminate the lice.

**Mites** are too small to be seen with the naked eye. The more common mites affecting equines are the sarcoptic, psoroptic, chorioptic (leg mange), and demodectic (rare in horses) mite. Collectively, these mites form a condition known as mange. The mange-causing mites result in thickening of the skin, dermatitis of the fetlock (known in draft horses as greasy heel), and lesions that may involve the entire body if not treated. The choroptic mite commonly affects draft horses, infesting the heavy-feathered fetlock area. Treatment for mange-causing mites is several oral doses of ivermectin or moxidectin and a dry environment. Since mites spread quickly between animals, all animals should be treated.

The straw itch mite *(Pyemotes tritici)* also affects equines. These mites feed on organic material, such as straw bedding and hay. Animals fed from a hay rack may have infestations on their faces. If fed on the ground, the legs and fetlocks may be affected. Signs include small pimples that may form a rash on the neck, face, and leg areas. Treatment includes steroid drugs and a dry environment.

**Ticks** are dangerous to horses and mules in large numbers, when they result in blood loss and general discomfort. Tick activity varies with the season. At times during the spring and summer I have hardly seen a tick, while other times it seems that all I do is pick ticks. Severe infestations run in cycles, particularly in the foothills of the Mountain States. During some years ticks are light, while in other years they affect nearly all equines in the area. Tick control is a two-phase process:

- Carefully inspect your animal after working in the woods or other tick-infested areas and remove any ticks you find.
- Proactively control tick infestation with insecticides approved for use on equines.

Lyme disease, caused by the spirochete *Borrelia burgdorferi*, may be transmitted to a horse or mule via the deer tick *(Ixodes scapularis)* on the East Coast and in the Midwest, and by the western black-legged tick *(Ixodes pacificus)* on the West Coast. The prevalence of this disease is directly correlated to the prevalence of infected ticks, which varies by location; the number of animals that test positive in a certain location varies from less than one percent in nonendemic areas to 68 percent in endemic areas.

Signs of Lyme disease include general lethargy and intermittent lameness, with recurring inflammation of the joints leading to chronic inflammatory arthritis. Irreversible complications of the central nervous system may also occur. Veterinary treatment may include an aggressive regimen of antibiotics and anti-inflammatories, with cold hosing and supportive standing wraps applied to ease swelling. Activity should be reduced or halted until the symptoms are gone.

Equine ehrlichiosis is caused by *Ehrlichia equi*, a rickettsial organism transmitted also by *I. scapularis* and *I. pacificus*. Equine ehrlichiosis causes symptoms similar to those of Lyme disease, but does not affect the central nervous system. Other signs include yellow mucous membranes and irregular heartbeat. Death is rare; with proper veterinary care, the affected equine should recover.

## Care of the Teeth

For many years, equine dentistry consisted of the veterinarian sticking a hand rasp into an equine's mouth, scrubbing it around a bit, and calling it good. Today, however, the subject of equine dentistry is understood as a routine measure for increasing the animal's quality of life. This change in approach to dental health has occurred both because horses and mules spend more time in confinement, with diets different from those of pasture or range animals, and because they are expected to perform at higher levels, requiring them to be in tip-top shape. Not only is good dental health important for adequate nutritional intake and overall health, but it can also improve equine behavior. If a horse's or mule's teeth or mouth hurt, the animal will not want to accept a bit and may resist having his head handled.

The equine jaw and teeth are made for chewing fibrous plants, and therefore the teeth (except the canine teeth) erupt constantly and are ground down by the action of chewing. The lower jaw is narrower than the upper jaw, allowing for the side-to-side motions necessary for chewing grass and other forage. If the horse is confined and fed a high-protein diet consisting of concentrated feed and hay, his teeth are inactive for much of the day and the offset jaws begin to present problems. Because the upper and lower teeth do not meet evenly, sharp ridges called hooks or points form over time on the inside of the lower teeth and outside of the upper teeth as the animal chews. These hooks must be rasped down in a process called floating.

If these hooks aren't floated, their sharp points can cause many problems, such as persistent sores or ulcers in the mouth, difficulty eating, and trouble with the bit. Feral horses consume enough abrasive material, such as sand and gravel, to minimize these points. On the other hand, you won't see many feral horses living for 20 or 30 years. Tooth fractures, retained caps, incomplete eruption, and overbite or underbite are other problems that can cause mouth pain in the draft horse or mule.

You will encounter horse and mule owners who have never had their animal's teeth examined or treated. They are, in fact, likely to laugh at the idea of equine dentistry as "just another ruse for the vet to get more of our money." Yet these same people work with animals that have obvious dental issues: a gelding that throws his head when bridled, a mule that dribbles feed when he eats and is in poor flesh, a young mare that resists turning to the left.

Instead of solving these problems by contacting the veterinarian, these teamsters push their animals. They lose money, time, and often their tempers — all of which could have been prevented by addressing the tooth fragment embedded in the gelding's jaw (perhaps the result of being kicked), the sharp points caused by the mule's imperfectly aligned bite, and the mare's two-year-old caps, one of which stubbornly refuses to come off, causing bleeding whenever pressure from the bit is applied to the left side of her jaw.

Yes, identifying and correcting equine dental problems costs money. But in the long run, you will save money because your animal will use his feed more efficiently, work more comfortably, and display fewer behavioral issues.

## SCHEDULING DENTAL EXAMS

While all horses and mules should have their teeth examined regularly, animals in certain age groups need particular attention. Young animals two and three years of age and animals older than fifteen are most at risk for dental problems and require more frequent examination than equines in other age brackets.

Two- and three-year-old equines are experiencing significant dental change. Between the ages of two and five, each animal will shed and replace 24 teeth. Until the horse or mule is five years old and tooth eruption slows, a dental exam every six months should be part of your routine maintenance program. Young horses are at risk for retained caps, premolar points that can cause sores, and pain associated with wolf teeth. Located in front of the second premolar, wolf teeth are small but can cause large problems. Since their roots are short, they do not sit firmly in the jawbone. Wolf teeth do not always cause problems, but are often routinely removed to prevent them from interfering with the bit.

Equines between the ages of five and fifteen are in the lowest-risk category. Still, they should have a dental exam every year to check for missing and damaged teeth, the formation of sharp hooks and points, and other problems. If you want your equine to live beyond his twenties, he must have an even bite plane established and maintained throughout the teen years, which can be done by correct floating of the teeth. Your veterinarian can determine the need for teeth floating during the oral exam.

Horses and mules more than twenty years of age tend to have smooth teeth from wear over the years. They are also likely to have missing and damaged teeth. These older animals should have a dental exam at least every year, preferably every six months.

Normal jaw

Overbite (parrot mouth)

Underbite

> ◀ **CAUTION** ▶
>
> ### Heading Off Dental Problems
>
> Contact a proficient equine veterinary dentist if you notice your horse or mule:
>
> - Losing weight or condition
> - Experiencing discharge or odor from the mouth or nose
> - Drooling excessively when eating
> - Packing feed in a cheek
> - Dribbling feed when chewing
> - Dropping partially chewed wads of feed
> - Chewing with an open mouth or noticeably offset jaw
> - Developing a lump under or on the side of the upper or lower jaw
> - Experiencing an abscess or drainage under the upper or lower jaw

### FINDING A DENTAL PRACTITIONER

Many states prevent anyone but a licensed veterinarian from treating equine teeth. The Veterinary Practice Law for many states says that a nonveterinarian equine dentist can practice under the direct supervision of a veterinarian, which leaves significant loopholes.

Before hiring someone to examine or treat your equine's teeth, make sure the person is working within the law. Make sure, too, that your veterinarian is qualified; not all veterinarians have experience or are competent to do equine dental work.

## Care of the Hooves

The old adage "No hoof, no horse" is too true; an animal with feet that can't support his body won't get far. Hoof-management practices vary dramatically among equine owners. Hoof care may be intensive or minimal, depending on management philosophy, draft use, and the animal's condition.

Many ranchers leave their draft horses barefoot (unshod) most of the year, putting shoes on only when required by a specific task, such as a long trot down the road. Most mules inherit the tough, elastic feet of their donkey forefathers and often don't require any shoes unless subjected to daily use on hard pavement or extremely stony conditions. Some draft horses, when trimmed and balanced correctly, adjust to a completely barefoot existence, even on rough terrain. More intensive hoof care may be needed for an animal with poor hoof conformation or problems such as laminitis or a predisposition to cracking, or for an animal engaged in jobs that are hard on the feet, such as city carriage work. To determine which management practices your horse or mule needs, consider the following questions:

**What type of maintenance is needed?** If your animal is on pasture most of the year, no shoes and a simple six-week trimming schedule will probably suffice.

**How often do you use the horse or mule?** If your animal is receiving daily work, you should maintain a consistent trimming schedule, and, if needed, shoe at least the front feet.

**Does your horse or mule require corrective shoeing and/or trimming?** Corrective work is necessary for animals with conformation problems or deformities, bad hoof cracking, laminitis, or lameness.

Your best source of information is a competent farrier. A good equine veterinarian should be able to provide you with the name of a local farrier who will help you to develop a plan for optimal hoof care.

Without proper care, equine hooves can become overgrown or cracked.

## PROPER TRIMMING AND SHOEING

Caring for draft horse hooves is different from caring for draft mule hooves. Mules, as a general rule, do not need the same frequent trimming and other careful attention to their feet required by the draft horse, and they require shoeing only in the toughest conditions. If your mule is on the road frequently or if you are packing over rough, rocky terrain, you may choose to shoe your mule.

Draft horse feet, because of their size and the great weight they bear, are susceptible to cracking, spreading, and bruising, and may need shoes to help prevent or correct these problems. If your pastures are stony and rough, shoeing is a good way to minimize sole bruising in a susceptible horse. A draft horse also needs regular hoof trimming every six to eight weeks.

For the beginner, having a professional do the job will help head off potential problems, such as quarter cracks and other cracks you may not notice until they have advanced alarmingly. You may, if you choose, eventually learn how to properly trim your horse's or mule's feet yourself. Attend a clinic,

Knowing how much to trim off can be more challenging than holding that huge hoof between your legs.

## Mule Hooves versus Horse Hooves

**Mule hooves differ** from horse hooves in many ways:
- The mule's hoof is denser than that of the horse, therefore less susceptible to quarter cracks.
- The mule's foot is longer and narrower than the horse's more rounded hoof.
- The mule's frog is proportionally larger than that of the horse and bears more direct weight.
- The mule foot is smaller than the foot of a horse of the same weight; the draft mule might wear a size 1 or 2 shoe, while the draft horse would wear a 4 or 6.
- The angle of the mule's hoof is steeper than that of a horse; the average angle for a mule is 60 degrees, compared to an average of 55 degrees for the horse.

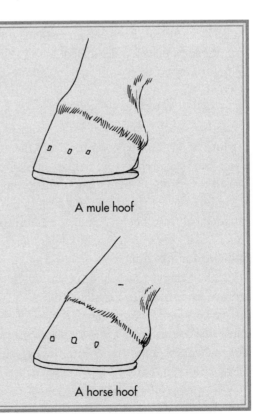

A mule hoof

A horse hoof

ROUTINE HEALTH CARE  **167**

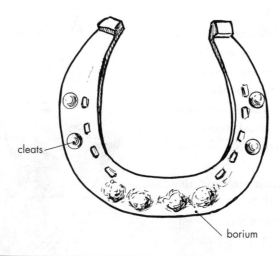

A shoe with cleats and borium for traction

read books, and watch videotapes before attempting to grab your draft animal's feet and start trimming. Some farriers are happy to show their customers how to maintain equine hooves between regular professional trims.

Take time to find a good farrier and develop an ongoing relationship. You may have difficulty finding someone willing to work on your draft horse or mule. Draft horses have enormous, heavy feet that can discourage even a dedicated farrier. Mules can be notoriously ornery with their feet, and some farriers shy away from them, as well. Training your equine to stand for hoof care will go a long way toward enticing a qualified farrier and encouraging him or her to keep coming back. Once you find a qualified and reliable farrier, he or she can help you to determine if regular trimming, shoeing, or a combination of the two is the best option for the maintenance and health of your draft animal's feet.

## Shoeing for Traction

Draft horses or mules pulling heavy loads can require extra traction, particularly when working on paved roads, ice, mud, and snow. A barefoot equine or one shod with regular shoes may slip and slide, where an equine shod with traction shoes can dig in and keep moving. A substance called borium may be applied to the bottoms of shoes to improve grip. Toe calks or cleats that screw into the shoes are often used to improve traction for horses that log or do heavy farm work.

Traction shoes are beneficial to the working equine, but can also become a deadly weapon. Traction devices increase the amount of damage that can occur to other equines or to you as a result of a well-aimed kick. A few of the draft horses on our farm wore traction shoes and all of our driving horses had traction shoes on the front. While we never experienced trouble with animals getting hurt in the pasture, any animal known to exhibit aggressive behavior should not be turned out with other animals while wearing traction shoes.

## Shoeing for Pulling

The shoes used in competition pulling depend on the rules of the competition. In so-called professional pulling, the competition can be fierce and traction shoes with calks or cleats are allowed. The primary purpose and training of animals engaged in professional pulling is to compete. Some competition pulling horses are also used for logging.

By contrast, horses engaged in what is called barnyard pulling are kept primarily for fieldwork or logging, and pull competitively for recreation. In barnyard pulling the horses or mules must wear flat shoes or go unshod. If shod, the shoes may have a small amount of borium applied to the bottom, but cleats and calks are not allowed.

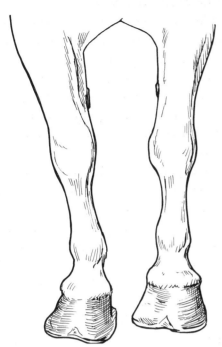
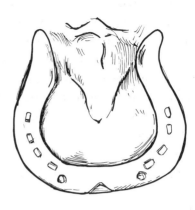

A Scotch bottom show shoe

## Show Shoeing

Shoeing the show animal can be quite an elaborate process. For show, the size of the hoof and the horse's gait are important. Show-shoeing techniques are often designed to enhance the size of the hoof, as well as the flashiness of the gait, by using Scotch bottom shoes that are weighted and shaped to alter the appearance of the hoof and the way the horse raises his feet. Most show shoeing methods encourage a flared, overgrown hoof that can lead to lameness and early retirement. Think seriously about subjecting your animal to the dictates of fashion.

### CORRECTIVE SHOEING

Corrective shoeing and trimming are necessary in cases of laminitis, as well as for hoof cracking, contracted heels, toeing in or out, feet that are asymmetrical, and improper hoof wall angles, all of which can cause lameness. Almost all of these issues, except laminitis (which is caused by an internal toxic reaction), result from inattention to hoof care, poor nutrition, or worse: improper and incorrect hoof care. The broad, mushroomed hoof with cracked sidewalls is almost synonymous with the neglected draft horse. Poor hoof conformation, which can be blamed on genetics, is also a factor. Hoof issues, if let go, can become expensive and time-consuming. Besides, a lame animal is not much use.

## The Importance of Grooming

Most horses and mules enjoy touch, and grooming is one way to productively touch your animal. Equines are extremely sensitive, however, and do not like to be tickled or slapped. A rhythmic rubbing motion is the most appreciated, as long as it is not in one of the supersensitive areas that include the flank, girth, legs, nose, ears, or belly. Although your horse or mule can become accustomed to your rubbing and cleaning these sensitive areas, never approach a new horse by rubbing these areas.

Grooming a draft horse or mule sometimes seems like a futile activity. The moment you turn the animal out to pasture, he finds the dirtiest corner and joyfully rolls around, covering all those careful brush strokes with a thick coat of dust. So why bother?

Appearance is one reason to keep your animals clean and looking good. Grooming also presents a daily ritual you can engage in with your equine, a time to connect and observe changes in body condition. Other reasons are more utilitarian and for the animal's physical well-being. Grooming helps remove external parasites and foreign objects from the skin and coat, and keeps clean those body parts that are most susceptible to irritation or disease. Remove all dirt, plant debris, and other material and smooth the hair before harnessing to prevent sores. After the harness is removed, groom again to straighten the hair before it dries in a sweaty mess, making it harder to brush the next time.

Aside from helping an animal look cared for and beautiful, daily grooming gives you the opportunity to notice subtle changes in appearance, such as an abnormal lump, enlarged joint, growths, wounds, or skin conditions you may miss by just looking from a distance. Use this daily routine also to notice changes in weight. You don't need to gauge your animal's weight daily, but if you begin to suspect changes, monitor weight more closely.

## ROUTINE GROOMING

Burrs, twigs, and other foreign objects lodge in your draft animal's coat, particularly in winter when the coat is the longest. A harness or saddle rubbing these objects can create discomfort and painful sores. Always brush your horse or mule before putting on any tack and after removing it. Regular brushing also keeps the coat free of mud that can compromise the animal's ability to shed rain and to fluff up his hair to keep warm.

Ticks and bots may be managed to some degree by hand removal during grooming. Don't rush over the neck and back and call it good. Be thorough. Ticks like to collect under the "armpits" (the area directly in front of the girth), up between the hind legs, under the jaw, and in other less-visible places.

Small stones can become lodged in the hoof, and a daily picking will decrease the chance of bruising. Pick your equines' hooves often if they are working on gravel or in excessively wet surface conditions.

### Care of the Mane and Neck

The equine neck experiences friction from the movement of the harness collar when the animal is hitched to an implement with a tongue. As the horse walks or trots along, the tongue keeps time with the step, rocking back and forth slightly. This movement in turn moves the collar, which can create sores if the collar does not fit well or a too-large collar is not padded. For the draft horse, the collar could be thought of as an athletic shoe. Would you want to run a marathon in sneakers that were too small and pinched or too big and sloppy?

Since long, matted hair moving back and forth under the collar creates an environment for the formation of hot spots, our family used to roach, or trim, the manes of our draft horses in the summer. We found that clipping the manes decreased the chance of lesions forming from the constant friction of the collar. With the mane roached, we could easily examine the area on the neck/withers where the collar sat, to ensure it wasn't getting sore. Mowing machines, in particular, can quickly rub lesions, since the tongue, in addition to rocking back and forth with the forward motion of the animals, can be quite heavy on the animal's neck. However, if the hair is not kept short enough, the collar can force the stubble into the skin, causing inflammation and tissue damage, so once a mane is roached, the roaching must be maintained as long as the animal is working.

### Cleaning the Genitals

The genital area can accumulate *smegma*, a secretion that causes irritation and has been associated with a type of skin cancer called squamous cell carcinoma. Cleaning your equine's genitals is an important part of preventing skin irritation and/or skin cancer and should be done at two levels:

- Routinely hose the area with a moderately forceful stream of water to help maintain good hygiene.
- At least once a year, conduct a thorough cleaning, particularly of male animals.

Cleaning the genitals of a mare can be annoying to her, but if she is restrained properly and is obedient, you can generally do it without sedation by gently rubbing or peeling off the waxy secretion that accumulates on the udder. The most invasive you should ever be with a mare is a stream of water on the vulva. *Never* insert the hose into the vulva or vagina.

Stallions and geldings require a much more involved cleaning than mares. Cleaning the penis and sheath is not a job for the timid or inexperienced — have your vet or a knowledgeable horseperson show you how to do it safely. The horse or mule must be willing to accept a stream of warm water over the entire genital area. Gently insert the hose into the sheath; alternate with reaching in with a gloved finger to remove accumulated matter from deep inside. A sheath-cleaning agent is available from your veterinarian or tack shop.

The end of the penis is highly sensitive but important to check; a hardened secretion called the *bean* can form at the end of the penis and obstruct urine flow. Unless the animal willingly drops his penis and leaves it dropped for cleaning, consider having your vet sedate the horse or mule.

## Grooming Tools

**M**any different tools are used for equine grooming. For your animal's comfort, know which one to use where.

**Rubber currycomb.** The rubber curry is a soft brush with rubber teeth. Used with moderate pressure, it works well for heavily muscled parts of the body to loosen dirt and debris. Never use a metal curry on your animal; the sharp, abrasive teeth are intended only for cleaning hair from grooming brushes.

**Medium brush.** The medium grooming brush is stiff enough to whisk dirt out of the coat and to help an unruly coat to lie flat. Use it on the ribs and legs.

**Soft brush.** The softness of this brush is appealing to sensitive parts of the equine: the face, lower legs, belly, and flank. The soft brush gives the animal an allover shine after the currycomb and medium brush have been used.

**Sweat scraper.** This metal or plastic tool removes water from the coat after a bath or a hosing down after a hard workout. Use it only on heavily muscled parts of the body; the pressure will irritate bony areas, such as the lower legs.

**Hoof pick.** Use a hoof pick to clean your equine's feet of any gravel or other foreign objects. Pick out hooves both before and after working an animal.

If you own equines for any length of time, you will no doubt find yourself dealing with a serious illness or injury at some point, no matter how well you care for them. Knowing some basic first aid is important and knowing when to ask for help is even more important. Although you can learn a certain amount of basic doctoring that you may do yourself, you also need to learn to separate the wheat from the chaff. Over the years I have run into plenty of home-brewed cures that made me scratch my logic-loving head.

For one old-timer, the cure for the thumps (an eastern Oregon term for laminitis, not to be confused with the other common definition of thumps, a hiccup or small muscle convulsion seen in the flank) was a simple procedure. To effectively cure a horse with the thumps, he said, all you have to do is take blood out of the hoof area and insert it into the chest. He had cured two horses this way, by George, and that was enough to convince him. I must have looked incredulous, because I could see he felt sorry for my ignorance.

Another "sure fire" treatment for deworming horses without medication is simple enough: tap rapidly on the left side of a horse's rib cage at regular intervals. Worms don't appreciate the rhythmic tapping and vacate. How the folks who advocated this method know that it worked is a mystery, as they never mentioned seeing any worms pass.

I am sure some home remedies do help, but anyone who has read James Herriot knows that many of those old-time potions were injurious to an animal's health. Yet not all treatments originate from the pharmacy, and if a home remedy may be explained physiologically, go ahead and try it. But remember that if an animal regains health after the use of one of these methods, chances are pretty good he recovered in spite of the remedy. You are almost always better off consulting your veterinarian when you are concerned about your animal's health.

Your best defense against illness and injury, particularly unexpected emergencies, is to develop a solid working relationship with a good equine veterinarian. If you want your vet to take you seriously, don't meet him for the first time when your horse has nearly severed his leg on a piece of wire. Not only is a veterinarian going to wonder if he will get paid (if you are an unknown client with an emergency), but he may not be able to accommodate your call. Many equine veterinarians have large enough and busy enough practices that they cannot afford to potentially lose time and money with well-established clients by tending to a new, completely unknown client with an emergency. Take the time to get to know your veterinarian. It could save your horse's or mule's life.

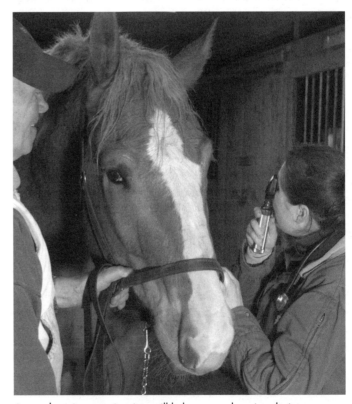

A good equine veterinarian will help you make wise choices about your animal's health.

## Dealing with Wounds and Injuries

Some of the more common conditions requiring first aid in your equine are cuts and bruises, some of which you will never find the cause for. Improperly fitting collars and harnesses can also cause sores that require treatment.

You are almost assured of one day finding your horse or mule with a deep laceration, a strange soft tissue swelling, or some other trauma-induced injury. These conditions may be life threatening and require immediate assistance. An obvious laceration, profuse bleeding, or puncture wound constitutes an emergency — call your veterinarian as quickly as possible. Your vet will likely ask you about:

- Signs of shock
- Bleeding
- What the wound looks like

Your ability to succinctly answer these questions will help the vet to assess the situation and decide how urgent it is. The vet may ask if you have a first-aid kit (which you should) and advise you on what steps you should take until he or she arrives.

A comprehensive first-aid kit contains the following items:

- Nonstick wound dressing pads
- Sterile and nonsterile gauze squares for bandaging
- Cling gauze
- Cotton rolls
- Adhesive tape
- Cohesive bandages (Vetrap or Coflex)
- Antibacterial ointment
- Antibacterial soap (Betadine or Nolvasan scrub)
- Antiseptic solution (Betadine or Nolvasan solution)
- Oral Banamine paste
- Sterile saline solution
- Bandage scissors
- Latex exam gloves
- Leg wraps
- Duct tape
- Permanent marking pen
- Hemostats and forceps
- Pliers
- Clippers
- Six-inch (15 cm) PVC tubing cut in 1½–4 foot (45–120 cm) lengths and cut in half vertically for emergency splints
- Rectal thermometer
- Stethoscope
- Flashlight and spare batteries
- First aid information
- Horse transportation phone numbers
- Your veterinarian's and alternative veterinarian's phone numbers

### TREATING COLLAR SORES

Improperly fitting collars and harnesses can rub nasty, deep sores in a surprisingly short time. The first step in solving this problem, obviously, is to ensure a good fit. Even a good-fitting collar can cause sores, however, if your animal has a tender (not work-toughened) neck. The rocking motion of a sickle bar mower is particularly notorious for causing sores on animals. Using collar pads can reduce the risk of collar sores; we always use collar pads when mowing. The obvious difficulty with healing

---

### Signs of Shock

**Any extreme trauma,** whether from an injury, severe pain, overexertion, or other dramatic change, can result in shock. This condition must be taken seriously and dealt with immediately. While waiting for the vet to arrive, keep the animal warm and quiet. Treatment generally involves specific intravenous fluids, oxygen therapy, antimicrobial therapy, and antiendotoxic therapy; steroids and surgical therapy may also apply.

Signs of shock in a horse or mule include:
- Shaking or trembling of part or the entire body
- Cold ears, muzzle, limbs
- Poor CRT (see page 155)

collar or harness sores is that the points will continue to rub to some degree each time the animal is worked. The animal may need to be laid off work for a few weeks in order to heal properly. Cleanse the sores daily with water and an antiseptic. A course of antibiotics will reduce the risk of infection.

## Dealing with General Lameness

Fortunately for you, draft horses and mules do not experience many of the lameness problems affecting highly athletic, lighter sport horses. However, just because draft horses and mules don't typically spin around barrels, fly down a racetrack, or soar over jumps doesn't mean they are immune to joint, leg, and hoof problems. Draft equines carry heavy weight on their legs and are still susceptible to lameness, which can result from any number of causes, including an injury such as a stone bruise or a puncture wound, laminitis, poor hoof conformation, joint problems, and disease.

Lameness can appear as a subtle hitch in a step, a dramatic, gait-altering, three-legged condition, or anything in between. The best way to prevent a serious or chronic lameness is to make a habit of carefully observing the way your animal normally moves. Once you recognize that he is not moving normally, you then have to determine if it is because of lameness. Understanding the origin of the lameness will help you decide how best to manage your animal to resolve the condition. Unfortunately, lamenesses can be hard to diagnose and difficult (and expensive) to treat.

### EVALUATING LAMENESS

If you notice your horse or mule moving differently from the way he normally does, he could be altering his gait to compensate for a painful limb. A definitive diagnosis requires a lameness evaluation by a knowledgeable person, typically an equine veterinarian with a good reputation. Not all veterinarians have the same amount of experience in diagnosing and treating lameness.

Unless the animal is three-legged lame or a fracture is suspected, a lameness evaluation usually begins with the vet examining the animal while he

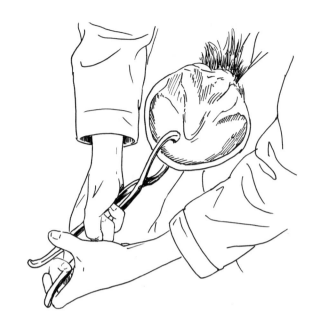

A hoof tester handled by a competent veterinarian or farrier helps to determine whether the hoof is in pain and causing lameness.

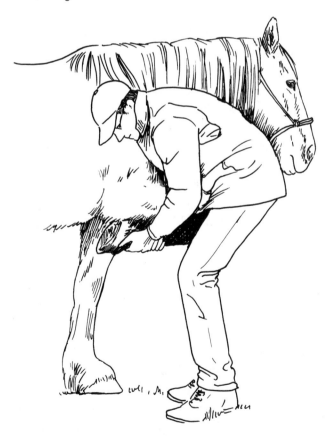

A flexion test by a competent veterinarian helps to determine whether a joint is the cause of pain.

trots in a straight line, on a loose lead and on a flat, smooth surface. Many draft horses and mules are unaccustomed to trotting in hand, which can make lameness evaluation extremely difficult. A slightly lame horse or mule will bob his head, dropping the head when the normal leg bears weight and throwing it up when the painful leg bears weight.

A lameness exam often includes evaluation with a hoof tester. This pincer-type device applies pressure to specific areas of the hoof to check for abnormal sensitivity. A thorough examination also includes looking for inflammation or heat anywhere along the leg.

A flexion test is usually administered as part of the lameness evaluation. The flexion test involves holding the leg with the suspected joint flexed (bent) for a few minutes. When released, the lameness will be exaggerated in the affected joint when the animal moves off immediately.

## DIAGNOSING LAMENESS

After conducting the lameness evaluation, your veterinarian may choose to use various diagnostic tools to isolate the lameness. Although pinpointing the exact site of lameness can be difficult and often frustrating, it is important for determining an effective treatment. Diagnostics such as nerve and joint blocks, radiographs, scintigraphy, magnetic resonance imaging (MRI), ultrasound, and testing of blood, joint fluid, and tissue samples are all tools your equine veterinarian may use to isolate the lameness.

**Joint and nerve blocks** are often administered in an attempt to isolate a potential site of lameness. Your vet will follow a system of ruling out various locations by chemically blocking the nerves via injection and thus reducing sensation to a particular joint until the lameness disappears. Nerve and joint blocks are common tools for isolating lameness problems.

**Radiographs (X-rays)** are helpful in determining changes and damage to bony tissues and should be interpreted only by an experienced equine veterinarian. Radiographs *do not* provide much information about tendons, ligaments, or intrajoint structures, which are often the source of lameness.

### Grading the Lameness

**Lameness has ranges** of severity, making evaluation, diagnosis, and prognosis difficult. To aid in its description and diagnosis, the American Association of Equine Practitioners has developed a lameness scale from 0 to 5.

**0:** Lameness is not perceptible under any circumstances (such as under saddle, while circling, walking on an incline or hard surface). This rating is considered "normal" for the animal.
**1:** Lameness is difficult to observe and not consistently apparent, regardless of circumstances.
**2:** Lameness is difficult to observe at a walk or when the animal is trotting in a straight line, but appears consistently under some circumstances.
**3:** Lameness is consistently observable at a trot under all circumstances.
**4:** Lameness is obvious at a walk.
**5:** Lameness produces minimal weight bearing when the animal is in motion and/or at rest, or the animal is completely unable to move.

**Scintigraphy,** also called nuclear scanning, is a procedure used mainly at referral or veterinary teaching hospitals. Radioisotopes are injected intravenously in concentrated amounts in the affected area. An image of the suspected site is then scanned with a gamma camera, which provides an image. Scintigraphy is often used to image bony changes, such as subtle fractures, that are not observable on radiographs. The affected areas show up as hot areas on the image.

**Magnetic resonance imaging (MRI)** is an expensive but excellent diagnostic tool with well-defined resolution and detail for evaluating soft tissue and bony problems, primarily of the lower leg. It can show structures that aren't revealed by radiographs. MRI for equines is not available everywhere, but as the need for and benefit of this technology becomes widely known, more veterinary teaching hospitals and referral centers are using it.

**Ultrasound** is another diagnostic tool for lameness used by many experienced equine veterinarians. This technique uses ultrasonic waves to create images of internal structures that radiographs fail to achieve.

**Lab work** is appropriate for a lameness that may be caused by infection, rather than by trauma or injury. To determine the presence or cause of the infection, blood, joint fluid, and/or tissue samples may be taken. These samples require laboratory analysis, which takes time.

## COMMON LAMENESSES OF JOINTS AND LEGS

Draft horses and mules of all ages may be affected by lameness as a result of a disorder of the joints and/or legs. Some of these disorders result from nutritional deficits and are preventable; others result from poor hoof maintenance and are also preventable. Still other lamenesses are a result of injury. Some develop slowly over the life of the animal, often because of a conformational fault. Lameness due to genetics should not be perpetuated; the animal should be removed from the breeding pool.

### Developmental Orthopedic Disease

Developmental orthopedic disease (DOD) is a condition that affects growing horses. The rapid growth experienced by many draft horses and mules makes them particularly susceptible to DOD. Young drafts may experience a host of problems relating to the balance between bone, tendon, and ligament growth, as well as problems with the conversion of cartilage into bone.

Reducing the risk of a DOD involves providing growing horses and mules with a well-balanced diet that is high in fiber and low in soluble carbohydrates (i.e., concentrated feeds, such as grain), has a proper calcium-to-phosphorus ratio, and provides a sufficient supply of the trace minerals zinc, copper, and manganese to support the growth of strong bones and healthy joints. Young, growing equines should have free access to sufficient space for exercise to reduce the risk of DOD.

### Degenerative Joint Disease

Bone spavin and bog spavin are two kinds of degenerative joint disease (DJD) affecting adult draft horses. Spavin often results from improper trimming or shoeing, which causes the animal's natural gait to be altered. Spavin may also be caused by excessive strain on the hock joints from overly hard pulling and from conformational defects, such as sickle or cow hocks. Many show and hitch horses are shod with shoes that exaggerate their gait, and many of these horses show signs of DJD later — and, unfortunately, not so much later — in life.

*Bone spavin*, a bony growth in the lower hock joint, is typically caused by improper hoof care and abnormal hoof conformation. *Bog spavin*, a fluid-filled swelling on the lower inside of the hock, is usually caused by unnatural strain on the hock joint, such as when an animal is required to pull a heavy load through a resistant force, like mud.

Spavin in the draft horse generally occurs because of the animal's inherent conformation. Although bog spavin rarely causes lameness, it is a conformational defect. Being cow-hocked (hocks too close together) or post-legged (too straight through the hocks), as many draft horses are, appears to predispose horses to bog spavin, bone spavin, and other hock disorders, making this a common problem.

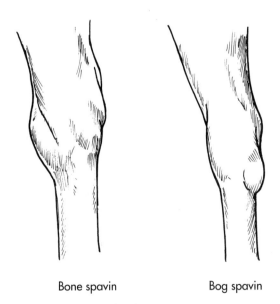

Bone spavin    Bog spavin

### Splints

Lameness associated with stress or trauma to the splint bones in the lower legs is known as *splints*. The nature of draft work — pulling, rather than high-speed twisting and turning — generally protects draft equines from this injury. The primary cause of splints in draft horses is banging the splint bone with the opposite hoof, which occurs in horses with wide chests that tend to paddle (move their feet in circles) when they move.

### Stocking Up

*Stocking up* is a term for swelling in the lower legs, typically in the tendon sheath. Stocking up usually results from inadequate circulation in the legs when an equine has been standing still for too long, such as during long shipping hauls or confinement to a tie stall. Pressure wraps help prevent stocking up. Violent movement that strains the leg can also cause excessive fluid to form in the legs. Activity usually reduces the swelling. If the fluid is a result of injury, however, activity may make the problem worse. If stocking up does not resolve with exercise, or you suspect injury, contact your veterinarian for a further evaluation.

## COMMON LAMENESSES OF THE HOOVES

The enormous feet of all draft horse breeds offer lots of surface area where things can go wrong, including sidebone, hoof cracks, sole bruises, canker, and thrush. Genetics, to some extent, plays a role in the toughness of a horse's feet. Percherons and Suffolks, for instance, were bred for work on compacted clay soils and therefore tend to have tougher feet than Belgians, originally bred to work where the soils are richer. Clydesdales and Shires have larger, flatter feet adapted to the moist areas where these breeds were developed and may therefore experience problems if worked on the type of hard or rocky land to which Percherons and Suffolks have adapted.

Draft mules are far less affected by lameness of the hooves than are draft horses.

### Sidebone

The hardening of the cartilages of the coffin bone is known as *sidebone*. The condition is common in both adult and growing draft breeds and may be the result of a genetic predisposition, as well as poor hoof conformation and improper trimming and shoeing. Sidebone does not always produce lameness, although the damaged cartilage can become infected, causing significant lameness. It is usually a progressive condition, seen more often in older animals than young ones. Treatment includes the use of anti-inflammatory drugs, as well as proper trimming and shoeing.

### Hoof Cracks

Hoof cracks are a common ailment of draft horses. Quarter cracks, occurring on the side of the hoof, are the most common hoof crack. Cracks result from poor trimming or failure to trim at all. Cracks may also result from a nutritional deficiency or heavy concussion such as from trotting on hard surfaces. Regular hoof trimming that produces a well-balanced hoof will help the cracks to eventually grow out.

Quarter cracks are common in draft horse hooves that are not properly trimmed.

### Sole Bruises and Abscesses

Other problems common in draft horses — perhaps as a result of their large, shallow hooves — are *sole bruises* and *abscesses*. Susceptibility to bruises predisposes a hoof to abscesses. Bruises and abscesses may result from dramatic changes in the contact surface of a hoof sole, such as going barefoot after being shod for a long period or moving from a soft, muddy pasture to a hard, rough pasture. Bruises often cause minor lameness, while abscesses can cause a three-legged lameness because of the extreme pain.

Treatment for abscesses includes opening and draining the affected area, a job you, your farrier, or your veterinarian can do. Some abscesses are stubborn and too far from the surface to drain, but may be treated by soaking the hoof in Epsom salts and antibacterial medications.

## Canker and Thrush

A condition often associated with painful feet is *thrush*, a bacterial infection of the hoof frog resulting from filthy bedding or persistently wet soil. Some animals seem to be more susceptible than others to chronic thrush, which is best prevented by keeping stalls and other footing conditions dry. Treatment for thrush includes cleaning the hoof with Betadine or any other specialized thrush scrub available from most feed and livestock supply stores. Your veterinarian may also prescribe an injectable antibiotic.

A similar but more serious infection is *canker*, medically known as necrotic pododermatitis, which affects not only the frog but also the sole and heels, and causes the death of infected tissue. The resulting odor is quite foul. Lameness may be slight at first, but quickly worsens as the condition escalates. Canker is caused by lack of proper hoof care for equines subjected to wet and bacteria-contaminated soil and/or bedding.

Treatment varies and may involve a one-time surgical removal of the dead and dying tissue while the animal is under anesthesia, or may involve a more routine removal of the superficial dead tissue, routine cleaning, and packing of the hoof with antibiotic medication. Injections of antibiotics may be needed to control this difficult-to-treat condition. Until the bacteria are completely eliminated from the hoof, tissue will continue to die and must be regularly removed. Eventually a healthy sole, heels, and frog remain. During recovery, the animal's bedding should be keep scrupulously clean to maintain dry conditions.

## Dealing with Laminitis

The prevention of laminitis should be part of every equine management program and a suspected case should always be treated as an emergency requiring veterinary attention. *Laminitis* is an inflammation of the laminae of the foot caused by the disruption of blood flow. The result is a degeneration of the attachments between the coffin bone and the inner hoof wall. If the laminitis is severe, this degeneration ultimately results in a complete failure of the attachments. Farriers refer to this failure and the subsequent rotation of the coffin bone as *founder*.

Because equines support 65 percent of their body weight on their forefeet, laminitis tends to affect the forefeet more often than the hinds, although in a severe case all four feet may be involved. Laminitis may be acute or chronic and can be life-threatening. Chronic laminitis is an acute case that lasts longer than two days and/or results in rotation of the coffin bone. In addition to general lameness, signs of acute laminitis include:

- Shifting weight from one leg to the other
- Standing with legs braced and extended (camped-out)
- Muscle tremors
- Bounding digital pulse
- Feet that feel warm or hot
- Refusal to pick up the feet
- Reluctance to move
- Reluctance to stand once lying down

Any of these symptoms are cause for concern and require immediate medical attention.

### COMMON CAUSES OF LAMINITIS

Because laminitis is prevalent in horses, horse owners must pay diligent attention to prevent or treat it. Laminitis is often the direct result of mismanagement. The exact cause may be difficult to pinpoint, but certain factors are known to increase the risk.

**Grain overload** is a common cause of laminitis. Metabolic changes in the digestive tract produce an endotoxin that causes an inflammation of the laminae. Excessive amounts of oats produce only a mild case of laminitis or sometimes no signs at all. Barley, wheat, and corn are the grains most frequently associated with grain overload. A horse that breaks into the grain room must be treated immediately by a veterinarian. Do not wait until symptoms advance, as a grain binge is almost sure to result in laminitis and can be fatal.

**Grass laminitis** is common in horses grazing on lush, rapidly growing pasture, and often occurs in the spring after a winter of little or no pasture. Legumes are implicated more than grasses, perhaps because of hormones in the legumes. Ponies and overweight horses of all breeds, often characterized by a heavy, cresty neck, are the most susceptible to grass laminitis.

**Medications,** such as corticosteroids and strong laxatives, have been linked to laminitis when used in high doses or for prolonged periods.

**Injury** of a limb, particularly a forelimb, creates increased stress on one or more of the other legs, and can cause laminitis in the overstressed limbs. Controlling the animal's weight, providing deep bedding, using an anti-inflammatory, and treating the injured limb all help minimize the chances of laminitis occurring in the healthy limbs.

**Black walnut shavings** produce a toxin that can cause acute laminitis. Do not bed your equine with black walnut shavings, mixed shavings, or shavings from unknown origins, as they pose too great a risk.

**Retained placental tissue** after a mare foals often causes an infection, leading to laminitis in the mare. This type of laminitis is serious and must be addressed immediately if you suspect retained placental tissues. A retained placenta must be removed quickly.

**Road stress** is a form of laminitis resulting from repeated concussion on hard surfaces, as occurs in carriage horses required to spend long hours on paved roads. Laminitis is frequently seen in draft horses as a result of sole bruising of their large, shallow feet. Road laminitis also may occur in horses hauled over rough roads in a trailer with inadequate springs or cushioning.

**High fever or acute illness** that causes metabolic disturbances, such as West Nile virus or Potomac horse fever, have the potential for causing laminitis.

## TREATING LAMINITIS

Consider acute laminitis an emergency, as permanent damage can occur if treatment is delayed. If you suspect your horse or mule has laminitis, contact your veterinarian immediately. The vet will administer medications to help control the pain and will suggest support for the soles of the feet, as well as careful removal of shoes if the animal is shod. *Do not pound on the hooves,* which will cause excruciating pain. Hosing the hooves with cold water, or standing the animal in a running stream, may help reduce the pain and inflammation while you wait for the veterinarian.

In a case of grain- or grass-induced laminitis, your vet may administer a laxative to speed up passage of the feed through the animal's system, as well as to neutralize the toxin. Treatment for chronic laminitis includes regular, diligent hoof care and other precautions to prevent the coffin bone from penetrating the sole of the hoof as it becomes separated from the inner hoof wall. To reduce the potential for further damage, you will need to consult both your veterinarian and your farrier.

A horse or mule affected with laminitis remains unsound for the rest of his life. Although with proper management an animal with a mild case of laminitis could be usable for extended periods, do not consider purchasing an animal with a history of laminitis if you intend to use him as a work animal.

A horse in pain from laminitis stands camped out and with his head up.

> ◀ **CAUTION** ▶
>
> ### Withholding Water
>
> Everyone has heard that laminitis occurs when an overheated equine consumes large amounts of cold water, and for this reason many horse owners withhold water from an overheated horse. Research has not shown that drinking water causes a hot horse to develop laminitis. Many equine professionals, however, insist that too much water drunk too quickly after strong exertion can cause colic, while others argue that withholding water from the horse can cause dehydration and impaction colic.
>
> To be on the safe side, in conjunction with cooling off your hot horse by putting him in the shade and giving him an easy walk and a good rubdown, offer him small amounts of cool (not icy) water as he cools down before letting him drink his fill.

## PREVENTING LAMINITIS

Certain equines have an increased risk of laminitis, which may be a result of conformation, genetics, or illness. A horse or mule has an increased risk if he is overweight, has a heavily crested neck, has a previous history of laminitis, or is an older horse with Cushing's disease. Managing weight to prevent obesity will reduce your chances of having to deal with laminitis.

Dietary risk factors include feeding large meals of carbohydrates and consistently feeding a fixed amount of concentrates to an animal that is worked erratically. Carefully monitor your draft equine's dietary needs. Introduce your animals to spring pasture gradually to avoid grass laminitis. Be vigilant in preventing access to grain and feed storage areas.

## Dealing with Colic

*Colic* simply means abdominal pain. It is not a disease, but an indication of illness. Nearly all horses experience some signs of colic at one time or another. Colic has many causes, and can range from mild with no lasting consequences to severe enough to cause torsion (twisting of the intestines) or other potentially fatal conditions. Most horses and mules will experience mild colic at some point in their lives; many such cases go undetected. However, distinguishing a mild from a severe situation at the onset of pain is almost impossible, and thus you should carefully monitor an animal with *any* signs.

The gastrointestinal (GI) tract, where many signs of colic originate, is a complex system that begins with the esophagus and stomach, loops through the small intestines (hanging from the mesentery, which is attached by one point to the abdomen), goes through the cecum and the large intestine, and ends at the anus. This huge maze of digestive complexity is attached to the abdominal wall in only a few places, folds over on itself multiple times, and can twist, contort, or do other potentially harmful gymnastics with frightful ease. Unfortunately, contortions may be difficult to diagnose without opening up the belly to take a look. Understanding general preventive measures, signs, common colics, and what to do in case your horse or mule develops colic will prepare you to deal with your equine's pain.

The first signs of colic may be so subtle that they are easy to miss. To head off further trouble, carefully note these symptoms:

- Lying down more than usual, or getting up and lying down repeatedly
- Moving restlessly
- Looking at or kicking at the belly
- Frequently standing as though to urinate but not producing urine
- Showing little or no interest in food
- Curling the lip in a flehmen response for no apparent reason

Other signs are intensely acute. Call your veterinarian immediately if you observe these signs:

- Pawing the ground repeatedly
- Biting or kicking at the belly
- Rolling repeatedly or more than normal
- Not producing manure
- Acting interested in water but not drinking
- Sweating without cause
- Breathing rapidly with flared nostrils
- Appearing depressed or listless

## COMMON CAUSES OF COLIC

Colic is so common that the cause can be hard to pinpoint. In most mild cases, you will never know the cause for sure. Some causes of colic, however, are well known.

**Gas colic** comes from a buildup of gas in the intestines, usually in the cecum and/or large intestine. A gas buildup that has no deeper underlying problem usually resolves easily. Walking your horse or mule may help ease further buildup and stimulate the passage of manure or gas.

**Lack of water** can lead to impaction colic. In the fall, when horses have an inadequate water supply and are consuming dry forage, most vets see a sharp increase in colic cases. Constant access to water is imperative for healthy digestive function. Impaction colic usually occurs at or near a junction in the GI tract, where feed becomes blocked and then builds up to form a solid mass. Impaction is usually easy to resolve with veterinary assistance, but is sometimes a sign of a more serious colic, such as torsion.

**Tapeworms** can cause a telescoping effect in the intestines (intussusception), leading to symptoms that may be mild to severe, and potentially requiring surgical treatment.

**Gastric distension or rupture** occurs when an animal eats significant quantities of grain or some substance that expands with moisture. Horses cannot vomit to reduce the stomach's contents, and a stomach that becomes overextended may rupture. A ruptured stomach inevitably leads to death. The only prevention for this type of colic is to call your veterinarian immediately if you suspect your animal has gorged on concentrated feed; the vet can quickly reduce the stomach contents using a nasogastric tube, and take precautions against laminitis.

**Twisting, torsion, volvulus,** or displacement of the intestine causes severely painful colic. In these cases, the intestine twists, moves to an abnormal position, or tears. The few points of connection for the GI tract and mesentery allow the intestines to become displaced, usually resulting in a complete blockage of the GI tract. Immediate surgery is the only option between your horse and death. These types of colic often reduce the supply of blood to tissues, so the longer treatment is delayed, the greater the risk that the tissues of the affected intestine will begin to die. Colic of this nature often begins with mild symptoms, but can quickly move to the severe stage. If the torsion is caught and treated early, surgery may save your horse, though equine surgery is costly and requires specially designed facilities.

**Sand colic** typically occurs in regions of the country where the soil is naturally sandy. Sand is an intestinal irritant, and the initial signs of sand colic can be treated with anti-inflammatory drugs. The weight of the sand, however, may lead to intestinal displacement or torsion, requiring immediate surgical intervention. Avoid feeding on sandy ground if at all possible; use a feeder instead. If you keep your horse or mule on sandy soils, check the feces for sand. If the feces contain sand, consider supplementing your equine's diet with psyllium, which helps eliminate sand from the gut. Mineral oil works for some kinds of colic, but is not helpful for sand colic because it will generally pass around the sand and not move it out of the GI system.

**Inflammation** of the small or large intestines can cause enteritis and colitis. Blister beetles in alfalfa hay are one cause of this type of painful colic. Another possible cause suspected by some veterinarians is a bacteria, although the specific bacteria has yet to be identified. This case is serious and should be handled promptly by your veterinarian.

## TREATING COLIC

If the signs of colic are mild, take action to prevent a more serious case. First call your veterinarian and describe the signs. Then place the animal in an area where you can closely observe him. Remove *all* feed, but keep water available except in a case of grain overload, when water must also be removed. Walk him to prevent him from rolling, but don't exhaust or stress him with too much exercise.

If the signs of colic are severe, your preliminary report of the animal's condition will greatly help your vet to evaluate the situation. When you call, be prepared with the following information:

- Specific signs of colic (indicating the severity)
- Pulse or heart rate
- Respiratory rate
- Rectal temperature
- Mucous membrane color
- Capillary refill time
- Digestive sounds (or lack thereof)
- Recent changes (pasture, feeding, exercise, accidental overconsumption)
- Bowel movements (or lack thereof)

If your animal is behaving in a violent manner, the situation is critical. Severe pain can indicate life-threatening colic. In most situations, relieving the pain, which your veterinarian can do with drugs, is the first step toward recovery. *Do not administer anything to your equine unless the veterinarian tells you to.*

While waiting for your veterinarian to arrive, try to keep the animal standing. If he lies down and rolls, the chance of injury increases. You are no match for a 2,000-pound (900 kg) beast in abject agony. If you cannot keep him from rolling, try at least to keep him from hurting himself while he is thrashing around. Place hay bales around the stall edges to keep him from crashing into the walls and move everything possible out of the way.

## PREVENTING COLIC

Taking steps to prevent colic is not a foolproof way to avoid pain. Horses and even mules get bellyaches from unknown causes. But by ruling out common causes of colic, you give your animal a better chance of remaining pain free. Although colic can sneak up on the most diligent equine owner, the following management practices will help you to reduce the risk of colic in your draft horse or mule:

- Maintain a rigorous parasite control program.
- Provide a safe place for your horse or mule to run and buck, important activities for the proper functioning of the equine GI tract.
- Keep your feeding program as regular as possible; the less variety in scheduling, the better.
- Do not feed hay, grain, or anything else that smells remotely of mold.
- Feed hay before concentrated feeds.
- Provide at least 60 percent of the digestible energy of your feed in forage (pasture or hay).
- Provide this forage throughout the day, rather than in one feeding.
- Make sure that your equine has constant access to clean water, particularly after hard work.
- Exercise your draft animal consistently.
- If you make any changes to diet or exercise, do so slowly to allow your animal to adapt.

## Equine Polysaccharide Storage Myopathy

Equine polysaccharide storage myopathy (EPSM) is a metabolic condition that appears to be hereditary and particularly affects heavily muscled equines, including draft horses and mules. The fact that it occurs in mules as well as horses indicates that

Walk your horse during a colic attack (unless he is in so much pain that you are in danger of being hurt).

the gene is dominant. This condition is incredibly common; more than half of the draft horses studied in North America have been diagnosed with some stage of EPSM. The condition may be effectively managed with a proper diet, and the affected draft horse or mule may be successfully worked for many years.

EPSM is apparently caused by the inability of certain equines to utilize the soluble carbohydrates found in grain and other concentrated feeds. Because the muscles cannot obtain sufficient energy for proper function, they begin to break down their own proteins for energy. As this process continues, the muscles symmetrically atrophy. In a severe case, the animal becomes little more than a skeleton. The onset of EPSM signs varies among animals and does not occur after a specific event or at a particular age. The condition is the result of high levels of carbohydrates building up in the muscle instead of being metabolized. The buildup causes an energy deficit that leads to the signs associated with EPSM.

Horses affected with EPSM can experience muscle weakness and severe muscle cramping in the hind limbs that makes exercise painful. Animals with EPSM may also be extremely thin to emaciated. The most common signs of EPSM include:

- Symptoms of colic following exercise
- Poor performance
- Weakness in the hindquarters
- Lack of energy
- Stumbling
- Stiff and awkward gaits in the hind end
- Trembling after exercise

### DIAGNOSING EPSM

The only way to definitively diagnose EPSM is to have your veterinarian do a muscle biopsy. A small amount of muscle tissue is taken from the rump and examined for abnormal carbohydrate storage and muscle damage.

Exercise testing also may be used to diagnose EPSM. Elevated blood levels of the muscle enzymes creatine kinase (CK) and aspartate aminotransferase (AST) after 15 minutes of exercise are likely indicators of EPSM. Severely affected horses have elevated CK and AST levels, even after weeks of stall rest.

Many owners diagnose their draft animals merely by putting them on an EPSM diet. If the signs of EPSM are reduced by a low-carbohydrate, high-fat diet, it's a good bet that the animal has EPSM.

### TREATING EPSM

Horses diagnosed with EPSM benefit from exercise conditioning, plenty of pasture turnout, and a diet that is high in fat and low in soluble carbohydrates (grain). Your veterinarian can help you develop a diet.

A horse with a severe case of EPSM will experience loss of appetite, may stop eating all together, and will become lethargic. Such an animal should not be worked until he gains back weight and strength. Some horses with severe EPSM never recover, but become so weak they must be euthanized.

---

## Monday Morning Sickness

**M**onday morning sickness was commonly recognized by old-time teamsters. After resting their draft animals over Sunday, they noticed that upon resuming work on Monday, some horses showed signs of severe muscle injury — thus the term "Monday morning sickness."

The condition — also known as tying up, rhabdomyolysis, black water, setfast, and azoturia — is a manifestation of EPSM. The same horses consistently show signs, while their stable mates on a similar exercise program have no problems. The signs include pain and swelling in the rump, reluctance to move, brown to port wine urine color, and difficulty or inability to rise after lying down.

# Travel Precautions

**Traveling is stressful for horses and mules,** no matter how many times they do it, and the stress can lead to more serious health issues. Long hauls are more stressful than short hauls; regardless of the distance, take precautions to ensure your animal arrives at the destination safely and in good health. Exercise, water, and feed must be continued during hauling.

Besides observing common sense precautions, you must also follow health regulations for shipping equines across state lines and for attending shows and other events. If you plan to haul your equine, contact your veterinarian about obtaining a current health certificate and negative Coggins test before going on your trip. Some states also require brand inspections.

**Creature safety and comfort.** Draft animals require larger accommodations than many slant horse trailers designed for light horses can comfortably provide. Drafts are therefore often shipped in stock trailers that allow for a bit more room and maneuverability. If you plan to do any hauling, ensure that your trailer is in good working order with a floor that's strong enough to handle the weight.

Shipping in the summer can result in dehydration or overheating. Make sure your trailer has adequate ventilation. If you are hauling a long distance, consider hauling at night when the temperature is lower.

If you are trailering only one animal, letting him travel completely loose is an option, but if you haul more than one animal and do not have dividers to separate them, you must tie. Tie as loosely as is safe — not so loose that an animal can put a leg over the rope or become tangled with another animal — so the animal can lower his head to keep his airways clear. Some animals prefer to travel facing the rear of the trailer, which may not be possible with your trailer configuration or if you haul more than one animal.

**Long-distance hauling.** To reduce the stress of long-distance hauling, plan on exercising your equine at least every six hours. After a long haul, refrain from hauling the same animal again for at least one week. Allowing the animal to rest will help reduce respiratory diseases and other common stress-induced illnesses.

Dehydration after hauling is common, so offer water regularly during transport. Weather conditions will dictate how often you need to provide your equine with water. In hot weather, offer water at least every four hours. Using a bucket that is familiar to the animal will increase his willingness to drink.

Provide plenty of hay when hauling, but *no* grain or other concentrated feed. Equines sometimes have trouble digesting concentrated feeds at any time and the additional stress of hauling will not improve their digestion. Feed hay as close to the floor as possible so your horse or mule doesn't have to reach up for it. Lowering the hay level reduces the chance of respiratory illness that can result from hauling.

**Health certificate.** Most states require a health certificate for legal entry of your animal. Almost all shows, auctions, and other equine events also require a health certificate. The health certificate is written by a veterinarian and certifies that your animal appears healthy and free from infectious disease or other debilitating conditions. It is a general inspection and guarantees only what your veterinarian observes.

**Coggins test.** This test for equine infectious anemia (EIA), a viral disease that may be carried by horses without symptoms, is required for hauling equines across state lines and to events. No treatment or vaccine is available for EIA, and by law, an equine with EIA must be isolated for the rest of his life or destroyed. Most states and shows require a negative Coggins test in the past year; some require a negative test in the past six months.

# EXPECT THE UNEXPECTED

**Jereld Rice, DVM**
**Vale, Oregon**

Type of Practice: Mixed, with emphasis on equine and bovine

Other equine experience: Starting colts, wrangling

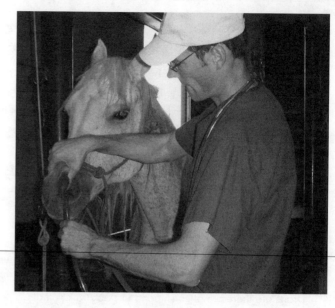

Jereld Rice grew up wishing for horses. When he was young he pored over an *Encyclopedia of the Horse* for hours, learning the markings, colors, breeds, and characteristics of horses. His father managed a busy university dairy, and the family had little time or money for horses.

The first occupation Dr. Rice chose, in elementary school, was that of a veterinarian. He loved animals and expressed it in the concern he showed for any animal in his care. While in college, Dr. Rice realized his dream of working with horses when he wrangled for a children's summer camp. Eventually he became the cowboy camp director and managed 60 head of riding and draft horses.

Chemistry was his nemesis in college, causing him to abandon his dream of veterinary medicine and instead choose a path in elementary education. However, one year in the classroom left him with the painful realization that he had missed his calling. A serpentine path led him to starting colts for a number of years, then to finishing his prerequisites for veterinary school, and finally to vet school at Washington State University, where he had the opportunity to work with one of the world's premier equine orthopedic surgeons. After earning his veterinary degree, Dr. Rice joined a rural practice that emphasizes bovine and equine care and also has a small animal component.

Because he serves a largely ranching clientele, Dr. Rice sees many emergency cases that require immediate action to save a life. Some of the more common equine emergencies in his practice are deep lacerations and colic, both situations in which prompt medical treatment is crucial. On a typically busy day he may perform a C-section on a first-calf heifer, surgically repair a horse's torn flexor tendons, and treat a dog for liver failure, along with whatever else may walk through the clinic door. The clinic offers night and weekend on-call emergency services, and Dr. Rice gets plenty of quality time with clients during these hours.

Dr. Rice's motto in treating emergencies is, "Stay calm, work quickly, and never give up." Adhering to this motto has earned him an excellent reputation with his clients, as well as success with some incredibly difficult cases. On one colic case, for which the clients could not afford surgery, he hospitalized the horse for six days on IV and nasogastric fluids, administering a total of 120 gallons of fluids. On the sixth day, the animal finally passed the grass clipping impaction and today remains a hard-working horse.

Dr. Rice seeks to educate his clients to avoid any circumstance that will create an emergency. Emergencies, however, are bound to happen, and he encourages all of his clients to maintain a fully stocked first-aid kit. The ability to act while the veterinarian is en route can often result in saving the horse's life.

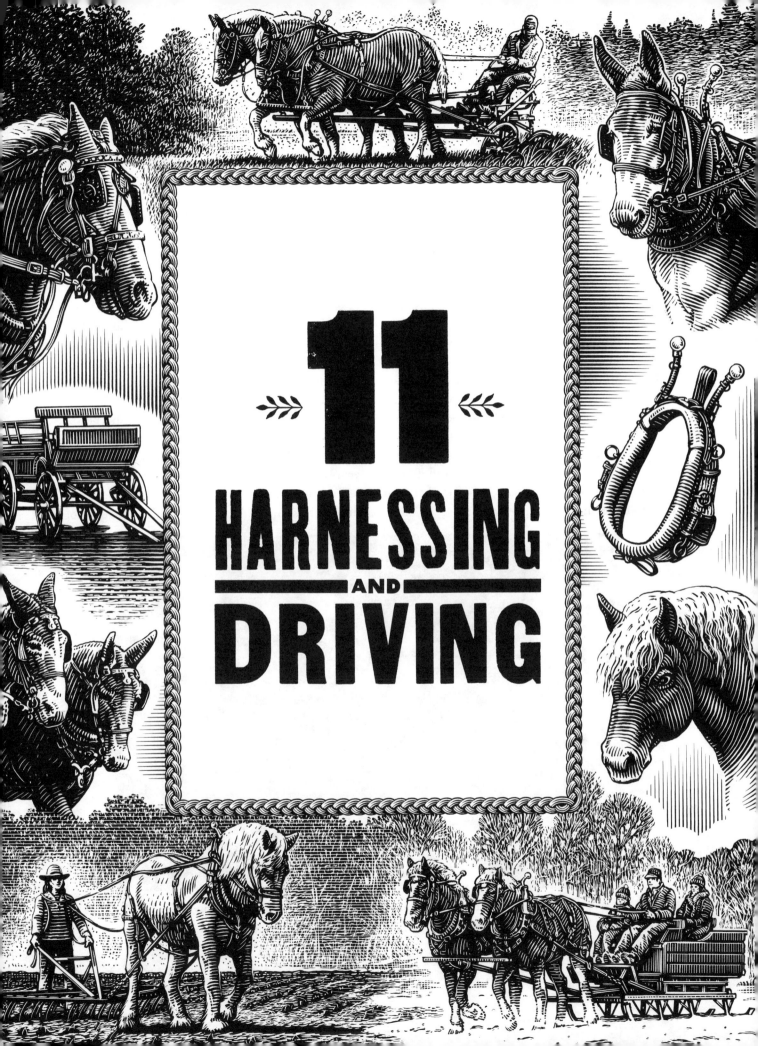

# 11
# HARNESSING AND DRIVING

My dad's criterion if I wanted to drive was that I must be strong enough to harness. I was eleven years old before I was able to lift a harness onto our Morgan driving horse. The sense of accomplishment that followed was second only to the pleasure of trotting down the road in the cart on a pleasant summer's day.

I have been driving draft horses since I was quite young and therefore have difficulty imagining *not* knowing the feel of the lines, of the strength in the horses, of the synergy between teamster and team. I remember the first time my dad allowed me to disc alone with three horses — the experience was powerful and liberating. Nothing else is like having the responsibility for guiding that much power to do the job you want.

On your journey toward using your draft horses and mules, you will need to acquire two important skills: harnessing and driving. Having beautiful animals and all the fancy equipment in the world is worthless without a driver who knows how to effectively communicate with the animals and a good harness to provide the physical connection between driver and team.

## Learn the Components of the Harness

The first step toward using your draft horse or mule is to literally harness its energy, using a physical harness that ergonomically fits the animal's body and uses the power of the chest, forelimbs, and hindquarters for pulling, turning, and stopping equipment. Nearly all teamsters and muleteers speak of their chosen harness style in tones of near reverence. The harness makes working their precious animals possible, and most good drivers treat their harnesses with as much respect as they do their animals. A properly maintained harness will provide many years of service.

To understand how the harness functions, you must first know its basic parts. Different styles of

Harnessing your draft horse or mule for the first time is likely to be a clumsy, awkward attempt, with straps going in all directions.

harness vary somewhat in their makeup, but all harnesses may be simplified into four main components: the communication (steering) parts, the pulling parts, the turning parts, and the backing and stopping parts.

## THE COMMUNICATION COMPONENTS

You will use both vocal commands and cues from the harness to effectively communicate your wishes to your draft equine. Because your voice may be imitated by anyone around you (and could influence your animal's movement without your consent), always ensure that your animal responds *first* to cues from the communication parts of the harness. These parts are the bridle, the bit, and the driving lines fastened to the bit.

These three parts may be likened to a tractor's steering wheel, gas pedal, and brake pedal. They form the link between the driver and the source of power. The communication parts of the harness allow you to direct your animal to back up, stop, go forward, or turn right and left. By the way, if you wish to be taken seriously in draft horse circles, it's worth remembering that the leather strips attached to the bit for communication purposes are *driving lines*. Reins are reserved for riding.

Because these parts of your harness are so vital to communication, they must always be 100 percent functional, free from defects, and well cared for. Failure in any of these important communication links could spell disaster.

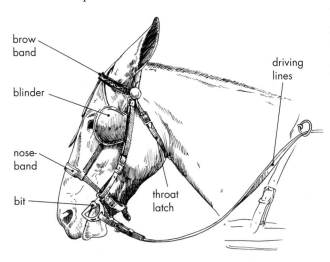

**PARTS OF THE BRIDLE**

> ## Blinders or No?
>
> **Most people working horses** and mules in harness use bridles with blinders (called blinkers on pleasure-driving harnesses). Blinders keep the animal's vision focused straight ahead, in binocular vision, effectively eliminating monocular vision, where the animal can see two pictures at once. Limiting the range of vision and potential distractions supposedly keeps the animal more focused and reduces the potential for reactions to unfamiliar sights.
>
> On the other hand, a horse or mule with the freedom to look around, move his head, and see the entire picture has the potential for being more at ease with his surroundings. Working your animal without blinders, however, means training him not to be fearful of his surroundings, the equipment he pulls, and any vehicles that pass. If you are willing to put in the time required to teach your animal to calmly accept you and his surroundings, you will have a superior team member, rather than one that plods through life in a tunnel.

## THE PULLING COMPONENTS

The majority of work accomplished by draft horses and mules involves moving a piece of equipment forward. Although the motion looks like pulling, in reality the load is moved when the animal *pushes* against the harness's pulling parts. A harness designed to pull a heavy load is called a *hame-style harness*, and the pulling parts are the collar, the hames, the hame straps, and the traces.

The hames, hame straps, and traces use the physics of the collar and its ability to harness the power of the equine's forelimbs and hind limbs. For pleasure driving, a *breast-collar harness* functions in a similar fashion, but uses a horizontal strap across the chest instead of the collar, hames, and hame straps. Because of the great stress and pressure exerted on and by the harness's pulling parts, these parts are generally stronger than the rest of the harness.

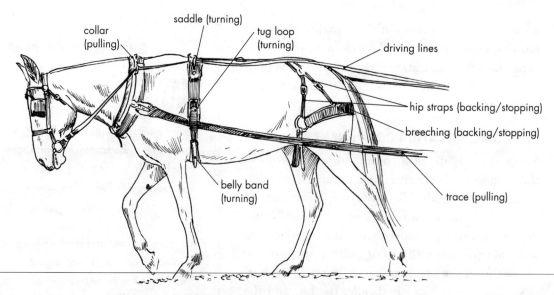

**BASIC HAME-STYLE HARNESS COMPONENTS**

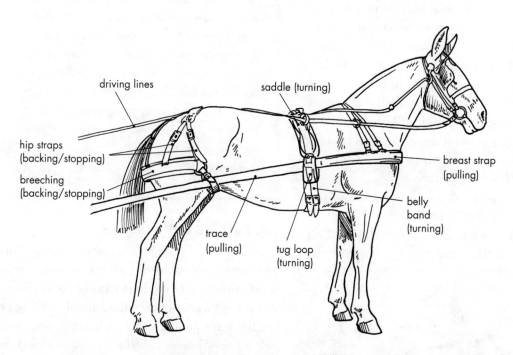

**BASIC BREAST-COLLAR HARNESS COMPONENTS**

### THE BACKING AND STOPPING COMPONENTS

Once your draft animal starts moving a load, at some point you'll want to stop or reverse the direction of travel. The harness must therefore include a way to stop and reverse the forward motion of the equipment being pulled. The backing and stopping parts of the harness are the quarter strap, the breeching, and the hip straps. These parts also hold the load back on a downhill grade.

Different styles of harness use variations of these parts, but all serve the same basic purpose. These stopping and backing parts of the harness work in concert with the neck yoke or hold-back straps of the equipment to back the load or keep the load from running over the animal.

## THE TURNING COMPONENTS

While you're watching your horse or mule in the pasture, notice that when he turns, the motion begins with the head, moves to the shoulders, and is followed through with the hindquarters. Under harness, the equine uses lateral movement in different ways, depending on whether he is hitched in a team or alone. The parts of the harness used for turning when the animal is in a team hitch are the collar, the hames, and the traces. The traces control the lateral movements of the animals and translate them to the lateral movements of the equipment via the neck yoke, which is directly attached to the hames and the tongue.

Vehicles or equipment operated by a single animal are usually equipped with shafts. The parts of the harness that move the shafts are the saddle, the belly band, the belly band billet, and the tug loop.

### How to Harness

After you know what each component of the harness does and where it goes on the horse or mule, you need to know how to put the harness on the animal. Applying a harness sounds quite a bit simpler than it may prove to be for the novice. If the harness is not arranged in the correct manner, you will wind up fighting straps, buckles, snaps, and hames.

This is how I go about harnessing a horse (for unharnessing, follow each of these steps exactly in reverse order):

1. Halter the animal and tie him to a secure ring.
2. Brush him down thoroughly.
3. After checking that the collar is clean inside, slide it over the animal's head; for this job you will have to unsnap the lead tying him to the ring. If the animal's head is too large to slide the collar over, open the collar at the top and put it on over the animal's neck. Check the collar for proper fit, as neck size changes with an animal's condition.
4. Go to the harness rack and first put your *right* arm through the breeching and under the saddle. Put these parts onto your upper arm/shoulder. Then grab the right hame with your right hand and the left hame with the left hand. Lift the entire harness off the rack in one fell swoop. You should be able to walk easily with the hames held in your hands and the rest of the harness resting on your upper arm/shoulder.
5. Approach the animal gently from the left side. Put the hames on the collar, one on each side. Ease the rest of the harness onto the horse's back with the right traces and right side of breeching falling on the right side of the horse.
6. Arrange the harness so the saddle is just behind the withers, the breeching is on the hindquarters, and the quarter straps hang down evenly.
7. Always begin fastening the harness in the front. First fasten the bottom hame strap. Make sure the hames fit snugly behind the collar rim with no gaps or looseness.
8. Buckle or snap the breast strap and the belly band, and snap the quarter straps to the tongue strap, which usually has a D-ring for the purpose.
9. Put the bridle on the horse.
10. Pull the driving lines through the hame rings and fasten them to the bit. Pull them to the left side and securely loop them over the hames.
11. Recheck everything piece by piece to make sure everything is right.

You are now ready to hook your animal to the implement.
BE SAFE!

# A LIFETIME OF WORKING HORSES

**George Hatley**
**Deary, Idaho**

Type of Farm: Timber, cattle, and horse ranch
Animals Owned: Two Belgian geldings

George Hatley was born in 1924 on a 400-acre wheat ranch in the Palouse hill country of northern Idaho. If anyone knows about driving horses, it is this man. All those acres of wheat were plowed with 12 head of horses. A highlight in young George's life was the day each year when 33 head of horses from around the neighborhood were first hooked to the ground-driven grain harvester.

While some people sought to make a fortune in the stock market, George sought to make his mark in north Idaho through the true stock (horse and cow) market and in real estate. George and his wife of 60 years, Iola, owned large ranches in north Idaho where they ran 350 mother cows, raised Appaloosa horses, made hay, cultivated beautiful gardens and orchards, and had a sizable silviculture (forestry) enterprise. George has never been a man to sit still.

For as long as most people have known him, George has kept a team of Belgian draft horses. These horses provide power for his beautiful array of horse-drawn vehicles, collected over the years and lovingly restored and preserved. Among his collection is a fully stocked chuck wagon that he used for years on his annual roundup and branding. He also owns a farm bobsled, a small cutter, a one-horse sleigh, a side-spring top buggy, a piano box buggy (made by World Buggy Company), a mountain wagon (made by Studebaker), and a 12-seater wagon.

A number of years ago, realizing that not all folks who enjoy driving horses and mules have somewhere to drive them, George and Iola began opening their ranch twice a year for the public to enjoy. Dozens of drivers and as many as 50 horses and mules come to trot along the miles of open roads. In the winter, the deep snow of north Idaho beckons teams with their merry jingle of bells and the swish of runners.

I have a strong personal connection with George and Iola. When I moved from Kentucky to Idaho to pursue graduate work at the University of Idaho, George offered me his team to drive any time I wanted, and I have had the pleasure of participating in the rides mentioned above. In 2004, George's horse-drawn wagon carted the bridal party to the wedding site where my husband and I exchanged vows, and George generously loaned me the vintage sidesaddle on which I rode to meet my groom.

Although he is getting up in years, George still takes time to show anyone who is interested his collection of horse-drawn vehicles, as well as his nice team of Belgians, Jake and Chub. As long as George, an inductee into the Idaho Hall of Fame, and Iola have anything to say about it, driving horses and mules for pleasure will always be an option in this beautiful part of the country.

George Hatley with Alina and Jereld Rice

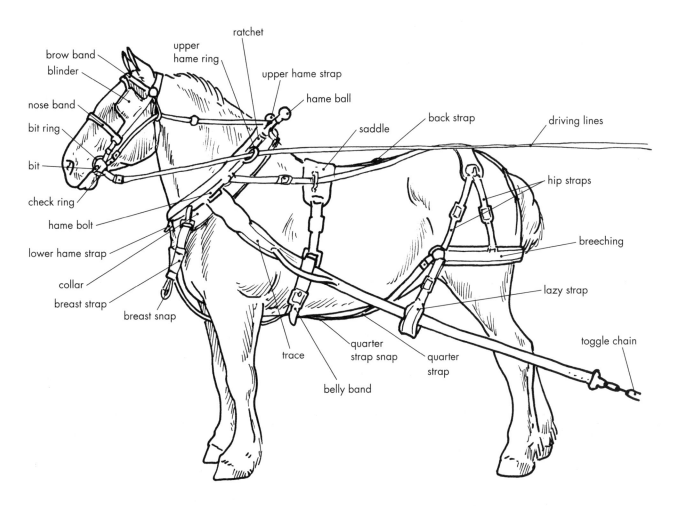

**BASIC WORK HARNESS COMPONENTS FOR A TEAM**

## The Importance of Ensuring a Good Fit

Take time to consult with well-informed teamsters before rushing out to buy the first harness and collar you see. Properly fitting the harness and collar to your horse or mule ensures that he can function at full capacity.

If you ever rode a bike that did not fit you, wore a backpack that rubbed in the wrong places, or hiked for miles in ill-fitting shoes, then you understand the necessity of fitting your harness to your draft horse or mule. He will wear this harness each time you set out to do draft work. An uncomfortable animal quickly associates the harness with pain and will make life miserable for you. Worse, if you continue to use this ill-fitting harness, the unnatural pressure exerted by the harness will result in sores, lameness, or other health issues and injuries that may have lifelong consequences.

**HARNESSING AND DRIVING** 193

| Typical Harness Sizes | |
|---|---|
| **HARNESS** | **FITS EQUINE** |
| Pony | up to 800 pounds (360 kg) |
| Horse | 800 to 1,400 pounds (360–630 kg) |
| Draft | 1,400 pounds or more (630 kg) |
| Custom | based on equine's size and weight |

## USING THE CENTER OF GRAVITY

To maximize your draft animal's efficiency and safety, his natural center of gravity must be in equilibrium with the load forces. The center of gravity on a working horse is typically located where the trace and surcingle intersect. If the trace is offset from the center of gravity, the efficiency of the harness cannot be maximized. The breeching of the harness should point to the center of gravity; if it is too high or too low, the efficiency of stopping or backing movements is undermined. On the other hand, when an animal is holding back a load, his body compresses, causing the breeching to drop lower than usual. To ensure that the breeching fits correctly when it is needed to hold back a large load, place the breeching slightly high (an inch or two above what looks right with a relaxed breeching), so that when the animal compresses his body the breeching won't drop too low to do its job.

To find the center of gravity on your horse or mule, draw an imaginary line from the point of the shoulder to the point of the buttocks. The center of gravity is the point where this line intersects the space between the twelfth and thirteenth rib.

## FITTING THE COLLAR

The collar must fit the specific draft horse or mule that it is meant for. If you have more than one animal, each should have his own collar. An improperly fitting collar quickly creates sores on the animal's withers and/or shoulders. A collar pad may be used to cushion the friction of the collar against the shoulders.

The show-ring practice of fitting a draft horse with an overly large collar just to make the neck appear larger is not a good idea. Similarly, draft horse and mule owners who compete to see who has the largest animal, may substantiate their claims by fitting their animals with collars too large for their necks, just to be able to say that their animal wears a size 26 collar, when in reality he might need a 24. If the animal is regularly worked wearing a too-big collar, his neck and shoulders will soon have sores to testify to the teamster's irresponsibility.

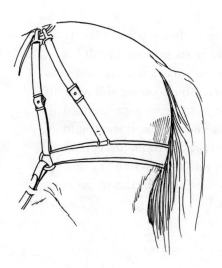

Correct placement of the breeching

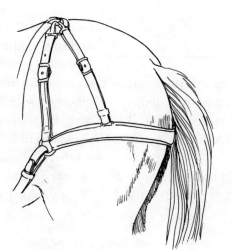

Breeching too high

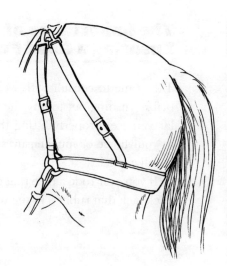

Breeching too low

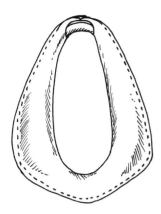 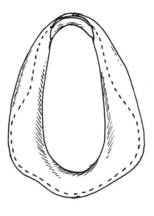 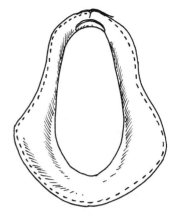

Regular collar        Half-sweeney collar        Full-sweeney collar

A regular or full-face collar fits an animal with a long, flat neck and might be used by a mule or light horse. A half-sweeney collar suits an animal with a medium to thick neck, such as a pony or an average draft horse. A full-sweeney collar is made for a heavy draft animal with a thick neck and steep shoulders.

A collar that is too large can cause friction and pressure on the shoulder joint, a bony area with no padding of muscle. It can also push on the animal's bony shoulder blade, or scapula, and press against a nerve that is largely unprotected by the light muscling. As the horse or mule leans to pull a heavy load, the collar slams against the scapula, crushing the suprascapular nerve. Over time, muscle atrophy in both shoulders results in an irreparable condition known as *sweeney*, a word derived from the Pennsylvania Dutch word *schwinne*, meaning wasting away or atrophy.

The word *sweeney* is also used to describe a collar's shape, as determined by the amount and placement of stuffing in the collar. Half-sweeney and full-sweeney collars have less stuffing in the upper part, making them wider and more rounded than a regular collar, which is sometimes called a full-face or straight-face collar. Draft horses with strong, thick necks often require a half- or full-sweeney collar. Draft stallions, in particular, have large, thick necks and require a full-sweeney collar for proper fit. Mules, on the other hand, have thinner necks and usually take a regular collar, which muleteers call a mule collar.

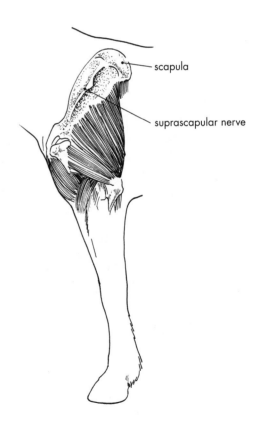

Proper collar fit is essential to protecting the scapula and the suprascapular nerve.

## Measuring the Correct Collar Size

**Collar size is measured** in inches from the inside top to the inside bottom. To determine what size collar your draft animal requires, measure the straight-line distance between the base of the neck and the top of the neck, directly in front of the withers; round up to the next inch.

This measurement is difficult to obtain precisely, so use it as a guideline and plan to try several different collars on your equine. A properly fitting collar is long enough that you can put one hand horizontally between the inside bottom of the collar and the animal's neck when the collar is pushed back against his shoulder. The collar should be wide enough that you can run your fingers up and down between the collar and the neck.

Hames come in even sizes only and should be one to two inches larger than the collar size. A size 24 or 25 collar, for example, requires a 26-inch (66 cm) set of hames. Hames that are the right size for the collar naturally fall into the area behind the collar's rim.

Like all of us, draft horses gain weight when they are less active and lose weight when they exercise. A collar that fits well at the beginning of the work season therefore might be too big by the end of the season. To fill in the gap and prevent the collar from rubbing and causing sores, use a collar pad under the collar.

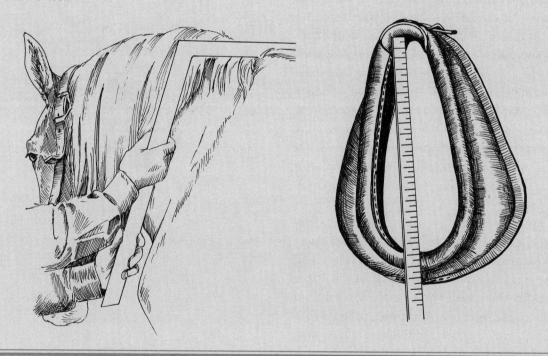

## Types of Harness Material

Good harnesses are made by independent craftsmen who guarantee an excellent product. You can buy cheaper, mass-produced, imported harnesses, but you won't get the same quality or safety features.

Harness shops sometimes offer used harnesses for sale. If you wish to acquire a used harness, have an experienced person examine it for sound construction. Using an unsound harness is not worth the risk, although sometimes a worn or rotted part may be replaced easily. A knowledgeable teamster or harness maker can tell you if a secondhand harness is usable or reparable and safe.

When you buy a new harness, you pay for the time spent constructing it, as well as the materials used. Three materials are commonly used to make harnesses: leather, nylon webbing, and plastic-coated webbing.

**Leather** has been the material of choice for years and continues to be for many show and field folks. If maintained regularly and kept in good repair, leather will last for the life of your horse or mule and longer. Nothing feels better to me than leather driving lines, and few things smell better than a leather harness. A well-made leather harness will cost more than a harness made of synthetic materials.

**Nylon webbing** has certain advantages over leather. It is easy to clean, doesn't rot, remains flexible in cold weather, and is strong yet lightweight. Despite its many advantages, some people don't like the looks of a nylon harness. Its chief disadvantage is that hair and dirt readily stick to it and cause chafing, and the cheaper versions contain stiffening agents that rub and chafe.

**Plastic-coated webbing** (brand names include BioThane and Ohio-Thane) looks like leather, but has some of the advantages of plain nylon webbing. It may be easily cleaned with a sponge and water, is impervious to moisture, remains flexible in cold weather, and resists scratches. It is also much lighter in weight, making it easier to handle. It is particularly attractive to the ecofriendly farmer who does not like the idea of killing a cow to harness his horse or mule. Its main disadvantage is that it is not entirely accepted in the show ring and might negatively influence a judge's decision. Although the material used to make most of the harness is coated with polyurethane, driving lines are usually made of vinyl-coated webbing (brand name Beta) that feels more like leather.

Although the maker, quality, and hardware influence the price, a new, stout leather harness costs the most, a new nylon harness will cost the least, and a new plastic-coated-webbing harness falls somewhere in between.

## Different Styles of Harness

In addition to being made of different materials, harnesses come in different styles, depending on the intended use. Today's draft horses and mules have two main occupations: driving and working in the fields. These activities require harnesses with different capacities and strengths. A driving harness allows the animal to move under a lighter weight, while a heavy work harness is built for strength and durability.

### TYPES OF PLEASURE-DRIVING HARNESS

The driving harness for a draft horse is usually heavier than the counterpart for the light driving horse. A show-draft-driving harness may be heavier and more elaborate than an average work harness. A pleasure-draft-driving harness, however, is typically lighter in weight than either the show-driving harness or the work harness.

If you drive a single animal, you will need a single driving harness; if you drive a pair, you will need a double driving harness. You can convert a single driving harness to a double driving harness with a few modifications, so if you sometimes drive single and sometimes drive a team, you won't need two completely different sets of harnesses. To drive single, you will need a harness with tug loops for the shafts of your vehicle. Driving harnesses come in two basic styles: breast-collar and hame-collar.

### Breast-Collar Harness

The *breast-collar harness* is a light harness that is not suitable for pulling heavy loads, but is often used for showing and is also quite functional for light vehicle

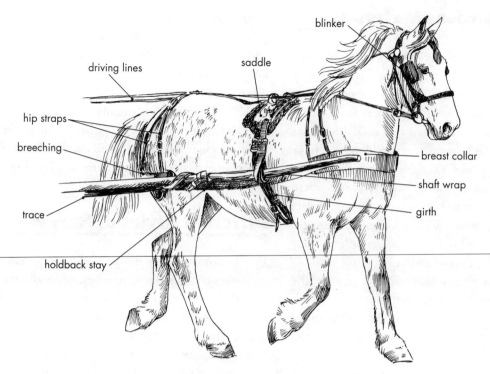

**BREAST-COLLAR HARNESS AND BRIDLE**

use. The breast collar must be properly padded to prevent rubbing. The breast-collar harness does not require the additional purchase of a collar, making it a more economical choice than the hame-style harness. The breast-collar harness is also lighter than the hame-style harness, making it a nice choice for women and children.

## Hame-Style Harness

The *hame-style harness* uses a collar in addition to the harness. Except for the collar and hames, the other parts are the same as the breast-collar harness. The hame-style harness typically weighs more and is more stoutly made than the breast style, and is appropriate for large loads pulled for long distances.

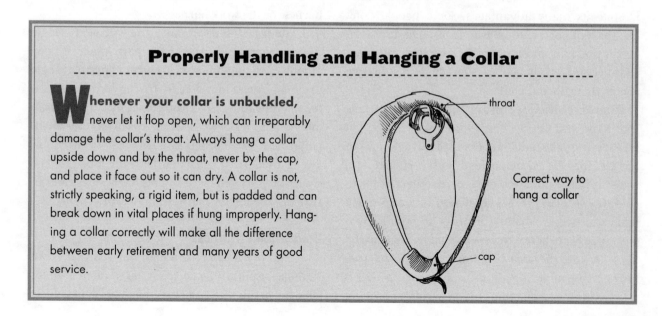

## Properly Handling and Hanging a Collar

**W**henever your collar is unbuckled, never let it flop open, which can irreparably damage the collar's throat. Always hang a collar upside down and by the throat, never by the cap, and place it face out so it can dry. A collar is not, strictly speaking, a rigid item, but is padded and can break down in vital places if hung improperly. Hanging a collar correctly will make all the difference between early retirement and many years of good service.

Correct way to hang a collar

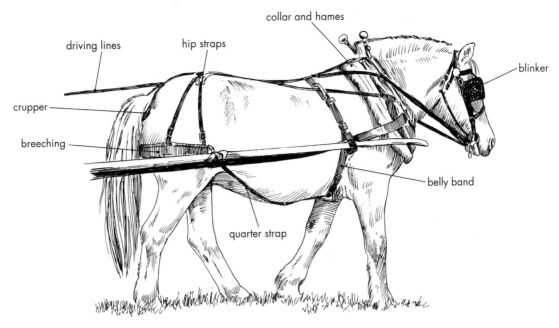

**HAME-STYLE DRIVING HARNESS**

Compared to the breast style, the angle of the traces from the hames is physically more ergonomic for the equine pulling a heavy load.

## TYPES OF WORK HARNESS

A work harness is heavier and stronger than a driving harness and has larger hames to handle heavy loads. Your work harness should be in top working order, and you must regularly inspect it for signs of damage or wear. I have been party to two accidents caused by a lack of attention to normal wear of workharness leather, and they were not only embarrassing but also potentially dangerous.

One accident happened when the breast strap broke in two where the buckle had caused regular rubbing. When the breast strap broke, the tongue dropped out of the neck yoke, which made steering impossible. The situation could have deteriorated into a disaster if the horses had been less well trained, but fortunately a combination of good horses, luck, and wagon brakes kept both humans and horses from getting hurt.

The other accident was caused by a quarter strap breaking, also where the buckle rubbed the leather. Unfortunately, this accident involved a mower. When the strap broke, the horses spooked and took off, and didn't stop until the cutter bar slammed into a tree. This accident was extremely ugly for the mower, but fortunately the horses and human suffered little damage other than increased heart rates.

So inspect your harnesses regularly and keep leather ones clean and well oiled. Oiling helps to keep leather from drying out, and drying out speeds deterioration and leads to breaks. Many harness shops have large vats of harness oil where the entire harness may be soaked. It is a rather messy affair, but an effective way to thoroughly oil your harness. Soak your harness a minimum of once a year; if you live in an excessively dry and dusty environment, soak it in the spring and fall.

All work harnesses are made with the same objective in mind: to provide an energy-efficient way to use the power of the draft horse or mule to move large, heavy loads. To accomplish this objective, three main styles of work harness have evolved: box breeching or Western-style, Yankee breeching, and New England D-ring. Teamsters, often grouped geographically, eloquently (and endlessly) argue the virtues of their chosen style.

## Western-style or Box Breeching Harness

When my family began working horses, we used the Western-style harness. Most of the draft horses and

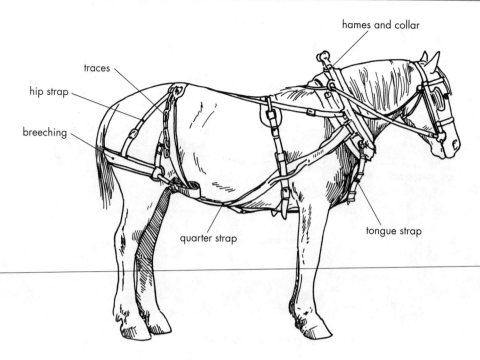

**WESTERN-STYLE OR BOX BREECHING WORK HARNESS**

mules worked today use this style. The western harness's functionality and simplicity make it an excellent option for the beginning teamster. The main functional parts of the Western-style harness are the hames, breast strap, quarter straps, breeching, pole (tongue) strap, and traces. Any work harness requires the addition of a collar, which adds to the cost; an adjustable collar that will accommodate up to three sizes may be worth the extra money.

## Yankee Breeching Harness

The significant difference between the Western and Yankee breeching harness is where the breeching lies. On a Western harness, when the equine is holding back a load, the weight is directed at the upper part of the hind leg, just below the point of the hip. On the Yankee breeching harness, the weight is directed above the tail, a part of the anatomy with considerably more bulk than the hind leg. Placing the breeching above the tail eliminates the chance for the hind legs to get swept out from under animals while they are holding back a large load.

The Yankee breeching harness also has a crupper, which helps to hold the breeching in place. Because less material is used in the breeching area of the Yankee breeching harness compared to the Western-style work harness, expect to pay a little less than for the Western-style harness.

After working draft horses for five years or so, my dad switched all his harnesses to the Yankee breeching style, because he felt it was more efficient at holding back hay and other large loads. Bringing in a massive load of hay to our barn was tense no matter how you looked at it, because the horses had to go downhill all the way from the fields, and then stop on a hill and hold the load until we chocked the wheels. With the Yankee breeching harness, the mares almost sat down to hold back the load, yet I still felt comfortable knowing the harness would hold together.

The only drawback I experienced was the height of placement of the snaps on the hip straps. With a Western-style harness, these snaps are about halfway down the hip strap. With a Yankee breeching, they are at the point of the croup, which, compared to Western style, is quite a bit higher and harder for a shorter person to access. Yankee breeching works great with team harness, but is not designed for use with shafts.

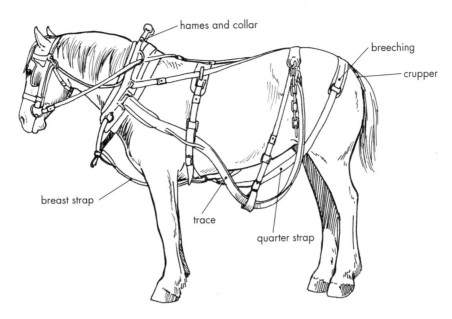

**YANKEE BREECHING WORK HARNESS**

## New England D-Ring Harness

The New England D-ring harness is used by a small but growing number of teamsters. I have never used it myself, but I know teamsters who refuse to use anything else. The differences between the D-ring harness and the other two work harnesses include an absence of quarter straps (no snapping below the belly, except for the belly band) and breast straps (instead, neck straps and side-backer straps do the job of putting the tongue where it needs to be, as well as holding back, slowing, and stopping the load).

The benefit of the D-ring harness is its ability to prevent tongue pressure on the top of the animal's neck. All the downward pressure is transferred to the saddle, a place that does not become as hot or experience as much friction as the collar. Anyone who has discovered a collar sore on a horse's or mule's withers after a long day of mowing hay will understand the benefit of this harness. The New England D-ring work harness costs more than the Western-style harness.

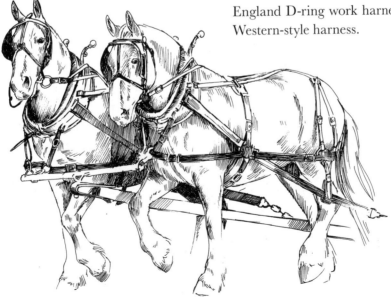

**NEW ENGLAND D-RING WORK HARNESS**

**HARNESSING AND DRIVING**

> ◀ **CAUTIONARY TALE** ▶
>
> ### The Danger of Incomplete Harnessing
>
> Today you are going to mow hay. You feed your mules and take them out for harnessing. Your mules are great animals — well trained, quiet, and obedient. You put the collar easily enough on mule number one, ease on the harness, and buckle up. Unfortunately, you are distracted by a cat that darts through the second mule's legs, causing a bit of a ruckus, and you forget to snap the second quarter strap on the harness of mule number one.
>
> You put mule number one's bridle on and then move on to mule number two, and repeat all the same steps. This time, however, you remember both quarter straps, but you don't tighten the hame strap enough because you fail to notice that the hames have slipped out of adjustment. You snap the driving lines on and hook the mules to the mower.
>
> On the first downhill swath, you notice that mule number one's breeching start tilting over to one side of his hindquarters. Before you can act, it's all the way over and he starts acting squirrelly. *Whoa!* Time to stop. You pull back on the lines, but now mule number two is having problems with his hames; one has completely slipped off the collar. Now you are really in trouble.
>
> Fortunately for you, the downhill isn't bad and you manage to stop your well-trained team. You are shaking as you fix the errant harness pieces. Your fault? Definitely — every single part of this problem was your fault.
>
> **Lesson learned:** Always double-check every part of your harness before leaving the barn. With less well-trained animals and a worse downhill run, you would have had a wreck.

## Learning Different Driving Techniques

Learning to drive horses or mules is the work of a lifetime; you will constantly learn better ways to drive. The initial experience may be frustrating and will probably involve failed attempts at communicating your wishes to your animals. Your best approach is to exercise patience with yourself and your animals.

I was fortunate to grow up driving horses, and was never aware of developing a feel for the lines. When I was young, I would put my hands in front of my dad's while he drove our powerful team of Belgian mares. As I grew older and gained enough weight to maintain the correct tension on the lines, I drove the animals myself. So my learning was gradual and almost imperceptible.

Theoretically, driving horses is as simple as driving a car. You maintain the correct tension on the lines, just as you maintain the correct pressure on the gas pedal and the brakes. In reality, though, driving a horse is not as simple as driving a car; the horse has his own mind and as a result, line pressure and tension need constant adjustment. In addition to the lines, well-trained horses understand and obey voice commands, which go hand in hand with proper line tension. Once you have the basics, you will learn more subtle ways to communicate with your animals through the bit.

No amount of reading or watching DVDs can teach you how to drive horses and mules. No words can teach you how to develop the intimate connection required between teamster and team. You can learn to communicate with your animal's bit via your driving lines only with much practice, trial, and error.

As a novice you *can*, however, prepare yourself for driving your horses or mules before you take the lines in your hands. You can understand the concept of single and team driving, as well as theories about tension, voice commands, and line driving. You can learn about general approaches to driving and practice some exercises that will give you confidence to climb into the driver's seat.

### SINGLE VERSUS TEAM DRIVING

Most driving concepts apply to both single and double driving. The basics of single and team driving are the foundation from which you can eventually move toward more complicated driving arrangements and techniques. You may one day wish to move on to more complex hitches, such as driving more than two abreast or driving four-up (two teams at once, one in front of the other), or even bigger hitches.

Driving a single horse is simpler than driving a team — two driving lines go straight to the bit, so you have one horse, one bit, and one driver. This setup has the fewest possible variables and is therefore the best way to begin learning to drive. Driving two horses side by side, as is needed for most team implements, requires two horses, two bits, two driving lines (that split to go to the two ends of the two bits), and one human. This set of variables introduces the complexity of driving two animals with two hands, a complication overcome by the splitting of the driving lines. Driving a team is similar to driving a single horse, except that you have double the horsepower and double the chance of trouble.

As the teamster, you communicate with your animals through the bit and with your voice. A well-trained animal will already understand voice commands. For successful communication, you need to correctly know and use those commands in conjunction with the bit, which is manipulated through line tension. Do not expect your newly purchased animals to be push-button perfect in either voice command or line-driving technique. Although a well-trained team should know the basics, they more than likely still need refinement, constant reinforcement, and daily practice to attain optimal obedience. Since different teamsters may use different commands, find out how your horses were trained and which cues they understand.

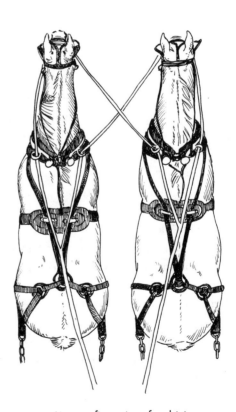

Line configurations for driving:
single (top) and team (bottom)

## MAINTAINING PROPER TENSION ON THE LINES

Driving horses and mules is all about maintaining proper tension on the lines. Correct tension lets your animal know you are still on the other end of the lines and allows smooth turning through curves. Be prepared to constantly change the tension because of such things as changes in the animals' energy, their reaction to stimuli, and shifts in the work load. If your tension is too tight, your horse will develop a hard mouth. If your tension is too loose, your animal may wonder if you are still back there and attempt to take over. Keeping the correct tension on the lines accomplishes a number of critical elements for a successful drive:

- You will be able to communicate your commands effectively and maintain control.
- The animal will understand that you are familiar with driving and will feel confident in your direction.
- You will be prepared for unexpected situations that may arise.
- Your animal will not become hard-mouthed from excessive pressure.

The *only* way you can acquire the sense of correct tension is to practice and practice and practice. A mentor can be helpful in letting you feel the correct tension and learn what is too little, too much, and just right. Eventually you will feel it. Until you do, prepare to be frustrated with your communication attempts. A well-trained animal may acquiesce to your clumsy attempts, but may also take advantage of them and do exactly what he wants. A green horse may become frightened by your lack of tension, which communicates your lack of experience, making a wreck almost inevitable. A well-trained and patient animal will be your best teacher.

## VOICE COMMANDS VERSUS LINE DRIVING

Teamsters argue articulately and verbosely on the advantages and disadvantages of voice commands versus driving-line communication. Some teamsters feel that voice should be the primary communication method, with the driving lines used for backup. This style of driving emphasizes training and obedience based on trust and true communication, versus the physical restraint of the driving lines.

Others feel that voice commands should not be used at all, primarily because you may find your animals doing things you did not intend. A common occurrence in parades, for example, is for a team to respond to a command made by the next teamster coming along behind, especially if the horses have been driven at some point by the other teamster.

I prefer to incorporate both voice and proper driving-line technique into my driving skills. An animal capable of responding to voice commands is a tremendous asset, particularly in an activity such as logging or operating a walking plow, when your hands may be full with another task. I also appreciate the skill involved in having correct line tension and lightness while driving, which I find imperative for refining any driving and draft animal.

So, in accomplishing the two critical aspects of driving — how to start and how to stop your animal or team — do you speak to them, or do you give them line cues? Using voice commands for starting and stopping has both pros and cons. The most significant pro is the ability to start and stop your animals without having to pick up the lines. A team that responds to voice commands should be completely obedient to the stop command. If you can halt your team by voice alone, you have a team you can trust, but don't be lulled into complacency. To ensure your and your animals' safety, always be ready to use the lines if needed.

For those of us who farm with draft animals, having a vehicle we can control remotely is quite handy. For example, if you are picking up hay by hand, being able to trust your animals to move up and stop the wagon at the next bale simply by telling them to comes in handy. The disadvantage is that this team may follow unintentional orders — for example, if a nearby teamster asks his own team to move ahead. Other factors — such as thunder, yellow jackets, or dogs — can get between you and your animals' obedience. A runaway team is not just an inconvenience; it can be fatal to you and your animals. To avoid a serious and potentially deadly accident, you must be constantly alert and in control.

## Using Common Voice Commands

Teamsters all have their own unique ways of addressing their animals. A few commands, however, are universal and are spoken in English, even to horses originally trained in Amish or Mennonite communities, where Pennsylvania Dutch is the mother tongue. The reason for these universal commands is to avoid confusion when a draft equine is bought or sold.

Despite what you see in the movies, voice commands need not be shouted out. Horses and mules hear extremely well and respond better to softly spoken commands rather than shouting, which can startle them. Watch a really good teamster and you will see his team's ears swivel while they listen to commands that you can barely hear.

### Universal Voice Commands

| COMMAND | INFLECTION | ACTION EXPECTED |
|---|---|---|
| Get up | Upward | Move ahead |
| Whoa | Downward | Stop immediately |
| Easy | Downward | Slow down |
| Gee | Flat | Move to the right |
| Haw | Flat | Move to the left |
| Back | Flat | Move backward |
| Over | Flat | Move to the side away from me |

## Using the Lines Correctly

Many teamsters feel that a quiet, collected approach to driving is better achieved when only the driving lines are used to communicate with the team. If the purpose of your animals is strictly driving, rather than farming, you might consider this approach. The advantage of using lines only is that you will have the opportunity to refine your animals to a beautiful degree, achieving the balance, lightness, and collection visible in many performance-driving horses.

When using only driving lines, the commands to start and stop are made without a sound. The clear disadvantage is that a driver is always needed to control the cues. To cue with lines only:

- **Start** by picking up one line (in the direction you will travel), which cues your animal(s) to move forward. Some teamsters take this a step further and train their animals to know that when both lines are picked up, they are to walk straight ahead. This method of driving is more common in the show ring than in the field. Because of the risk that any line movement, intentional or not, may cue the animal to move, I use a double signal for my horses — when I pick up the lines I also give them an audible cue to move forward. If they don't hear the intended signal, they are not to move. Using a voice cue removes some of the risk involved with line movement as you enter a vehicle, or if an animal's own head movement causes him to think the line has been moved. Always treat your lines with caution and take care not to move them unless you intend for your animals to move.
- **Stop** with a slight increase in pressure on both lines.
- **Turn** by maintaining even pressure on both lines and then putting slightly more pressure (about as much as you would to stop) on the line in the direction you want to travel while releasing pressure on the other line. Because each animal (in a team) has a driving line attached to each side of his bit, both animals feel the same pressure at the same time in the same direction. Remember to release pressure on the other line, which allows the animals to bend as they turn.

## DEVELOPING CORRECT FORM

Although much of learning to drive involves simply doing it, having correct form is important. Correct form is the way you use your body during driving — exactly how you hold the lines, and how and where you position your body while driving. How you hold the driving lines may seem like a small part of the process of using draft equine, yet can make the difference between an average day with your team

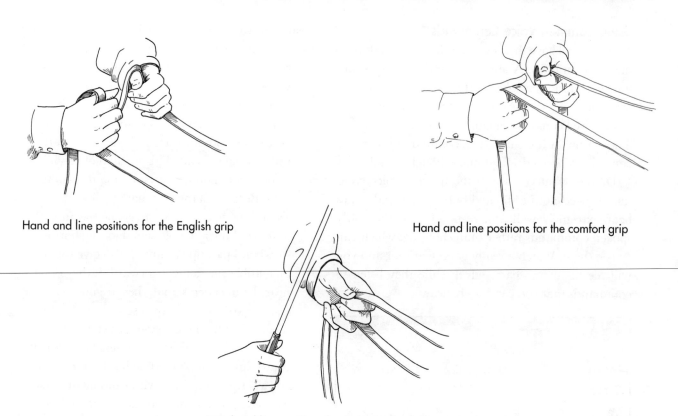

Hand and line positions for the English grip

Hand and line positions for the comfort grip

Hand and line positions for the Achenbach grip

and a disastrous wreck. Lines are generally held in one of three ways:

**The English grip** is my preferred way of driving. The lines are held in a secure grip (imagine your hand held with fingers pointing straight ahead and thumb pointing up; now curl the fingers into the palm) running from the bit under the pinky finger, through the palm and looped over the thumb. The line is gripped between the thumb and index finger. The thumb may then be used to gather up or release line as necessary for adjustment.

This method allows a strong, secure grip that may be quickly adjusted to compensate for the lengthened line needed by a horse pulling uphill with his head down, or shortened as the horse holds back going downhill. The strength of this grip comes from the double L formed by the fingers as the line is fed through the hands.

**The comfort grip** is, in my opinion, just a bit sloppy. Each line simply runs straight from the bit to between the index finger and the thumb and is gripped in the other four fingers as it runs through the palm. This grip offers little leverage. It is, however, quite comfortable, allowing the driver to drive with relaxed hands and easy grip. But that relaxed manner can all change in a matter of seconds; you may lose your lines if the animals jerk their heads suddenly and you aren't ready. Feeding the lines to achieve tension is also slow and cumbersome.

**The Achenbach grip** is a formal grip developed by the German equine driver Benno Von Achenbach. With this style, the lines are held in the left hand and the whip is held in the right hand. Imagine that the left hand is held with the fingers pointing forward and the thumb pointing up. Curl the fingers in toward the palm. The left driving line runs from the bit over the top of the index finger and is gripped with the thumb. The right line runs from the bit to between the two middle fingers. Both lines lay in the palm and form an L when they hang down to exit the palm.

When my hands tire from holding the lines in the English grip, I switch to the Achenbach grip. It allows more agile and versatile handling of the lines. But because the grip is less secure, you have to be more vigilant than with the English grip.

# HARNESSING

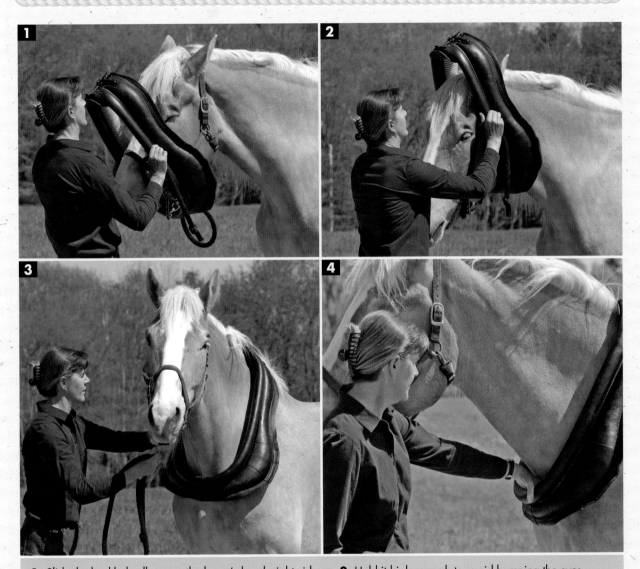

**1.** Slide the buckled collar over the horse's head, right-side up. **2.** Hold it high enough to avoid bumping the eyes. **3.** Settle it on the shoulders. **4.** Check for proper fit.

OR

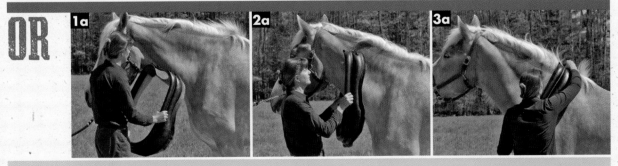

**1a.** If the horse's head is too large, hold the collar open. **2a.** Slide it over the narrow part of the neck. **3a.** Buckle it.

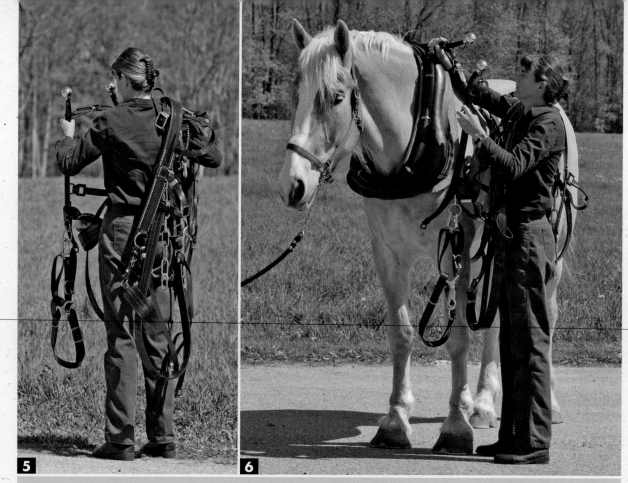

**5.** Carrying the harness is easy if you first put your right arm through the breeching and under the saddle and hoist it onto your shoulder. Hold the right hame in your right hand and the left hame in your left. **6.** Slide the hames into place on the collar. **7.** Ease the rest of the harness onto the horse's back. **8.** Make sure the saddle is just behind the withers, the breeching is on the hindquarters, and the quarter straps hang down evenly.

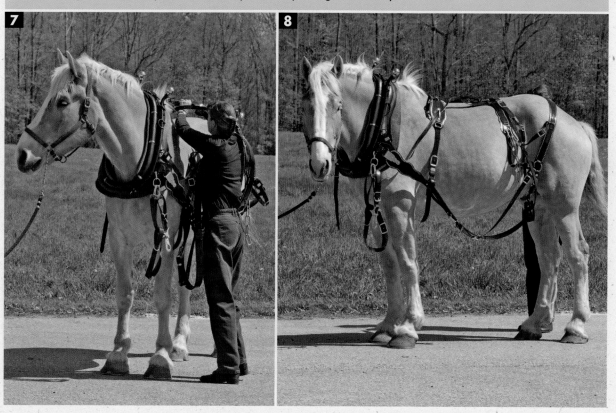

**9.** Slide the breeching into position. **10.** Begin fastening the harness from the front, starting with the bottom hame strap. Make sure the hames fit snugly onto the collar. **11.** Buckle the crupper.

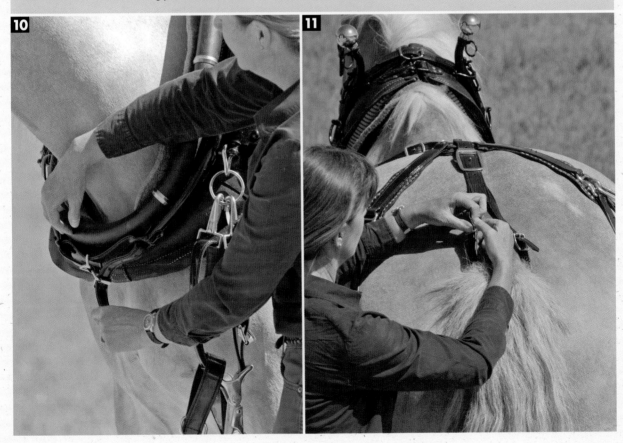

## OVERVIEW

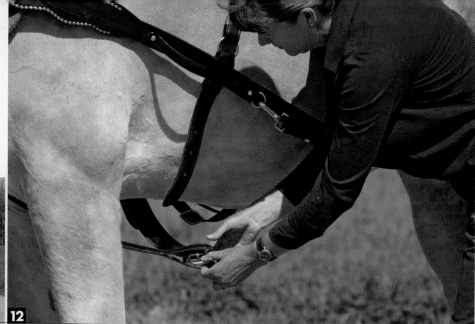

**12.** Buckle or snap the breast strap and slip the belly band through the pole loop.

**13.** Fasten the belly band.

**14.** Clip the quarter straps into place.

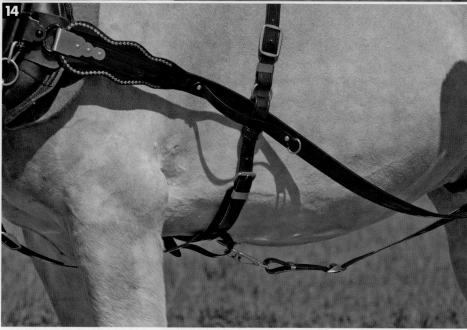

210  HARNESSING

**15.** After buckling, always check again that the breeching fits correctly.

**16.** Check that all other components, such as the breast strap, are lying flat and are properly snapped.

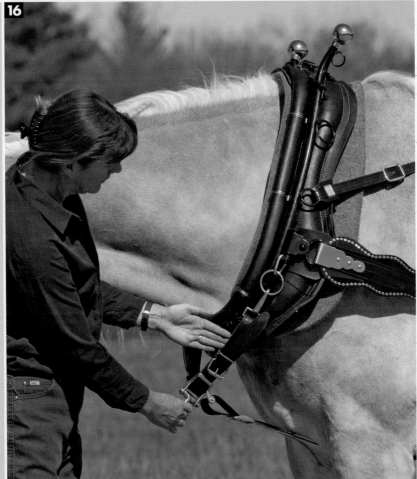

HARNESSING 211

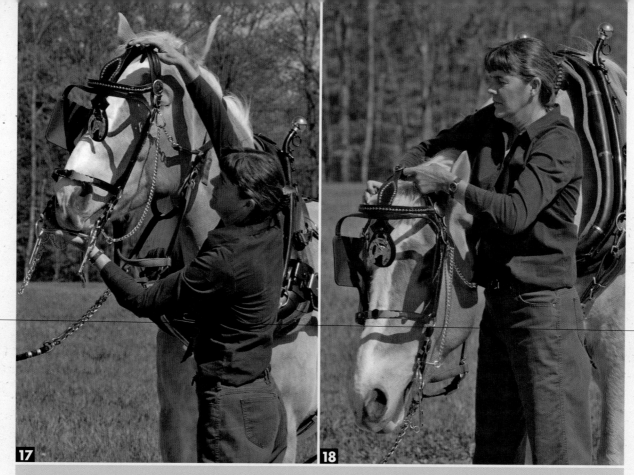

**17.** Slide the bit into the horse's mouth. **18.** Bring the crownpiece over first the right ear and then the left. Training your horse to drop his head makes this process much easier. **19.** Put the driving lines through the loops. **20.** Slide the lines through the hame rings before fastening to the bit.

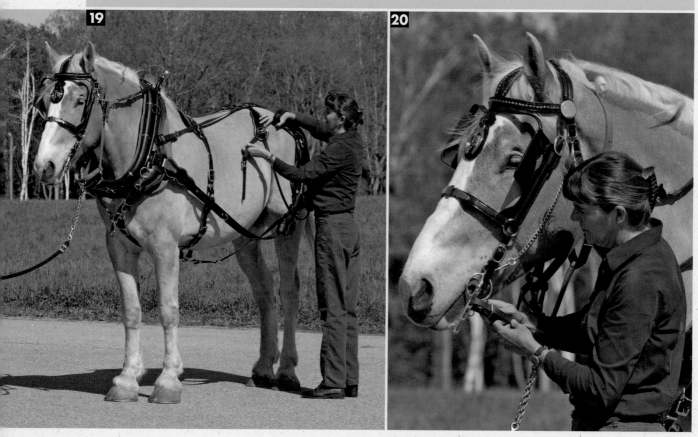

**212 HARNESSING**

All harnessed up and ready to move on to the job at hand.

# HITCHING

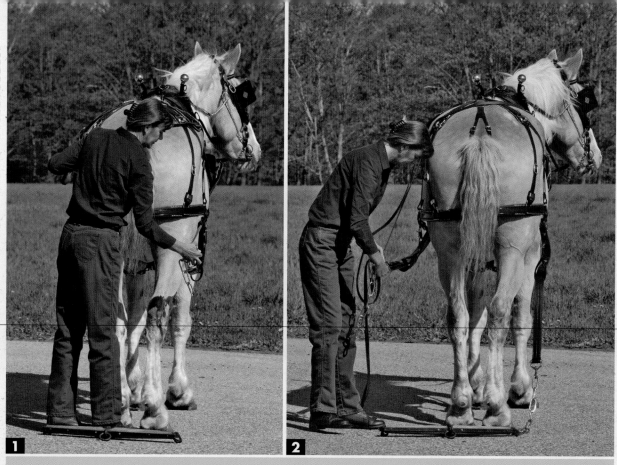

**1.** To fasten your harnessed horse to a singletree, bring the first trace chain down the hook and fasten it. **2.** Leaving the singletree on the ground, fasten the second trace. **3.** Now you are ready to hook up to your equipment.

Special thanks to our models, Granite State Barney, owned by Robin Lovejoy, and Elizabeth Bridges, and to Ruth Meader and Meader's Supply Corp., Rochester, New Hampshire, for providing our location.

## Positioning Your Body

Driving novices tend to seize up like posts, becoming stiff and immobile. Relax. The animals that you are driving can feel your tension, which will undermine your credibility. If you feel yourself getting tense, roll your shoulders and take a few deep breaths. Allow yourself to move with your animals, to enjoy them, and to communicate with them through the driving lines. Proper position also allows you to work more efficiently and more comfortably over a long day of driving.

How you position yourself will be dictated by the implement you are driving. You may need to stand, sit on a single seat, or sit on a seat designed to be shared. If the implement has one seat — such as those found on a manure spreader, disc harrow, or rake — the decision of where to sit is simple: You sit in the seat.

If the implement is a farm wagon, you will probably stand and drive your horses chariot style. Standing while driving is a skill achieved with practice, but is much simpler than it looks. Don't attempt it unless you have good balance, however. Depending on the purpose of your wagon, you may use front, back, and sideboards, or no boards at all. When driving only with the bed, you won't have anything to grab if you lose your balance.

If the implement is a spring wagon, a forecart with bench seat, a buggy, a sleigh, or any other vehicle designed for transportation, you will likely sit on a wide seat designed to be shared. The correct place for you, as the driver, is on the *right* side of the vehicle. If you drive a car, sitting on the right while driving a team of horses or mules may seem counterintuitive, but it is where tradition dictates that the teamster sits. Although the animals don't seem to notice which side you drive from, sitting on the right allows you a clear view of the road's shoulder.

## PRACTICE EXERCISES WITHOUT A TEAM

The two biggest challenges of driving with proper tension are learning to hold the bit lightly and keeping the correct tension through a turn. To spare your horse or mule your clumsy first attempts, practice these two concepts using nonequine models. You may feel a bit silly doing these exercises, but they will help you to develop a better understanding of how to handle the lines and maintain the proper feel of the bit. Your first equine driving candidate will thank you for having the forethought to practice.

### Learning Correct Tension

To learn correct tension, you must be able to feel the point at which the bit in the animal's mouth begins to exert too much pressure. You can simulate this feel by doing a simple exercise using springs. For this exercise, you need two 18-foot (6 m) pieces of driving line or rope and two electric gate handles or other springs with similar tension and hooks.

Fasten the gate handles to a solid point, such as a fence rail. Space them approximately shoulder-width apart. Fasten the lines (or rope) to the free ends of the gate handles. Stand at the far end of the lines and apply tension on them until you feel the

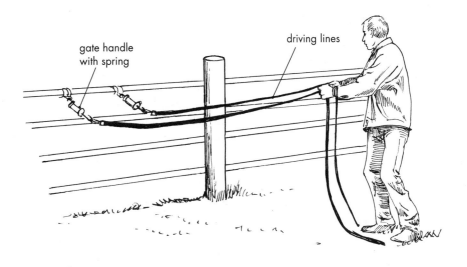

Practicing your driving line tension with spring-loaded electric gate handles attached to a fixed object such as a closed gate is a great way to practice without harming your animal's mouth.

# WORKING DRAFT HORSES FOR THE FUTURE

**Liza Howe**
**Lena, Illinois**

Type of Farm: Dairy and beef cattle, draft horses, and a few chickens for eggs

Animals owned: One Suffolk gelding, two Suffolk mares

Liza Howe grew up in Lena, Illinois, where her parents have draft horses and a dairy farm. As a young girl, Liza vastly preferred working the horses to working in the dairy. Gradually she began acquiring her own horses. All three of the horses she now owns she got as foals from her parents' Suffolk mares. Liza has been a part of their lives since the beginning, training them, showing them, pulling with them, and operating many different pieces of farm machinery with them.

Liza has been harnessing and driving horses since she was four years old, when her parents bought their first team of Belgians. Since then, the family's draft horse menagerie has changed color and shape — they first owned Percherons — then switched to the rarer American Cream and Suffolk. Liza takes every opportunity she gets to drive horses, and as a result has driven almost every team configuration imaginable from a one-horse rig to a twelve-horse hitch on a four-bottom plow to a unicorn hitch to a six-up in the show ring.

Because she is only five feet, three inches tall and because she competes in pulling events, Liza needs light, strong harnesses for her mares. She has found the nylon harness to work best for her, although she insists on leather driving lines. She prefers harnesses with hames over the breast-collar harness, mainly because she already owns the hame harnesses for her two mares working as a team and can't see a good reason to buy a breast collar harness. Besides, her horses "look tougher" in a hame harness.

Liza's experience in driving Suffolks, American Creams, Belgians, and Percherons has taught her that the Suffolk is an excellent choice for the beginner. Their rare-breed status, however, often makes acquiring them prohibitively expensive. Her advice for the beginner is to buy a dead-broke older team of Belgians that have been around the field more than just a time or two.

Liza has chosen to raise, show, and work with Suffolks because of their versatility. Although they will never compete in the big show rings with large hitches of Belgians and Percherons, her Suffolks can compete in smaller shows in many different categories and are great lightweight pullers. Her mares always have one foal on them and one in them, and the foals sell better than most Belgian foals, thus bringing Liza closer to her dream of becoming a veterinarian. The foals are helping fund her undergraduate education and she hopes they will help her through vet school if she is accepted.

Summing up her experience with draft horses, Liza says: "Working draft horses is like college — you get out of it what you put into it."

springs in the handle begin to stretch. Practice until you can anticipate the point when the springs begin to stretch by how the lines feel in your hands.

Tension is not static, but must be constantly adjusted as you drive. The goal of every teamster should be the ability to anticipate and maintain the correct tension. The ever-changing conditions you will encounter while driving will require you to feed or release line to avoid putting too much pressure on the bit while keeping extremely light tension. If you don't maintain enough tension, your team will not understand your signals. Too much tension can make your animal's hard mouthed, and your arms extremely tired.

### Learning to Turn

The equine turns first with his head, then with his shoulder, and finally follows through with his buttocks. To achieve a smooth turn while driving, you must retain correct tension on both lines. The line on the inside of the turn will be shorter than the line on the outside of the turn, because you apply slight pressure to the inside line and the animal responds by turning his head in that direction. Once the horse or mule begins to turn, you must feed out some line on the outside and pick up some slack on the inside line to maintain correct tension. As the turn is completed and the team straightens out, you must release the tension on the inside line while gathering the line on the outside.

A great way to experience the process of turning and the tension required is to use a tricycle or a similar device; it needs to have handles and to roll without needing to be balanced. Attach a rope to the stem of the handlebar and convince a helper to pull the trike while obeying your voice commands. Practice turning while the trike is in motion, maintaining the tension you practiced in the above exercise on *both* lines. Have your pulling partner go at different speeds, and you will experience to some degree what it is like to regulate tension on the driving lines of a horse or mule through a turn.

I well remember numerous attempts at instructing new drivers in the fine art of tension and using it correctly through a turn. One of those attempts

Having a helper pull a tricycle outfitted with driving lines attached to the handlebar grips lets you practice learning to turn with correct line tension.

involved a young man I will call Joe, who visited our farm to learn how to drive horses. He knew how to ride and therefore assumed driving could hardly be different. I assured him that, although both required good hands, driving and riding were two entirely different activities.

To simplify the exercise, I hooked a single horse to our spring wagon. We started off pulling uphill, and Joe's control was fine, particularly since the horse was pulling, naturally creating tension on the lines. As soon as we started going downhill, however, Joe started running into trouble. I was prepared for the trouble to escalate, since the hill ended in a sharp curve.

As the horse held back the wagon while traveling downhill, the traces and driving lines developed considerable slack. Joe was not prepared and fumbled trying to gather up the slack. The sharp curve was fast approaching, and still he was trying to reestablish tension. The horse was picking up speed, the lines were still slack, and right before we went into the curve, I made an executive decision and grabbed the lines. After we caught our breath, Joe's rueful comment was, "I guess it's a little harder than it looks."

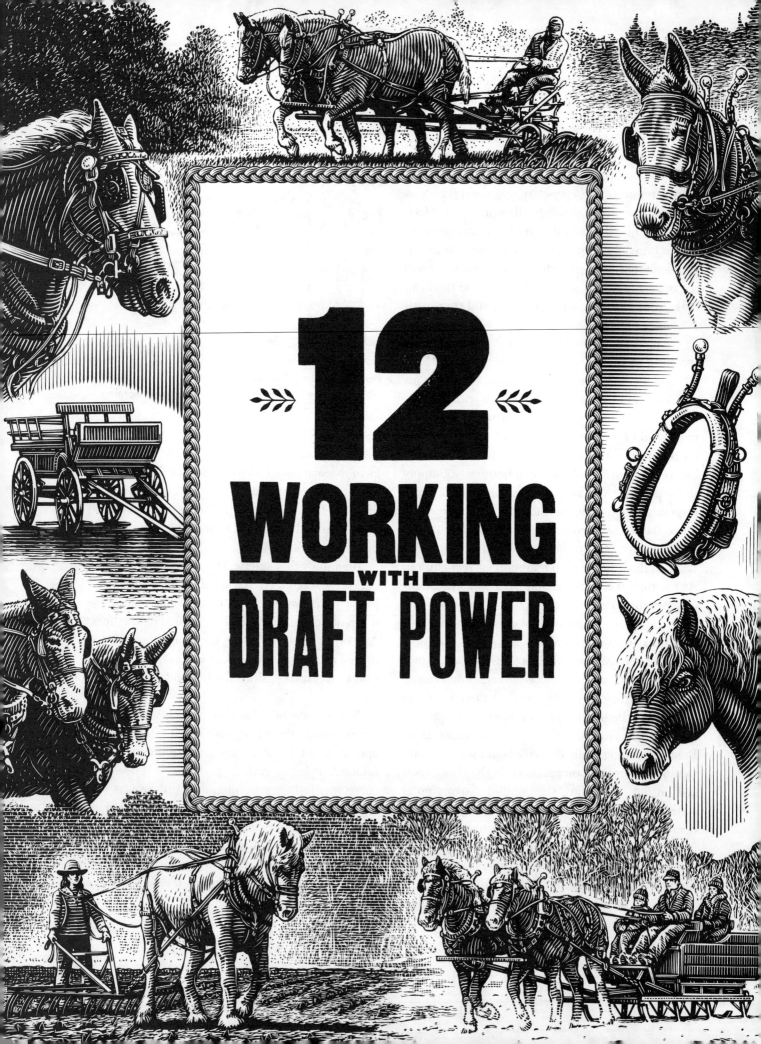
# 12
## WORKING WITH DRAFT POWER

On an early summer morning, dew still hanging heavy on the grass, I walk to the barn, a thousand thoughts running through my head. Today I plan to mow hay for the first time on my own. My team and I have spent time learning from my dad how to correctly operate the mower, how to turn corners, how to mow a straight swath, what to watch out for. We had our lessons while mowing pasture and other less valuable fields, but today I will guide my team down long, straight rows of much-needed timothy hay. I have greased the mower. I have made sure the sections are sharp and the guards are tight. The machine is ready. Am I? Are my horses? Today I will find out.

Having already finished their morning feed, my horses stand patiently in their stalls. I groom each one and then straighten my shoulders. I have harnessed these animals so many times I could do it in the dark. So why the tension? Because I realize the step I am taking today is significant. Today I go from doing minor work on my father's farm to doing a job with economic consequences. I am helping my family to realize our collective dream.

I harness my horses, snap the driving lines into place, and drive the team to the waiting mower. The off horse steps over the tongue and with a slight pull on the line to accompany my back command, the team backs into position. I fasten the neck yoke, slip it onto the tongue, and fasten the trace chains. I am ready to start.

I climb onto the seat and chirrup to the team. They know what to do and easily step out. By now, the dew has mostly burned off the hay field; the mower guards should slide smoothly and quickly through the hay. When I reach the field, I step off the mower and drop down the cutter bar. Then I slip back into the seat, engage the pitman arm and cutter bar, and cluck to the horses. The racket of the rapidly moving sections jars the morning, but the horses are calm and accepting; they recognize this noise. I swing the mower into position to cut our first round, let the bar down to the desired position, and watch the freshly cut hay fall in great, tall swaths into the grass-board and then out into a beautiful row. I am mowing hay — hay that will feed these very horses come winter. And *my* horses are powering the mower.

## Understand, Then Act

You no doubt have been dreaming about all the things you will do with your horses or mules, counting the days to when you can harness them, work with them, and enjoy their power. Even at this stage of the game, reality and dream can become blurred. You may have initial dreams such as driving down a country road, logging your own firewood, cultivating a garden, or some such thing. And yet you know that so many other activities exist for using draft equines. You have watched people work their horses and mules, and maybe even participated. Surely you can do it, too.

How difficult could it be? The level of difficulty you experience working your first team is influenced by a few variables. The most important of these variables is your skill level as a teamster. "Learn before operating" should become your mantra for *every* activity you do with your draft horses or mules.

A teamster looks to make sure that his furrow is even.

Not only will you be prepared, but you will also know what to anticipate and be ready to act when the need arises.

As a teamster, you must know your animals — their individual personalities, their idiosyncrasies, their strengths, and their weaknesses — and be prepared to act on your knowledge. Above all, you must project confidence no matter what situation you find yourself in, even if you are scared spitless. Never cater to a weakness in your animals by allowing them to choose what they will and will not do, and always be aware of the potential for danger. When you have a keen understanding between yourself and your animals, the time you spend with them is efficient and productive. In short, you and your animals must have a communication system that all of you understand, recognize, and respect.

## What Type of Equipment Do You Need?

In chapter one we explored some of the broad ways you could use your draft horses. Obviously, in addition to purchasing the right team for your needs, you will need to buy the right equipment. As you move in draft horse and mule circles, you will encounter many different styles and types of implements intended for many different uses. The differences may be large (a grain combine versus a stationary threshing machine) or small (a John Deere mower versus a McCormick Deering mower). How do you know what equipment will work best for you? Unfortunately, the only way to answer that question is to try a number of different options. Many products on the market are functional, efficient, and designed to be pulled by animals. The best way to learn about them is to attend public functions utilizing draft horses and mules, and ask the teamsters about their equipment and how it works.

## Basic Components of Draft Equipment

Before looking at any machinery, you must understand the basic elements of all draft equipment: the parts that control stopping, turning, and pulling. Your safety and the safety of your draft horses or mules depend on your understanding of these elements. They are fundamental, must be respected, and must always be considered when operating any implement. No matter what its purpose, most pieces of equipment that are designed to be operated by draft animals have two major components: the *tongue*, also called a *pole*, and the *evener*.

---

### The Hidden Side of Driving

**P**art of the job of being a teamster or muleteer is to be an equipment mechanic, as well. You cannot expect your equipment to run forever without a hitch. At some point, you will get greasy and dirty and intimate with your machinery. You must be willing to spend hours struggling to understand your equipment and then more hours loosening tight, rusty bolts, making adjustments, and fixing problems.

Being a mechanic does not come easy for me. I often have difficulty visualizing the physics of a piece of machinery. For many years, I plowed with our sulky plow and three horses. I knew exactly where to set the lever, how to hitch the traces, what the length of the jockey stick should be, and a number of other factors for successful plowing. But I never knew exactly why they needed to be that way.

When I learned that the clevis could be adjusted to maximize the efficiency of the horses' strength, that the evener could be adjusted to keep the plow from pulling too much one way or the other, and a number of other handy facts, I began to realize how little I knew about equipment. My dad had always set it up for me, then put me to work. When I began to get involved with equipment modification, restoration, and adjustment, my ability to fix problems myself increased exponentially, as did the ease of my day.

# A YOUTHFUL TEAMSTER

**Justin Nowell**
**Monroe, Tennessee**

Type of Farm: 104 acres (42 ha), diversified use includes corn and hay crops

Animals Owned: Belgian and Percheron mules, Belgian draft horses, grade draft horses

Justin Nowell is not your average 16-year-old high school student. Where most teenagers are immersed in the popular culture of the iPod, computers, and the latest video games, Justin stands out as an iconoclast. He has little use for computers and does not have an e-mail address. When he is not attending high school (where he is involved with Future Farmers of America), he is on the farm putting in quality time with the horses and mules his family uses to work their land.

Justin lives in a family that appreciates and uses draft horses and mules in a real way. His father, Mark, came to appreciate equine power through the mule his grandfather owned and worked and also through various friends and mentors who encouraged him on this journey. At one time Mark operated a horse logging business. Today he is a physician in rural Tennessee and uses the horses and mules to relieve the stress inherent to his occupation.

Justin already understands things that many people twice his age long to learn about draft equines. He has chosen the rewarding and challenging job of starting young horses and mules as a way to improve his equine skills. He has developed these skills by working with his father and with various friends and neighbors who have willingly shared their knowledge with him. The 32 head of horses and mules on his family's farm give him ample opportunity to hone his training skills. Justin enjoys training both riding stock and harness stock, often combining the disciplines to produce a well-rounded animal.

Justin has some interesting insights on the differences between horses and mules. Although he acknowledges that both animals have their unique challenges, he would rather work a mule in the field than a horse. "Horses," he says, "don't know how to regulate their energy, and tire out quickly. A mule can work right next to them and hardly be sweating, even though the horses are blowing and sweating hard."

Although Justin does not want to use solely horse power on his own farm, he is sure of one thing — he wants to farm. He does not see a formal college education in his future, but plans to pursue courses at a technical college where he can learn practical skills he will use on his farm.

## THE FUNCTION OF THE TONGUE

Equipment with wheels (such as a wagon, mower, or rake) usually has a tongue for steering and holding back. An exception is an implement designed for use with a single animal, which instead uses shafts for steering, turning, and holding back. Another exception is the sulky plow, which does not always have a tongue. As a general rule, though, an implement with wheels has a tongue.

When you communicate your intention through the driving lines and the animals respond accordingly, their movement is transferred to the neck yoke. The neck yoke in turn transfers the movement to the tongue, which controls the implement's stopping, turning, and backing. Neck yokes may be fastened to the tongue in a number of different ways.

**The standard stop** allows a neck yoke with a large ring in the center to slip on and off the tongue with ease — and that's the problem. The neck yoke is not fastened to the tongue in any way, but rests against the stop on the tongue, where pressure is exerted when the load is stopped or backed. The only thing keeping the neck yoke on the tongue is tension on the traces. Operator error can easily cause the neck yoke to slip off of the tongue. A safety chain attached to both the neck yoke and the tongue inhibits the neck yoke from slipping off of the tongue and is cheap insurance against a wreck.

**The safety stop** allows the neck yoke to slip onto the tongue, but not to inadvertently slip off. A piece of metal welded to the end of the tongue keeps the neck yoke from slipping off.

**The bolted-on tongue end** is the safest method of attaching a neck yoke. But removing the neck yoke from the tongue requires more time, making the neck yoke more difficult to transfer from one piece of equipment to another.

## THE FUNCTION OF THE TRACES

The traces are the components of the harness that connect the animals to the equipment. Each animal's traces originate at the collar, pass through the animal's center of gravity, and attach to a singletree. The team's two singletrees are then attached to an evener. On equipment with a tongue, the evener is attached to the far end of the tongue.

For single-horse equipment requiring steering, the traces are attached to a singletree at the end of the shafts. For equipment such as a harrow, which requires no braking or quick turning, the traces function as the pulling and turning power through the singletrees and evener. The evener is attached by a *clevis* to the main body of the equipment.

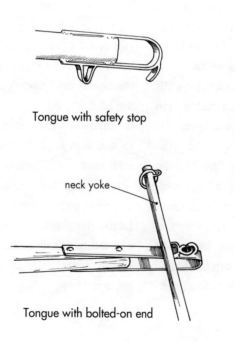

Tongue with safety stop

Tongue with bolted-on end

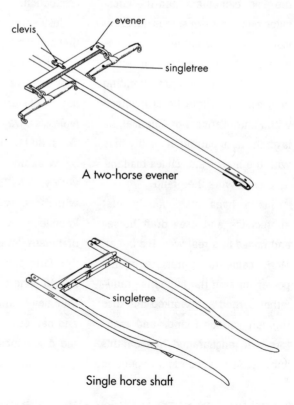

A two-horse evener

Single horse shaft

# A Forecart Adds Efficiency

**A forecart is** exactly what it sounds like — a cart made to go in front of another implement. Forecarts are designed to pull another implement (generally to increase the efficiency of the other implement), to provide a comfortable place for the teamster to ride (as when pulling a harrow, though using a harrow cart is safer), and to pull an implement manufactured for a tractor and thus lacking a tongue. A forecart may also be used to exercise or move your team, giving you a place to ride and your team a light vehicle to pull if you go any distance.

A basic forecart has wheels, a platform to stand on, and usually a seat for the teamster. More complex models are ground-, gasoline-, or diesel-driven forecarts capable of running a power takeoff (PTO) implement such as a hay baler. Power forecarts are the subject of much contention; purist horse and mule users detest the noise and pollution of the fuel engine and can't see any advantage in the extreme draft of the ground-driven forecart.

A logging cart is a specialized kind of forecart built for transporting logs. Logging carts increase the efficiency of a horse- or mule-powered logging operation by lifting the front ends of the logs off of the ground and thereby reducing draft.

A two-wheeled forecart

A power forecart

A logging forecart

## Selecting a Vehicle

Once you have an idea how the movement of horse-drawn implements is controlled, you are ready to look at equipment. You can enjoy a number of different activities with your draft animal or team, including transportation, cultivating crops, and logging. One of the most enjoyable activities is traveling behind your draft mules or horses, watching their tails blow in the wind and their heads held high while you control the lines. Traveling with real horse power is nothing short of thrilling. The freedom, the quiet, the grace are all attributes of equine transportation.

With shallow hooves in relation to their bulky bodies, draft horses are not suited for fast-paced, long-distance travel on pavement. If you want to move fast, you will need a lighter breed or perhaps a lightly built draft cross. For leisurely drives, however, the draft will steadily trot along.

Many vehicles are available, designed for various types of equine-powered transportation. They may be organized into light, heavy, forecart, business, show, and winter travel categories — all serving specific functions, some specialized and others multipurpose.

### LIGHT VEHICLES FOR TRANSPORTATION

Light forms of transportation such as a cart, light buggy, or surrey are all easy ways to move about with draft animals. These vehicles do not require the strength of two animals, thus allowing the owner of a single animal the freedom of equine-powered travel. These lighter vehicles, which can have two or four wheels, are meant almost strictly for transporting people, not for hauling goods or heavy loads. One light buggy is referred to as a courting buggy, which implies that it was meant for making an impression and for speed, rather than for moving cumbersome loads. The doctor's buggy is another example of a light buggy meant to transport a person quickly and efficiently.

I first became enamored with horse-drawn transportation when I was 10 years old and my dad got a Morgan gelding and a cart. My consuming passion was to prove to my dad that I was capable of driving this horse and cart on my own. I finally convinced him when I was 11, and oh, the joy of independence and freedom! My younger sister and I spent many hours jolting along in that little cart behind Prince, our willing horse.

In selecting the right vehicle for your horse or mule you need to consider many variables, including the size of your animal and what activities you wish to do.

A well-made farm wagon will serve many purposes from hauling hay or lumber to transporting guests.

## HEAVIER VEHICLES FOR WORK

Wagons come in different sizes, so consider the intended use before you buy one. If you plan to haul loads of, say, lumber, you will need a vehicle that can handle the bulk and weight, and you will probably need two horses or mules to do the pulling. A spring wagon with a small bed is a handy conveyance for smaller jobs. A single draft equine is an excellent option for the spring wagon, as long as your shafts are wide enough to accommodate the animal's bulk. Depending on whether the spring wagon is pulled by a team or a single horse, it can handle loads weighing from 300 to 900 pounds (135–405 kg).

For heavier jobs, such as hauling grain from the feed mill, you'll need a general farm wagon. A stout Amish- or Mennonite-made farm wagon can be an expensive but useful addition to your farm, and will last for as long as you keep it in good repair. A farm wagon is at least 15-feet (4.5 m) long — enough to haul a good load of hay (with hay racks), a stout load of wood, or even a load of pumpkins. In addition to removable hayracks, it should come equipped with front, back, and side boards. A good farm wagon should be able to haul a load weighing 1,500 to 2,000 pounds (675–900 kg). To haul a load this size up or down any grade, you will need three horses or mules.

## VEHICLES FOR A TRANSPORTATION BUSINESS

People will happily pay to be moved via horse or mule power. Carriage businesses in busy cities cater to tourists who enjoy the slower pace and the rhythm of hoofbeats on pavement. A conveyance for hauling paying customers needs to be safe, comfortable, and light. Carriages drawn for hours over hot pavement usually roll on rubber tires and generally have tops that protect clients from the sun.

If you are thinking about providing horse-drawn transportation for special events, a vis-à-vis is a conveyance often used to lend dramatic flair at weddings. In this intimate and elegant mode of transportation, the passengers sit facing each other. People tend to turn nostalgic for funerals, as well, and a horse-drawn hearse can add a solemn, elegant, and personal touch to this last rite.

## VEHICLES FOR WINTER TRAVEL

Nothing is more exhilarating than moving smartly behind a team of spirited horses or mules, bobsled runners squeaking through a snowy land of wonder and quiet. The bobsled may be used for pleasure travel or, as on large ranches in the West, a means of feeding hay to cattle.

A bobsled is basically a wagon box on four runners. The two front runners are connected to the

## Safety Rules for Horse-drawn Vehicles

**Transportation of any kind** offers challenges. Operating horse-drawn vehicles is no exception. While driving in public, always keep these safety factors in mind:

- Wheeled vehicles can travel much faster, will gain more momentum, and are harder to stop than implements without wheels. If you travel in hilly country or will be stopping frequently, have good brakes on your vehicle.
- *Always* have a slow-moving vehicle (SMV) triangle attached to the back of your vehicle when you drive on a public road; it is the law in every state.
- If you drive equines on a busy highway, drivers of motorized vehicles may have difficulty seeing you. To improve your visibility, attach a high, bright orange flag that may be seen from a distance. Such flags are available where bicycles are sold.
- When driving at night, equip your vehicle with lights, not so much to help you see as to help other people see you.
- Litigation is possible if your passengers are hurt because of a malfunctioning vehicle, even if they sign a waiver. To reduce the chance of being sued for negligence, inspect your vehicle before every trip to make sure the tongue is sound and not cracked, the wheels are all tight, the spokes are all there, the lug nuts are tight, the brakes are fully functional, and the grease zerks are full.

Winter travel by horse power is an exhilarating experience.

tongue for steering purposes. Other vehicles used on ice and snow are the sleigh, the sleigh's stylish cousin the cutter, and the sled. Sleighs and cutters operate on two runners, while a sled typically has no runners at all, but is simply dragged across the ground on its bottom surface.

## Transporting Goods without a Vehicle

Packing goods on the back of a horse or mule has been a method of transportation for thousands of years. In fact, long before the wheel was invented, people used donkeys as pack animals. While pack animals are still used around the world, in North America they are largely used to transport hunters and sightseers into remote wilderness country. In the 1920s and '30s, packing was invaluable to the United States Forest Service. Strings of pack mules regularly serviced remote fire lookout tower personnel during the long summer months of fire season.

A truly versatile horse or mule is able to work under harness, riding saddle, or packsaddle. However, medium-size animals in the 1,200- to 1,500-pound (545–680 kg) range make the best pack animals. Larger animals are too tall to hoist panniers onto and too bulky to navigate narrow trails. Smaller animals may be more agile, but can't handle heavy loads. Mules, with their strong sense of self-preservation, are renowned for their packing sense. The preferred pack mule is short and stocky.

Learning to pack correctly may seem easy, but it takes time. Try to imitate the effortless diamond hitch of the seasoned packer after he has slung the awkward panniers up and over the packsaddle, and you will understand what is involved.

As with any other form of activity using equine energy, pack animals need a harness of sorts. For the packer, this harness is called a saddle and consists of a stout arrangement of a wooden tree with pads, leather straps, and panniers to carry equipment and supplies. A well-built packsaddle will cost you several hundred dollars, but will last a lifetime if properly cared for. Many tack shops, particularly in the West, sell packing equipment.

Like teamsters with their harnesses, packers can wax eloquent on the virtues of each packsaddle type. Common packsaddles are the sawbuck and the Decker. Both styles serve the same purpose — carrying panniers. The panniers are what you transport your items in. They may be made of plastic, wood, leather, or canvas.

**A sawbuck packsaddle** has wooden crossbucks to hold the panniers and lacks the half-breed (the padded canvas cover with back boards used by the Decker). Most sawbuck packsaddles are double rigged, meaning that they have two cinches. Some packers prefer double rigging to single rigging for load stability. The rigging on a sawbuck packsaddle is riveted into place, which does not allow for adjustment. A disadvantage to the sawbuck saddle is that the crossbucks are made of wood and are susceptible to breaking if the animal falls or catches the saddle on a limb. A sawbuck packsaddle weighs around 20 pounds (9 kg).

**A Decker packsaddle** has metal hoops in place of the wooden crossbucks of the sawbuck. Many packers consider the Decker to be a stronger saddle because the panniers sit on the metal hoops instead of wooden crossbucks. The half-breed and

A sawbuck packsaddle

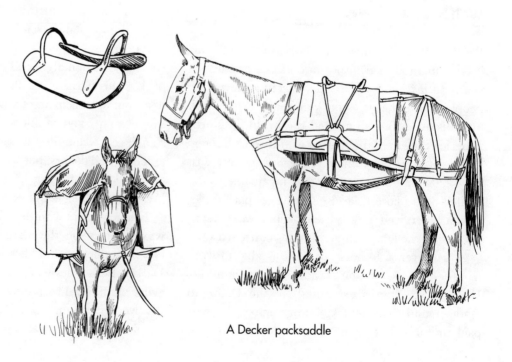

A Decker packsaddle

pack boards that are a part of the Decker help to evenly distribute the load. The Decker saddle is single rigged. The rigging on a Decker is buckled to the tree under the half-breed and is adjustable, allowing you to vary the position of the cinch and reduce the chance of sores. The Decker is often used with Decker hooks — four metal hooks that attach to the pannier hanger straps and may then be easily hung on the metal hoops. A Decker packsaddle (including the half-breed and pack board) weighs around 30 pounds (14 kg).

# Using Equipment for Cultivating and Farming

Cultivating the soil is a primal and ancient ritual, and farmers have partnered with draft animals for millennia to work the land and produce crops. Today, many farmers are choosing to return to the older ways and manage their land in a cycle of self-sufficiency. To involve your team of horses or mules in this beautiful but sweat-filled rite is to know fulfillment and purpose. Cultivating the soil is not just breaking and taming the sod; it is also planting the seeds, tending the plants, and harvesting the fruits of hard labor. The cycles involved in cultivation may be completed only through dedication to the task, but the results are well worth the hard work.

### THE PROCESS OF PLOWING

Plowing is the ultimate horse-drawn activity, as witnessed by numerous plowing bees and contests held each year throughout the country. It is both physically and mentally challenging and requires incredible skill of the plow operator and the animals pulling the plow. Successful plowing involves fit, well-trained animals, a well-adjusted plow, and limitless patience. Since plows come in a variety of shapes and sizes, you may find a higher level of satisfaction with your first plowing experience if you do a trial run on someone else's plow to find a good fit for you and your animals.

The plow's many parts are all necessary for a good plowing job. The *bottom* is the soil-turning part of the plow, and includes the *share* (which cuts the soil) and the *moldboard* (which helps turn the soil that's cut by the share). The plow bottom is where all the action takes place, and when a bottom is operating correctly, it is said to scour well, meaning that soil falls free from the moldboard, the furrow slice turns nicely, and the furrow is clean and straight. To cut efficiently and correctly, the plow bottom must be free from rust, and the share must be sharp and not worn down.

Most plows, both walking and riding, have a *coulter* — an independent disc that rolls ahead of the share and does an initial cut of soil and vegetation. The coulter also helps to ensure that vegetation is completely covered after it has been cut and turned. Coulters are often used alone on a walking plow, but on most riding plows are used in concert with a *jointer*. A jointer is basically a miniature plow that runs directly ahead of the share and behind the coulter, throwing trash (such as cornstalks, if you were plowing an old corn field) out of the way and ensuring good furrow turning.

Both the coulter and jointer must be correctly adjusted. For normal plowing conditions (not rocky or rooty), the coulter should ride directly above the share point. To correctly position the coulter, move the coulter shank clamp up or down the beam. To cut through trash and vegetation, the coulter must be kept sharp. If a jointer is used, the coulter may be set an inch or two ahead of the share and the jointer an inch or two behind.

### TYPES OF PLOWS

Plowing may be done two ways: riding on the plow or walking behind it. Both methods accomplish the same goal of turning the soil to an established depth, but in different ways.

**A single-bottom plow** turns only one furrow at a time. The common one-way plow plows only in one direction, to the left or to the right, not both ways. A two-way plow will plow either to the right or to the left, depending on which way it's set, and is designed to help the hillside farmer keep the soil on the hill and not in the river. Gang plows are multiple-bottom plows and require more horse power than a single-bottom plow.

A single-bottom riding plow is called a sulky plow. Riding plows are made with or without a steering

Plowing with more than one bottom requires a dramatic increase in horsepower.

tongue (pole). The tongue allows for rapid correction of direction to the share, but also makes the plow prone to tipping over if it is turned too tightly.

**A walking plow** is easier to maneuver than a riding plow, but is more physically demanding on the teamster and requires more skill and experience. Most walking plows are single bottom, although walking gang plows are still occasionally to be found.

Turning the soil creates tremendous draft, requiring strong, well-conditioned draft horses or mules to pull the plow all day. Multiple-bottom plows, such as two-gang or four-gang plows, require more horsepower. As a general rule, if you plan to plow all day with a 14-inch or larger bottom in a soil of medium consistency, such as silt loam or clay loam, figure three strong animals per bottom.

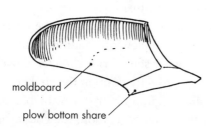

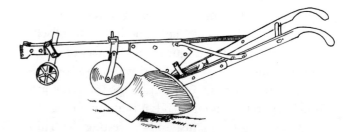

Walking plow with rolling coulter

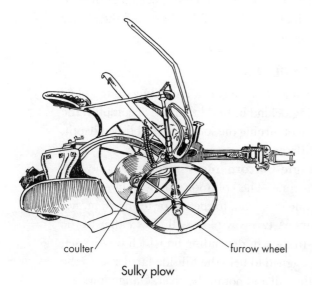

Sulky plow

A two-horse walking plow should be no larger than a 14-inch (35 cm) bottom.

## Common Plowing Problems

**Plowing straight** ensures that even turning occurs over the entire field, rather than one area getting skipped or turned shallower. A well-adjusted, smoothly cutting, straight-pulling plow is a beautiful sight, but a difficult one to obtain. Incorrect hitching of a walking plow is one of the most common problems associated with uneven furrows and results in hours of fighting the plow to keep the share in the ground.

Achieving a correct line of draft for your team is the solution to many of these problems. The line of draft controls the plow's front beam, from which the horses pull, which in turn controls the depth as well as the direction the plow takes. To control the sideways direction of the plow, hitch so the beam points straight forward when pulled by the horses. If you are plowing with three horses and a right-hand plow (the furrow turns to the right), offset the beam a little to the left.

The line of draft is largely controlled by the clevis, the device attaching the evener and the horses to the plow. The clevis attaches to the hitch, which has different holes to allow you to make the clevis lower or higher. If the clevis is too high, the beam will point down and the bottom will dig too deep. If the clevis is too low, the beam will lift up and you will be forced to fight the plow to keep the share in the ground.

Riding plows have three wheels, and you will know the vertical adjustment is correct if all three wheels carry approximately the same weight when the bottom is on the ground. Side draft — the tendency of a plow to wander to one side or another — in a riding plow is just as frustrating as in a walking plow. To correct for side draft, adjust the front furrow wheel toward the furrow wall by adjusting the lever controlling the furrow wheel.

### PREPARING THE SEEDBED

Once the soil is turned, it must be prepared for planting. Seedbed preparation is typically accomplished with a disc harrow, a device that cuts clods, smooths the soil, and jounces the living daylight out of you. After the freshly plowed earth has been cut and smoothed by the disc harrow, a spring-tooth or chisel-tooth harrow is applied to further refine the seedbed if necessary, depending on the type of soil.

**Disc harrows** come in an incredible array of styles, brands, and capacities. For the small garden, you may want a single-action disc harrow, which can maneuver and turn easily in tight spaces. For the larger produce patch, a tandem disc harrow (also called a double-action disc harrow) may be more appropriate, but requires the pulling power of four to eight horses or mules. Your local plowing bee is a good place to get a feel for the size of equipment you need for your specific operation.

A disc harrow is hitched to either the forecart or the evener by a clevis. Hitching is uncomplicated and does not require excessive adjustment. Simply make sure that the point of pulling is in the middle of the harrow and, if the angle of the draft is correct, all should go well.

**Spring- and spike-tooth harrows** work the soil with a series of teeth rather than discs. The spring-tooth harrow functions just as it sounds: The spring in the teeth keeps the piece of equipment moving smoothly across the ground, minimizing the bouncing of the harrow. The less the harrow jumps around on bumpy ground, the smoother your field will be. The spike-tooth harrow has aggressive teeth meant to rip the ground. It is typically used to drag manure in a pasture, to smooth a seedbed, or to aerate soil.

Both spring- and spike-tooth harrows are hitched by attaching an evener or the forecart to

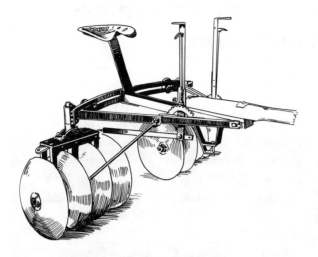
*A single-action disc harrow*

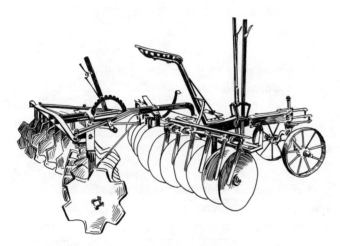
*A double-action disc harrow*

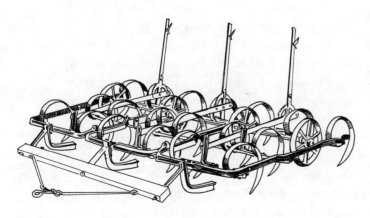
*A spring-tooth harrow*

the middle point of the harrow via a clevis. Hitching to the middle will keep the harrow moving in a straight line.

You can drive your animals hitched to the spring- or spike-tooth harrow in either of two ways: You can walk to the side or ride behind on a forecart or harrow cart. The method of locomotion you choose depends on the size of your field. I have done both for miles. Riding saves on leg power, but in a small field, turning with the forecart can be difficult. Turning a harrow is tricky enough anyway; you have to take care that it doesn't flip over, which can be extremely dangerous.

Harrowing is an excellent way to put good muscle tone on your horses and to help a spirited or willful animal remember what the harness is about. Although harrowing does not require the precision and concentration of plowing, it still requires great physical exertion, making rest periods imperative.

> ◀ **CAUTION** ▶
>
> ### Disc Harrow Safety
>
> Many of us teamsters have spent hours standing on a moving disc harrow — an absolutely unsafe practice. I have seen horrendous accidents result from this foolish behavior and I'm thankful that I was never injured in all of the hours I spent riding a disc. If your disc harrow does not have a seat, pull it with a forecart.

## PLANTING YOUR CROP

Planters have been made to accommodate many different seeds and seedlings, as well as operation sizes. If you are planting a hayfield, a grain drill will give you even distribution of the fine seeds. If you are planting corn or beans, a corn planter will carefully drop the seeds in neat, straight rows. Other common horse-drawn planters include potato planters and tobacco planters.

Planting with horses requires a good, steady team that will move methodically up one row and down the other. Nothing is more embarrassing than having crooked rows winding about the field in plain view of the highway and amused passersby. For this reason, many a beginning teamster has perfected planting technique on the back forty.

## MANAGING WEEDS

Many small-scale farmers growing corn, beans, and other row crops choose not to use herbicides, instead preferring regular weed cultivation. Tilling up the bare earth between the rows so no weeds remain is one of the simplest ways to manage weeds while allowing valuable crops to grow. Constantly digging up the soil, however, will result in the oxidation and loss of organic matter. Mulching with weed-free straw or other organic material is an alternative weed-management system that works well for small-scale enterprises. Another option is to use a green (living) mulch, such as hairy vetch or clover.

If you are choosing the cultivating method for weed removal, you have two options. Cultivators, like plows, may be either riding or walk-behind. Using a walk-behind cultivator presents some challenges. A walking cultivator may not come as close to the crop as a riding cultivator. A steady horse or mule pulling the cultivator will walk down the row carefully without wandering over your crop, but if the animal needs guidance, someone will have to drive your animal while you guide the cultivator. One of the first jobs I did on our farm when I was young was to ride the horse while Dad operated the one-row cultivator in our garden. I spent countless hours sweating on the back of a harnessed horse, guiding her up and down rows. When I got big enough to handle the cultivator, my younger sister rode the horse.

A riding cultivator straddles the row, so you cannot manage tall or wide-row crops with this equipment. The riding cultivator is often used for early corn, beans, onions, carrots, and other shorter row crops. The level of skill required to keep from knocking plants out of the row is quite high, since the operator controls not only the animals pulling the cultivator, but also the gangs where the cultivator shovels are attached, and can push them independently to the left or right with foot levers, depending on the correction required. Because of the number of levers, operating this implement demands great coordination.

### Cultivating Tips

**B**efore working with live plants, set up a row of cans and practice cultivating a straight line using all the levers. For good cultivation:

- Make sure that your shovels are evenly spaced.
- Keep your shovels tight and well adjusted to minimize wandering.
- Move slowly; a rapid pace causes shovels to throw soil and cover young plants.
- Use only steady equines for cultivation.

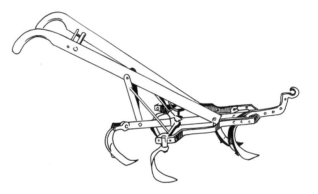

A walking cultivator

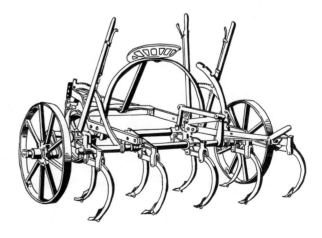

A riding cultivator

## HARVESTING YOUR CROP

Harvesting crops is the final step of cultivation. The way in which a crop is harvested depends largely on the hand labor a farmer is interested in investing. For example, some farmers own corn pickers, while others pick their corn entirely by hand. Some farmers plow their potatoes and then pick them up by hand, while others have a potato harvester that digs, sorts, and bags them. Because harvesting is done in such varied ways, covering the multitude of available harvesting methods would require a book of its own.

Most harvesting equipment, such as grain binders, threshing machines, and corn harvesters, require so much horse power that they are not practical for the average small-scale farmer. The old-order Amish and Mennonite people successfully use such equipment because they have sufficient collective manpower and horse power among their community members.

If your crop consists of garden vegetables, harvesting will continue for most of the season. Or perhaps you are interested in growing a few acres of pumpkins, sweet corn, tomatoes, or the like. In either case, you may use your horses or mules to transport your produce (yes, you will load them by hand onto the wagon), for which you will need a sturdy farm wagon that you don't mind muddying or denting and a quiet, well-trained team.

## The Art and Science of Making Hay

As you well know, horses and mules eat hay, and *a lot* of it. What better way to tighten the circle of sustainability than to harvest your own hay? It's as simple as cutting, raking, curing, and harvesting. Right? Wrong. To produce good-quality hay, you must make a science of haymaking. When pasture forage is not available, hay will constitute most of your animal's diet, and growing your own makes good economic sense.

It is a satisfying job to put up hay with the horses that will consume it.

**234** DRAFT HORSES AND MULES

## Mower Safety

**Mowing hay requires** a well-adjusted mower and constant attention to safety. The following tips will help you have a smooth-running mowing experience:

- Keep knife sections sharp and properly lubricated to increase cutting efficiency and decrease draft.
- Always keep the cutter bar in the full upright position when hitching and unhitching.
- When cutting lodged or tangled grass, tilt the cutter bar forward with the tilting lever.
- To prevent clogging of the guards or knife, try to keep the cutter bar out of previously cut grass. If you pick up any grass or other trash while the knife is disengaged, clean it thoroughly before beginning to cut again.
- Always disengage the gears when getting off the mower.
- Always get off the mower to the left and back; never dismount in front of the cutter bar.
- Never stand in front of the cutter bar.
- Never walk along the right side of the team; always stay on the left side.
- Keep your hands behind the cutter bar and out of the guards when cleaning the cutter bar, and use a sturdy stick to clean out plugs of grass.
- When transporting your mower, raise the cutter bar.
- To turn a corner, run the cutter bar just past the last standing grass. As soon as the last piece has fallen, stop the mower and lift the cutter bar with the foot lift. Swing your team around the corner and, if necessary, back into position to start a clean, straight swath. Ask your animals to move ahead, then engage the knife and drop the cutter bar just before it hits the first pieces of standing hay.

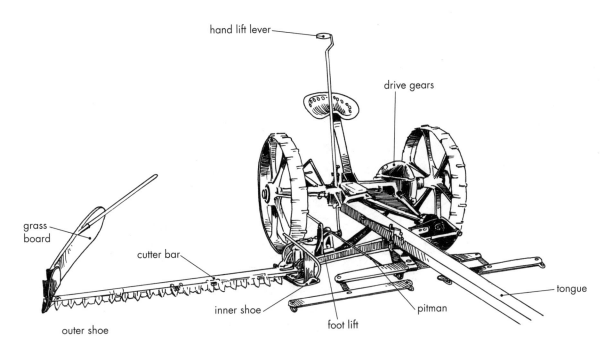

A SICKLE-BAR MOWER

Like any other crop, hay must be planted. You cannot mow a field of weeds and expect your animals to thrive during the winter on sticks and stems. You must plow and disc harrow your soil as for any other crop and then seed your field with the hay of your choice.

On our farm, we grew mixed grass and alfalfa. The grass, such as timothy and orchard grass, grows thick and tall to provide abundant biomass; the alfalfa provides quality and necessary protein. Oat hay is an excellent nurse crop, keeping the majority of weeds at bay while the new hay becomes established. The oat hay may be cut in the milk stage (when the grain is still soft and has a milky-white color), before the alfalfa matures, by which time the alfalfa will have grown enough to hold its own. The heads of oat hay are nutritious, but the stems are not, so expect to waste a lot when you feed it.

## MOWING HAY

Mowing hay is a beautiful and mesmerizing experience. To sit on the swaying seat of a sickle-bar mowing machine, hear the cutter bar whirring and the pitman arm clattering, watch the horses' round hindquarters powerfully pulling, and smell the clean, fresh scent of cut grass early on a new day is near heaven.

Mowing hay is also a noisy and potentially dangerous experience. The knife consists of many sharp mowing sections that move back and forth (powered by the turning of the wheels) within the cutter bar by the action of a pitman arm. When the wheels stop turning, the pitman arm ceases to move the knife. Guards, attached like 4-inch (10 cm) fingers to the cutter bar, protect the sections from rocks and other hidden objects. The cutter bar, once lowered from its upright transport position, may be set for cutting and raised for turning or avoiding stumps by a foot lift. It is engaged or disengaged by another foot lever.

Once engaged, the cutter bar does not stop moving until it is manually disengaged or the team stops. You have to be ready to hit that lever quickly in case an animal jumps across your path. Horrible accidents can result from a runaway team and a driver catapulted into the path of the scissor-sharp blades, and many a dog has lost a leg from being in the wrong place at the wrong time (a good reason to confine dogs on mowing days). You and your horses must know exactly what is going on and what to do if something goes wrong.

Most proponents of draft power today use one of two kinds of mower. A wheel-driven mower, to which the animals are hitched directly, is powered

A board and block may be used to counterbalance the weight of the mower.

by the turning of the mower's wheels and will stop cutting when the wheels stop turning. The same is true of the PTO on a ground-drive forecart. A tractor mower attaches to a forecart with a PTO that powers the cutter bar. If the PTO is engaged, the cutter bar is in motion, even if the mower is not.

The basic idea behind their operation is the same for both kinds.

Most mowers do not have a tongue truck (two wheels that hold the tongue up under the evener). The tongue therefore places incredible weight on the animals' necks, often leading to bad neck sores

## Determining the Moisture Content of Hay

**K**nowing when your hay is dry enough to cut and store is tricky. Everyone's impulse is to jump the gun and put it up before it rains, and many times before it is completely dry. Hay that is stored before it is properly cured presents a double hazard. First, it may produce enough heat from decomposition to spontaneously combust, potentially burning your barn to the ground. Second, it is likely to develop mold, which may lead to colic, respiratory diseases, and/or airway inflammation in your horses or mules.

For safe storage, loose hay and large round or rectangular bales should contain no more than 18 percent moisture; small rectangular bales should contain no more than 20 percent moisture.

Learning to recognize properly cured hay is almost a science. One way to learn is to get an old-timer to show you how to estimate the moisture level by feel. Another method is to weigh a sample and dry it in a microwave oven. In addition to the microwave, you will need a scale, such as a postage or kitchen scale, that can weigh at least 100 grams (3 oz), a paper plate, an 8-ounce (224 g) glass of water, a pencil and paper, a pair of strong scissors, and a clean, dry container in which to collect the hay samples.

To collect the samples, grab a minimum of 10 handfuls (several pounds total) of hay from various windrows at various depths. Fold each handful in half and use the scissors to trim off the outer edges. Cut each handful into pieces no longer than two inches (5 cm). Mix all of the samples together in the container. Select and accurately weigh 100 to 150 grams from your samples to obtain the wet weight.

Spread the weighed sample on the paper plate, leaving an open area in the center. Put the 8-ounce glass, about three-quarters full of water, in a back corner of the microwave. This water is needed to protect the oven; replace it as needed to keep the glass three-quarters full. Put the paper plate with the hay sample in the oven and turn it on medium or high. If the original sample was obviously moist, microwave it for 3 minutes; if it was fairly dry, microwave it for 2 minutes.

Remove the sample from the oven, weigh it, and record the weight. Mix the pieces together and reheat for 30 seconds. Weigh again. Be careful not to include the weight of the paper plate. If you weigh the moist hay on the plate, be sure to weigh the dry hay on the same plate. Continue weighing, mixing, and reheating the sample for 30 seconds at a time until it weighs the same twice in a row or until it begins to char. If the sample chars, use the last recorded weight, which is the dry weight.

To calculate the moisture percentage, subtract the dry weight from the wet weight, divide by the wet weight, and multiply by 100.

**Example:** If the wet weight is 150 grams and the dry weight is 111 grams:

$(150 - 111)/150 \times 100 = 26\%$ moisture

which is too moist for safe storage.

under the collar. To reduce the weight, wire a stout board under the seat and mower frame and hang a heavy block from the board. The weight of the block will counterbalance the mower.

## TEDDING, RAKING, AND HARVESTING

Depending on what part of the country you live in, you may need to do more to your mown hay than rake it into windrows. If you are in a humid climate, you may first need to ted your hay, or turn it over, after it has been drying in the sun for a day or two. Tedding may be done with a tedder or with a side-delivery rake set so that the teeth push out instead of pull in. Once the hay is turned, allow it to sit for another day before raking it.

When the hay is properly cured, it is raked for easy pickup. Three general kinds of rakes each having a different function and purpose, are available for use with horse or mule power.

**The side-delivery or reel rake** is the one to use if you are going to pick up the hay with a baler or hay loader and therefore need the hay raked in rows. The side-delivery rake is versatile because it can both ted hay (by reversing the direction of the reels) and sweep it up into windrows.

**The dump rake** or sulky rake does what its name implies — collects hay and dumps it into piles. It may be used to create windrows by lining up piles across the field for a hay loader or buck rake. A distinct benefit of the dump rake is that a single horse can power a small one. A wider rake requires the power of two or three horses. If you will be forking your hay onto a wagon by hand, use a dump rake.

**The buck rake** is used mostly in the West, often in conjunction with a dump rake. A buck rake is pushed along in front of two horses or mules to collect hay from windrows (made from a side-delivery rake) or piles (made from a dump rake) and transport it to a loose stack for storage.

Once the hay is raked into piles or windrows, it is harvest time. The method you use to put your hay into storage will depend on the size of your operation and the number of helpers you manage to round up. Storing hay loose often reduces the moisture content (and thus the potential for mold) of the hay, but storing it in bales is more space-efficient. A hay baler may be pulled by a team of horses on a powered forecart. The windrows created by a side-delivery rake will accept the teeth of either a baler or a hay loader.

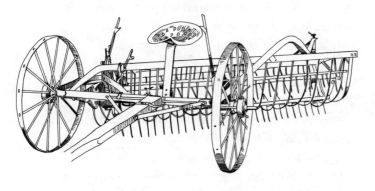

For nice, straight windrows, use a side-delivery (reel) rake.

A dump rake piles hay nicely for hand forking.

A buck rake gathers large quantities of hay and moves it to a stacking site.

The beaver slide stacker is used to make large haystacks.

**WORKING WITH DRAFT POWER**

In the West, hay was commonly stacked in large outdoor piles using a beaver slide and/or overshot stacker. Today the availability of balers or folks willing to bale hay for hire has cut down on the number of overshot stackers/beaver slides still in practical use. The overshot stacker and the beaver slide stacker both do essentially the same job: take hay off the buck rake and swing it through the air to stack it in a pile that may grow as tall as 20 feet.

On my family's farm, we put up hay loose using a loader, wagon, and stout team of horses. Occasionally if we were putting up hay in a hilly field and wanted to get really large loads, we pulled the wagon with three horses. After we loaded the hay from the field into the wagon, we had to load it from the wagon into the loft of our barn. Putting up hay is a brutal job. Whether you bale your hay or put it up loose, the task is dirty, hot, and merciless, and separates the hobbyist from the professional.

## Using Your Team for Logging

Logging is not a job for the slow of foot, faint of heart, or weak of arm. It requires a teamster who is quick-witted, strong, and agile, and a team that will pull with all their might and stop on a dime. If you and your team qualify, selective logging is a good way to manage your woodlot, teach your horses to pull dead weight, and stock a supply of winter firewood. A number of methods and opinions exist on the correct way to move a log; two of the most common are ground skidding and using a logging cart.

Ground skidding is easily adapted to working with a single horse or mule, provided the logs are not too heavy. For ground skidding you need a chain with a hook on it to attach the log to your singletree or doubletree, and a peavey (a stout wooden handle with a hook on the end) to roll and maneuver the log. Although this method requires the least equipment, it can be the most dangerous. Because the logs are simply pulled along the ground, you have no brakes; on a steep downhill slope, the traces become slack and the log cannot be steered.

A logging cart adds expense, but gives you a safer and more effective way to haul heavy timber out of the woods. The logging cart, which is generally pulled by a team, suspends the front end of a log off the ground, thus reducing drag so the team is able to move heavier loads, whether consisting of heavier logs or more of them. Some logging carts have powerful brakes, which help the team maneuver down steep hills more safely than with ground skidding.

## Purchasing Draft Equipment

The periodicals listed at the back of this book carry many advertisements for companies that produce brand-new horse-drawn equipment, as well as dealers in used or restored equipment. The annual Horse Progress Days, an event held in the East and Midwest, showcases horse-drawn equipment and manufacturers. Attending this event is an education in itself, particularly for anyone wishing to acquire new equipment.

If you don't want to fork over the money necessary for brand-new equipment, consider attending a draft-horse equipment auction or searching your newspaper for used equipment for sale. A third option, for those with the necessary time and skills, is to ferret out old equipment to restore. My dad restored all of the horse-drawn equipment that we used, and painting equipment is one of my favorite wintertime in-shop memories. If you are mechanically handy, restoring old equipment makes a great winter project. It is a time-consuming but rewarding endeavor, and when you are done you have exactly what you want.

# Logging Safety

**The forest is fraught** with danger from all directions, requiring the teamster to be on constant alert. Since your vision is obstructed in the woods, be extra careful about moving yourself or your equine. Here are some of the many things you must watch out for:

- Horses and mules pull hard when moving heavy logs, so make sure they have a clear path and destination.
- Logs roll without giving notice. Always stay on the uphill side and keep your eyes on a moving log as well as on your team.
- Stay on the outside of a turn as the log cuts the corner or, as horse loggers are fond of saying, stay out of the bight.
- Watch out for saplings, low branches, or bushes that may snap you or your animals in the face as you move logs.

- Avoid steep uphill or downhill skid trails, particularly when ground skidding. The dead weight of a log often requires great force to move over level ground; pulling uphill may be more than your animal can manage. On a downhill run, the log may build enough momentum to move faster than your animal — a potential disaster.

Before attempting to hitch your draft animals to a log many times their weight, educate yourself thoroughly on proper methods and safety precautions. A good place to start is to apprentice with an experienced horse or mule logger.

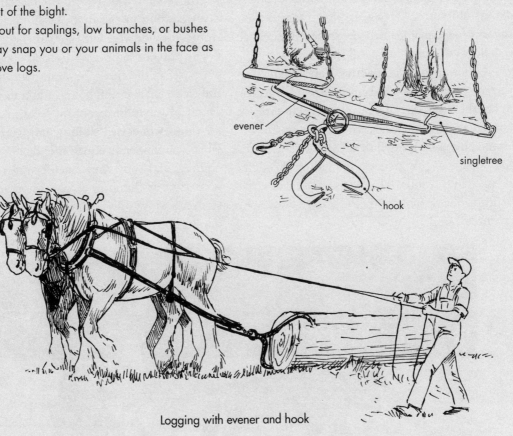

Logging with evener and hook

# A FINAL NOTE

Draft horses and mules have a distinctly romantic side attached to the pragmatic, practical, and self-sustaining reasons we support their use. Although we could offer a host of reasons to support, sustain, and encourage anyone interested in using draft horses or mules, perhaps the following scenario will suffice.

On a bitter-cold February day in northern Idaho — the kind of day where nature seems to stand still and wait for a hint of warmth from the sun — some 20 teamsters and their horses gather to enjoy the four-plus feet of snow with our horses, mules, bobsleds, and sleighs. Cold fingers grow uncooperative as teamsters finish hooking their eager horses to the vehicles.

When the passengers have all clambered aboard, the drivers chirrup to the teams, which take off in a spray of snow accompanied by the joyful ring of many full sets of sleigh bells. All the teams eagerly move out after a winter of inactivity, and the teamsters have their hands full keeping their animals headed in the same direction, in line, and in order.

What a wonderful winter scene — the long line of rigs, some pulled by spotted drafts, some by lighter driving horses, one by mules, all winding through the dense white woods. The melody of the sleigh bells adds to the festivity of the day, and every face wears a broad smile.

On your own road to becoming a successful teamster, you will encounter some bumps, but as long as you double-check every piece of your harness, make sure your driving lines are correctly snapped and triple-checked, see that your equipment is properly adjusted, and keep your brain fully engaged, you will have a pleasant journey down that road.

As you trot along, stop now and then to take time to remember how important those dreams were when you began your search for the perfect draft team. Remember the excitement you felt when you made the decision to purchase your animals. Remember the trepidation you felt the first time you stood next to your team and realized how big they were.

Above all, remember all the reasons you wanted draft equines in the first place: You thought they would provide you with an excellent way to slow down, practice some of the lost arts of our forefathers, get closer to the earth, be a bit more self-sustaining, use clean energy, and connect in a meaningful and practical way with beautiful creatures. Remember all those wonderful visions and dreams you had when you first entertained the idea of owning draft equines, and know they can all come true. We have witnessed firsthand many terrific teamster success stories, and we fully believe you are on your way to joining their ranks.

# GLOSSARY AND RESOURCES

## Glossary

**abreast.** Side by side and facing the same direction.

**Achenbach grip.** A method of holding both driving lines with one hand, by running the lines through the fingers, so the driving whip may be carried in the other hand. See also *comfort grip; English grip*.

**adjustable collar.** A collar for either work harness or driving harness that adjusts to three sizes.

**American Belgian.** A modern-style leggy Belgian.

**American Cream Draft.** A breed of heavy horse consistently cream in color, and the only draft breed still in existence to originate in the United States.

**apprenticeship.** A training program in which a novice learns a trade under the guidance of skilled tradesmen, often exchanging labor for learning; also called internship.

**barefoot.** Wearing no shoes on the hooves.

**barrel.** The section of the equine body between the forearms and the loins.

**base narrow.** A conformational fault in which the feet are closer together than are the gaskins or forearms.

**base wide.** A conformational fault in which the forearms or gaskins are closer together than are the feet.

**bay.** Hair coat color that is dark red (known as blood bay) to deep brown with black points.

**bean.** Hardened pellet of smegma sometimes found at the end of the penis.

**beaver slide.** A system of pulleys, ropes, poles, and basket that raises hay on rails to stack it in a pile up to 30 feet high, typically operated with horse or mule power.

**Belgian.** One of the most popular breeds of heavy horse.

**betadine.** An iodine-based antiseptic.

**binocular vision.** The ability to simultaneously use both eyes to produce one image.

**bit.** A metal piece on the bridle that goes in the equine mouth and is used to guide the animal's movements.

**blinders (blinkers).** A pair of small leather patches attached to the bridle to prevent the animal from seeing sideways and behind.

**blind staggers.** Acute selenium toxicity.

**blinkers.** See *blinders*.

**blow.** Take short, quick breaths after a period of exertion.

**bobsled.** A vehicle similar to a wagon, but having runners instead of wheels to glide over the snow.

**body condition.** The general appearance, weight, body fat, fat-to-muscle ratio, and health of an animal.

**bog spavin.** A conformational fault in which a fluid-filled swelling on the lower inside of the hock indicates inflammation.

**bone spavin.** A bony growth within the lower hock joint.

**borium.** Tungsten carbide composite welded to the bottom of equine shoes to increase traction and reduce wear.

**bottom.** The part of moldboard plow consisting of the share, the moldboard, the landside, and other parts that are essential for correct plow operation.

**bounding pulse.** A strong digital pulse that can be easily felt over the sesamoid bones. See also *digital pulse*.

**bowlegs.** A conformational fault in which the knees or hocks have too much space between them.

**box breeching harness.** Work harness with a breeching that comes over the croup and rests between the hock and the point of the croup, using quarter straps and breast straps to hold back the forward motion of a vehicle or implement; also called Western-style harness.

**box stall.** A stall with four sides and a door, in which the animal is allowed to walk about freely.

**Brabant Belgian.** A massive Belgian of the original type.

**breast collar.** A collar that fits over the pectoral and chest muscles, instead of the shoulders.

**breeching.** Parts of the harness that rest over the flank and croup to aid in backing and stopping.

**bridle.** Headgear consisting of a headstall, bit, chin strap, and driving lines.

**broodmare.** Mare used for breeding and raising foals.

**buck knee.** A conformational fault in which the knees bend forward when viewed from the side.

**buck rake.** A rake that collects hay as it is pushed in front of draft animals, often used in conjunction with a beaver slide or overshot stacker.

**buggy.** A small, light, four-wheel carriage pulled by a single horse or mule.

**calf knee.** A conformational fault in which the knees bend back slightly when viewed from the side.

**calk.** A projection welded or screwed to the bottom surface of an equine shoe to improve traction.

**calorie (kilocalorie).** A unit of measurement used to describe the energy stored in food.

**camped out.** A conformational fault, or sometimes an indication of hoof pain, in which the hind legs extend beyond the point of the buttock.

**camped under.** A conformational fault in which the hind legs are too far under the body.

**canine tooth (tush).** One of four relatively small, pointy teeth between the incisors and premolars of the upper and lower jaw; more often seen in males than in females.

**canker.** A bacterial infection of the hoof affecting the frog, the sole, and the heels, characterized by dying tissue.

**cannon bone.** The long bone between the knee or hock and the fetlock.

**carriage.** A four-wheel passenger vehicle pulled by two or more equines.

**cart.** A light two-wheel vehicle pulled by a single animal.

**cast.** Situation in which an equine rolls or lies down and becomes caught with his legs against a wall, fence, or other object and is unable to stand up.

**cecum.** A blind ending pouch at the junction of the small and large intestines, containing microbial organisms that digest cellulose; part of the hindgut.

**center of gravity.** The point in an equine body that may be used to represent the body's weight.

**check line (overcheck stay).** A piece of harness running from the bit to the hame on a work harness, or stay on a driving harness.

**chestnut.** A hair coat color ranging from reddish gold to deep red with mane and tail of similar color or lighter (never black); also a horny growth on the inside of the forearm and hindleg.

**cleat.** A removable calk.

**clevis.** A U-shaped piece of iron or steel with holes in the ends through which a pin is passed to connect one object to another, such as an evener to an implement.

**Clydesdale.** A breed of heavy horse originating from Scotland, made famous as an advertising symbol by the brewery industry.

**coated webbing.** Strong belting made of a synthetic material, such as nylon or polyester, coated with another synthetic material such as polyurethane or vinyl, and marketed under such brand names as BioThane, Beta, and Ohio-Thane.

**coffin bone.** The lowest bone in the hoof.

**Coggins test.** A test to determine the presence of the viral disease equine infectious anemia.

**coldblood.** A heavy horse, characterized as being hardworking while having a calm and gentle temperament.

**colic.** Any pain in the abdomen that may indicate illness, but is not itself a particular illness.

**collar.** The part of a harness that fits around the equine's neck, used to increase the efficiency of the shoulders in pulling heavy loads.

**collar cap.** The top part of a collar.

**colt.** An intact (uncastrated) male horse that has not yet reached physical maturity, typically less than three years of age.

**comfort grip.** A method of holding the driving lines that involves running each line through the palm of one hand and gripping it with the thumb and forefinger. See also *Auchenbach grip; English grip.*

**concentrates.** Feed products such as grain, fortified feeds, and pellets that, compared to forage, contain concentrated amounts of digestible energy.

**conformation.** The form, structure, and shape of an animal.

**coronet.** Area of soft tissue above the horn of the hoof.

**coulter.** A small disc that runs in front of a moldboard plow bottom to cut trash and vegetation.

**cow hocks.** A conformational fault in which the feet angle out and the hocks angle in when viewed from the rear.

**cresty neck.** Equine neck characterized by a high ridge along the top.

**cribbing.** A vice, commonly associated with windsucking, that involves biting or chewing wood and other objects.

**croup.** The part of an equine's rear from the front of the pelvis to the root of the tail.

**crupper.** A piece of harness that fastens around the tail.

**cryptorchid.** A male with one retained testicle.

**Cushing's disease (hyperadrenocorticism).** A hormonal disorder causing abnormal hair growth and shedding.

**developmental orthopedic disease (DOD).** A collective term describing all orthopedic problems in a growing foal, including metabolic, joint, and growth plate issues.

**diamond hitch.** A rope pattern, roughly in the shape of a diamond, that firmly attaches panniers to the pack saddle.

**digestible energy.** The energy remaining after the energy lost in feces is subtracted from the total available energy in a feed source.

**digital pulse.** The pulse felt in the midpastern or fetlock at the level of the sesamoid bones. See also *bounding pulse.*

**disc harrow.** An agricultural implement with a number of rotating blades used to cut and lightly turn the soil for seedbed preparation.

**dish face.** A concave facial profile as viewed from the side.

**dock.** To remove a horse's tail, including a significant part of the tailbone.

**double rumped.** Description of a draft animal's hindquarters in which a groove runs down the center, creating a heart shape when viewed from the rear.

**doubletree.** Two singletrees attached to a bar, called the evener, that keeps the pulling of two animals balanced and also attaches them to an implement. See also *singletree*.

**driving harness.** Light harness used for pulling a light transportation vehicle such as a surrey, buggy, sleigh, or cart.

**dump rake (sulky rake).** A rake that collects hay by scraping it up and dumping it into piles.

**easy keeper.** An animal that stays in good body condition without needing extra feed.

**eight up.** Four teams hitched one in front of the other.

**endophyte.** A fungus living in another plant.

**endotoxin.** A toxin released by bacteria or other microorganisms when they die.

**English grip.** A method of holding the driving lines that involves running one line through the palm of each hand, with the loose end coming up and over the thumb and the bit-end running under the pinky. See also *Achenbach grip; comfort grip*.

**enterolith.** A large, smooth, stonelike object formed in the intestinal tract.

**equine infectious anemia (EIA); (swamp fever).** A viral disease characterized by fever, depression, weakness, weight loss, edema, and anemia.

**equine polysaccharide storage myopathy (EPSM).** A muscle-wasting condition of equines.

**equine protozoal myeloencephalitis (EPM).** A disease of the central nervous system caused by exposure to opossum feces.

**estrus (heat; season).** The five to seven days during the estrous cycle when the female is most receptive to the male for breeding.

**evener.** A bar to which two or more singletrees are attached, which in turn attaches to an implement and balances the load that each animal pulls.

**ewe neck.** An equine neck that appears concave when viewed from the side.

**false rig.** A gelding with both testicles removed that still exhibits stallionlike behavior.

**far side (off side).** The right side of a team or a single horse or mule.

**fescue toxicosis.** A condition of pregnant mares caused by the consumption of endophyte-infected fescue and characterized by little or no milk production, prolonged gestation, and difficult birth.

**fetlock.** A joint of the leg just above the hoof between the cannon bone and the pastern.

**filly.** A female horse that has not yet reached physical maturity, typically less than three years of age.

**fistula.** An abnormal connection or tunnel between tissue.

**fistulous withers.** A chronic inflammation of the withers.

**flank.** The area between the rib and hip on the side of the equine.

**flehmen response.** A behavior in which the equine curls back his lips in response to an unusual or sexually enticing scent, or in response to pain.

**flexion test.** A test used to determine the location of a lameness, involving exaggerated flexing of an affected joint followed by immediate movement to see if the lameness increases.

**float.** To remove sharp edges from the teeth with hand or power tools.

**foal.** A juvenile equine less than one year old; also the act of giving birth.

**forage.** Plant material consisting of grasses, forbs, and small shrubs used as livestock feed.

**forb.** An herbaceous flowering plant other than a grass or grasslike plant.

**forecart.** A two-wheel cart used to pull a piece of farm machinery and provide a safe seat for the driver between the team and the equipment.

**founder.** Rotation and sinking of the coffin bone as a result of laminitis.

**four up.** Two teams hitched one in front of the other.

**free choice.** Feed and water offered without restricting the amount consumed.

**frog.** The fleshy pad in the central area of the hoof.

**furrow.** The cut and row of turned earth made by a moldboard plow.

**furrow horse or mule.** The animal that walks in the furrow while plowing and is responsible for creating a straight cut.

**Galvayne's groove.** A V-shaped groove appearing at the gum line of the corner incisor as an equine approaches the age of 10 years.

**gang plow.** A plow with two or more bottoms that operate at the same time.

**gaskin.** The muscular part of the hind leg between the hock and the stifle.

**gee.** A voice command meaning to turn to the right.

**gelding.** A castrated male equine.

**grade.** A nonregistered or mixed breed animal.

**green broke (started).** A horse or mule with minimal training.

**ground drive.** A system that uses the turning force of an implement's wheels to create enough power to run the implement; to guide an animal by walking and manipulating the driving lines.

**ground skid.** To drag logs along the ground.
**Haflinger.** Breed of small draft horse originating from the Tyrol region of Austria.
**hame.** One of a pair of steel arms that fit around the collar and to which the traces are attached.
**hame-style harness.** A harness that uses hames and a collar on the shoulders.
**hand.** A unit equaling 4 inches (10 cm) used to measure the height of a horse from the top of the withers in a straight line to the bottom of the hoof.
**hard keeper.** An animal that consumes large quantities of feed but fails to stay in good condition for nutritional, dental, conformational, or other reasons.
**harrow.** An implement used to smooth the soil; to use such an implement.
**haw.** A voice command meaning to turn to the left.
**hay belly.** A conformational fault in which the belly sags, often accompanied by swayback.
**hay loader.** An implement pulled behind a wagon to collect windrowed hay and drop it onto the wagon.
**hay net.** A knotted bag used as a hay feeder in a stall or trailer.
**head pressing.** A sign of illness in which the equine presses his head against any available object.
**headstall.** The band on a bridle that fits around the equine's head.
**heat.** See *estrus*.
**heavy horse.** A large horse of stocky build, generally used for draft purposes.
**hindgut digestion.** Digestion of fiber occurring in the large intestine.
**hinny.** The offspring of a jennet (female donkey) and a stallion (male horse).
**hitch.** A group of animals harnessed together; the method used to attach one animal or more to a vehicle or implement.
**hitch class.** A show class in which the animals are displayed driving in different team configurations such as single, one team, four up (two teams), six up (three teams), and so forth.
**hitch horse.** A leggy draft horse used in large hitches for show purposes.
**hock.** Joint in the rear leg between the stifle and the fetlock.
**horse mule (john).** A castrated male mule.
**horse or mule pull.** An event in which horses and/or mules compete to pull the heaviest load.
**hotblood.** A light equine breed, such as the Thoroughbred or Arabian, that tends to have a more reactive character than draft breeds.
**hyperadrenocorticism.** See *Cushing's disease*.
**in at the knees (knock kneed).** A conformational fault in which the knees bend toward each other.
**in hand.** Handled with a lead line, as opposed to being ridden or driven.
**intact.** Not castrated.
**intussusception.** The inversion of one portion of the intestine into another.
**jack/jackass.** An intact male donkey.
**jennet/jenny.** A female donkey.
**john.** A castrated male mule.
**jointer.** A sharp blade that runs in front of the coulter and plow bottom on a moldboard plow.
**junctional epidermolysis bullosa (JEB).** A rare but fatal genetic condition of Belgians and American Creams; also called hairless foal syndrome.
**kilocalorie.** See *calorie*.
**knock kneed.** See *in at the knees*.
**laminae.** Layers of interlocking sensitive and insensitive tissue connecting the hoof wall and the coffin bone.
**laminitis.** Inflammation of the sensitive laminae between the coffin bone and the hoof wall.
**landside.** The part of a moldboard plow bottom that runs against the furrow wall.
**line drive (ground drive).** To guide an animal while the handler walks behind and manipulates lines attached to the bit.
**lockjaw.** See *tetanus*.
**logging cart.** A cart used to elevate the front of a log to improve the efficiency of movement.
**loin.** The muscular area between the back and the croup.
**longear.** A mule, hinny, or donkey.
**Lyme disease.** A bacterial disease characterized in equines by stiffness or lameness.
**manure spreader.** An implement with gears, beaters, and chains designed for the effective spreading of manure and/or soiled bedding for fertilizer.
**mare.** A female horse.
**mare mule (molly).** A female mule.
**megacalorie.** 1,000 nutritional calories (kilocalories).
**milk teeth (temporary teeth).** Baby teeth that are shed as a foal matures and are followed by permanent teeth.
**moldboard.** The curved extension of a plow bottom.
**moldboard plow.** An implement used to turn the soil.
**molly.** See *mare mule*.
**monocular vision.** The ability to see separately with each eye.
**monogastric.** Having one stomach for digestion.
**muleteer (mule skinner).** A person who drives mules.
**multiple hitch.** More than two animals hitched together.
**near side (nigh side).** The left side of a team or a single horse or mule.
**neck yoke.** A bar that attaches to the collars of two animals and is used to hold back a vehicle's or implement's forward motion and to direct the tongue in turning.

**New England D-ring harness.** A style of work harness with no quarter straps or breast straps, but instead using neck straps and side-backer straps to hold back the forward motion of a vehicle or implement.

**nigh side.** See *near side*.

**Norwegian Fjord.** A small draft horse breed originating from Norway.

**off side.** See *far side*.

**osteochondrosis dissecans (OCD).** A degenerative joint disease (DJD) affecting the joints of young growing equine as a result of nutritional deficiencies.

**overcheck stay.** See *check line*.

**overo.** A coat color characterized by uneven splashes of white, with no white crossing the back.

**overshot jaw (parrot mouth).** Overbite of the upper jaw.

**overshot stacker.** A system of pulleys, ropes, poles, and basket that tosses hay onto a pile to stack it up to 30 feet high, typically operated with horse or mule power.

**pack.** To use animals for moving gear over a trail.

**pack saddle.** A piece of tack that holds panniers and other items for transport by an animal.

**Paint.** A breed of horse with pinto markings.

**pair.** See *team*.

**pannier.** A container fastened to a pack saddle to transport goods.

**parrot mouth.** See *overshot jaw*.

**pastern.** The area between the hoof and fetlock joint.

**peavey.** A tool consisting of a spike at the end of a lever, used to position logs.

**Percheron.** A popular heavy horse breed originating from the Le Perche region of France.

**pigeon footed.** See *toed in*.

**pinto.** A coat pattern with white patches.

**pitman.** A bar that connects a wheel to a cutter bar knife and converts the wheel's rotary motion into a back-and-forth cutting motion.

**plowing bee.** An event at which teamsters and their teams gather to collectively turn the ground of a field for fun.

**point.** The forward portion of a plowshare that penetrates the soil.

**point of release.** The moment an animal realizes he has been released from discipline, correction, or pressure.

**pole (tongue).** A length of wood or metal extending from a vehicle or implement, used to guide the direction of the vehicle or implement.

**post legged.** A conformational fault in which the hind legs are excessively straight.

**power forecart (PTO cart).** A forecart that uses a motor, or is ground driven, to operate a power take-off (PTO).

**power take-off (PTO).** A revolving shaft that uses power from an engine or a ground-driven device to drive the mechanism of another implement.

**proud cut.** A common term describing a true gelding that exhibits stallionlike behavior, erroneously attributed to the incomplete severence of the epididymis.

**purge deworming.** Administering large amounts of drugs a few times a year to control parasites.

**quarter crack.** A crack in the side of a hoof.

**reel rake (side delivery rake).** A rake that collects hay using a series of teeth moving in a circular motion to gather the hay into a windrow.

**retained cap.** A cap on a milk tooth that fails to fall off as the permanent tooth emerges.

**roach.** To trim the mane of a horse or mule to a few inches or less so the hair stands erect along the crest.

**roach back.** A conformational fault in which the back appears convex between the wither and loin.

**Roman nose.** A common feature of draft equines in which the facial profile appears convex.

**sand colic.** A pain in the abdomen as a result of ingesting sand.

**sesamoid bones.** The small bones at the rear of the fetlock joint.

**shafts.** Two parallel poles used for steering and braking, between which a single equine is attached to a vehicle or implement.

**share.** The replaceable cutting edge of a plow bottom.

**Shire.** One of the largest heavy horse breeds, originating from England.

**sickle-bar mower.** A mowing machine with a cutter bar that has a knife with sections for cutting grasses, legumes, and other small forbs.

**sidebone.** Hardening of the cartilages of the coffin bone.

**side delivery rake.** See *reel rake*.

**single.** A horse or mule hitched alone.

**singletree.** A bar to which the harness traces attach so a harnessed animal can pull a vehicle or implement. See also *doubletree*.

**six up.** Three teams hitched one in front of the other.

**skidding logs.** To move logs by pulling them along the ground; also called snaking or snigging.

**sleigh.** A vehicle with runners for sliding instead of wheels for rolling; used for transport on low-friction surfaces such as snow, ice, and sometimes grass.

**smegma.** A thick, foul-smelling substance formed when oily secretions from the skin accumulate under the foreskin of the penis.

**sorrel.** See *chestnut*.

**splay footed.** See *toed out*.

**spike-tooth harrow.** A harrow with sharp metal spikes that can rip up or smooth out soil.

**splint.** A bony swelling on the cannon or splint bones rarely seen in older animals but more common in a horse or mule less than three years of age as a result of overwork.

**splint bones.** The two small bones on the sides of the cannon bone.

**Spotted draft horse.** A heavy horse, typically of part Percheron lineage, with a spotted coat pattern.

**spring-tooth harrow.** A harrow with flexible spikes formed in a C-curve, typically used to smooth the soil.

**spring wagon.** A vehicle with seats in front and a bed in the back, pulled by one or two animals.

**stallion.** An intact male horse; also called a stud.

**stallion mule.** An intact male mule; also called a stud mule.

**started.** See *green broke*.

**startle posture.** Alert position with head up, tail slightly elevated, and muscles tensed.

**stay apparatus.** A system of muscles, ligaments, and tendons that allow an equine to remain standing during light sleep.

**straight behind.** See *post legged*.

**stifle.** The joint between the hip and hock, sometimes incorrectly called the knee.

**Suffolk.** A stocky horse bred for agricultural work, originating from England.

**sulky.** A light two-wheel cart.

**sulky plow.** A riding plow with one bottom.

**sulky rake.** See *dump rake*.

**suprascapular nerve.** The nerve that runs over the front of the shoulder blade.

**surrey.** A light carriage with two forward-facing seats.

**swamp fever.** See *equine infectious anemia*.

**swath.** In a hay field, the width cut in one pass with a mower.

**swayback.** A conformational fault in which the back appears concave from the wither to the loin when viewed from the side.

**sweeney.** Atrophy of the shoulder muscle and damage to the suprascrapular nerve resulting from an improperly fitting collar.

**tandem.** Two horses or mules hitched one in front of the other.

**team (pair).** Two horses or mules hitched side by side.

**teamster.** A person who drives horses or mules.

**ted.** To turn hay over to ensure proper drying.

**temporal fossa.** The hollow above the equine eye.

**temporary teeth.** See *milk teeth*.

**tetanus (lockjaw).** A condition caused by the bacteria *Clostridium tetani* producing painful, rigid muscle contractions, especially in the neck and jaw.

**threshing bee.** A community event at which teamsters use horse-powered steam engines and other old-style equipment to harvest grain.

**throatlatch.** The point at which the head and neck meet; the strap on a bridle that goes under the throat to secure the bridle.

**thrush.** An often painful bacterial infection resulting from poor bedding conditions or persistently wet soil conditions.

**thumps.** A hiccup seen in the flank; a colloquial synonym for laminitis.

**tie stall.** A stall in which an animal stands tied; typically not large enough for the animal to lie down in.

**tobiano.** A coat color characterized by regular patches of color, with white over the back.

**toed in (pigeon footed).** A conformational fault in which one or both toes point slightly inward.

**toed out (splay footed).** A conformational fault in which one or both toes point outward.

**tongue.** See *pole*.

**tongue truck.** Two wheels under the tongue, near the position of the evener, used to reduce the tongue weight of a sickle bar mower or other implement.

**trotting in hand.** Trotting consistently on a lead in a straight line with the handler on the ground.

**tush.** See *canine tooth*.

**tying up.** A form of metabolic muscle stiffness generally caused by irregularity in feed and work schedules.

**unicorn.** Two horses hitched side by side with a third horse hitched in front of them.

**vice.** A fault, bad habit, or other negative, often learned, behavior, such as wood chewing or stall kicking.

**vis-à-vis.** Vehicle in which the passengers sit facing each other.

**voice driving.** Guiding the movement of equines solely with voice commands.

**wagon drive.** A community event in which teamsters and their teams gather to drive a specific trail or road for a purpose, such as historical reenactment.

**warmblood.** The result of crossing a heavy horse (coldblood) with a Thoroughbred or Arabian (hotblood), or any horse that is not 100 percent coldblood or hotblood.

**Western-style harness.** See *box breeching harness*.

**windrow.** Hay or straw that has been raked into a long, narrow row; to rake hay or straw into rows.

**wind sucking.** A vice, commonly associated with cribbing, in which an equine bites an object, flexes his neck, pulls back with his teeth, and gulps in air.

**wither.** The highest point of an equine's shoulder.

**wolf teeth.** Small premolars that may erupt in the upper jaw in front of the cheek teeth.

**work harness.** A harness designed for equine pulling or holding back heavy loads, particularly in agricultural or transportation capacities.

**Yankee breeching harness.** A work harness similar to a box breeching harness, except that the breeching goes diagonally from the stifle to the point of the croup.

**yearling.** A horse or mule of either sex that is between one and two years of age.

# Resources

## BOOKS

Attar, Cynthia. *The Mule Companion: Celebrating the Mule.* Martinsville, IN: Airleaf Publishing, 2005. A guide to understanding basic training techniques and other things that sometimes baffle humans who work with mules.

Bowers, Steve and Marlen Steward. *Farming with Horses.* St. Paul, MN: MBI Publishing, 2006. Not a study in farming, but excellent, well-illustrated advice on harnessing, harness fitting, driving techniques, and hitching.

Brown, Christopher M. and Joseph J. Bertone. *The 5-Minute Veterinary Consult: Equine.* Baltimore, MD: Lippincott Williams and Wilkins, 2002. Written by veterinary clinicians, this technical book contains information on clinical signs, treatment, care, and prognosis for almost any disease affecting equines.

Burger, Sandra. *Horse Owner's Field Guide to Toxic Plants.* In collaboration with editors of Breakthrough Publications and Anthony P. Knight. Emmaus, PA: Breakthrough Publications, 1996. Although not illustrated well enough to be an identification book, this volume does provide good information on plants that could be harmful to your horse or mule.

Butler, Doug and Jacob. *Principles of Horseshoeing III.* Crawford, NE: Butler Publishing, 2004. This careful, thorough study of horseshoeing used by most farrier schools provides a wealth of information on shoeing and trimming the equine hoof.

Butler, Doug and Frank Gravlee. *Laminitis & Founder.* Cherokee, AL: Life Data Labs, 2007. Clear explanations and effective treatments for laminitis and founder written for hoof health care professionals, but of importance to all horse owners.

Cater, Don G. *Just How Old Is Your Horse,* 4th edition. Clarksville, AR: Tangleshoe, 2006. This home-produced book explains and illustrates how old-timers examined a horse's teeth to estimate his age, divided into 26 time periods ranging from one-to-three months up to 30 years.

Clay, Jackie. *Build the Right Fencing for Horses: A Storey Country Wisdom Bulletin, A-193.* North Adams, MA: Storey Publishing, 1999. Introduction to horse-safe fences and selecting the most suitable fence for a large or small pasture, corral, paddock, or training pen.

Cuffey, Robyn and Jaye-Allison Winkel. *The Essential Guide to Carriage Driving.* Buxton, ME: Trot-Online Publishing, 2003. This practical introduction to carriage driving covers all aspects from recreation and parades to competition.

Dutson, Judith. *Storey's Illustrated Guide to 96 Horse Breeds of North America.* North Adams, MA: Storey Publishing, 2005. This book details history, use, numbers, and many other helpful facts about the common and not-so-common breeds of horse found in North America.

Ehringer, Gavin. *Roofs and Rails: How to Plan and Build Your Ideal Horse Facility.* Colorado Springs, CO: Western Horseman, 2002. This detailed guide to planning and building horse housing and fencing does not specifically address draft horses, but comes closer than any other volume and contains lots of practical information that's easily modified to accommodate drafts.

Equine Research, Inc. *Equine Photos & Drawings for Conformation & Anatomy.* Tyler, TX: Equine Research, Inc., 1999. Everything you have ever wanted to know about equine anatomy and conformation lies between the covers of this carefully illustrated book, with attention to draft horse breeds where special note is called for.

Foreyt, William J. *Veterinary Parasitology: Reference Manual,* 5th edition. Ames, IA: Blackwell Publishing, 2001. This academic text geared for the veterinarian provides in-depth information on parasites and their hosts, life cycles, and treatment.

Gaton, Doris. *Breaking and Training the Driving Horse.* Hollywood, CA: Wilshire Book Company, 1984. This introduction to driving covers harness, bits and bitting, starting a green colt, working in poles, types of vehicle and harness, and hitching a cart or buggy; companion book to the DVD by the same name.

German National Equestrian Federation. *Principles of Driving.* Boonsboro, MD: Half Halt Press, 2002. An official instruction handbook explaining Germany's famous Achenbach system of training horses and drivers for competitive or leisure sport.

Heymering, Henry. *Hoof Care for Horses: A Storey Country Wisdom Bulletin, A-277.* North Adams, MA: Storey Publishing, 2001. An introduction to reasons for shoeing and how to keep a horse's feet healthy.

Hill, Cherry. *How to Think Like a Horse.* North Adams, MA: Storey Publishing, 2006. This book explains why horses do some of the things they do and helps the equine owner understand how to establish communication.

Hodges, Meredith. *A Guide to Raising and Showing Mules.* Fort Collins, CO: Lucky Three Ranch, 2003. Colorfully illustrated and worded, this book provides information mainly geared for the saddle and show mule, but the training techniques and general approach to mulemanship apply to any mule.

———. *Training Mules and Donkeys: a Logical Approach to Longears.* Crawford, CO: Alpine Publications, 1993. Directed more at the saddle and show mule than the draft mule, this book helps the muleteer understand how to approach the mule for a successful training experience.

Hutchins, Betsy and Paul, revised and edited by Leah Patton. *The Definitive Donkey: A Textbook of the Modern Ass.* Lewisville, TX: American Donkey and Mule Society, 1981, revised 1999. Although too brief in places, this book includes helpful information any mule lover will enjoy reading, from history to uses.

Jackson, Jaime. *The Natural Horse.* Fayetteville, AR: Star Ridge Publishing, 1997. This book, though not specifically about draft horses and mules, explains the social structure of the feral horse, thereby helping any equine owner understand many of the habits that are so ingrained in equines.

James, Ruth B. *How to Be Your Own Veterinarian (sometimes).* Crawford, CO: Alpine Press, 2nd ed., 2007. You will refer to this book again and again for information on everything from colic to lameness, presented in easily understood yet detailed and medically sound terms.

Kreling, Kai. *Horses' Teeth and Their Problems*. Guilford, CT: Lyons Press, 2005. Written by a German veterinarian, this thorough and extremely well-illustrated volume explains in plain English what every horse owner should know about the construction, function, and diseases of a horse's teeth, including mouth problems that result in behavioral issues.

Marten, Marty. *Problem Solving: Preventing and Solving Common Horse Problems*. Colorado Springs, CO: Western Horseman Magazine, 2003. While geared for the riding horse, this book offers the equine owner creative, thorough, and helpful solutions for such topics as trailer loading, hard-to-catch horses, barn-sour horses, herd-bound horses, and other common problems.

Miller, Lynn R. *Training Workhorses, Training Teamsters*, Sisters, OR: Small Farmer's Journal, 1994. This book offers philosophical and practical advice for the teamster just beginning to work with draft horses.

———. *Work Horse Handbook, 2nd edition*. Sisters, OR: Small Farmer's Journal, 2003. This complete and well-illustrated guide to horse power shows how to hitch singles, teams, and multiples and discusses the principles of driving for work, the dynamics of draft, and hitching up to various implements.

Mischka, Robert A. *It's Showtime! A Beginner's Guide to Showing Draft Horses*. Whitewater, WI: Heart Prairie Press, 1998. A beginner's guide to show preparation, from shoeing, trimming and bathing, braiding and decorating, to showing in halter or harness and what the judges look for.

Moore, Sam. *Implements for Farming with Horses & Mules*. Cedar Rapids, IA: Rural Heritage Magazine, 2006. If you own any horse-drawn equipment, if you plan to buy horse-drawn equipment, if you are just dreaming of getting horse-drawn equipment, you must have this informative and valuable resource.

Ramey, Pete. *Making Natural Hoof Care Work for You*. Fayetteville, AR: Star Ridge Publishing, 2003. Written in down-to-earth, plain language with loads of illustrations, this book documents unsound horses, including draft and carriage horses, that have been made sound through natural trimming.

Rooney, James R. *The Lame Horse*, 2nd edition. Neenah, WI: Russell Meerdink Company, 1998. Plain language explanations of the mechanics of a horse's legs and the causes, prevention, and treatment of every kind of lameness a horse is likely to encounter.

Rusbuldt, Anke. *First Aid for My Horse*. Guilford, CT: Lyons Press, 2005. Step-by-step and often entertaining instructions on how to prevent injuries, accidents, and emergencies, but should one occur, how to assess the situation and react carefully and skillfully.

Schwabe, Alison. *Your Horse's Teeth (Allen Photographic Guide)*, London: J.A. Allen, 2000. A photographic guide to examining a horse's mouth, including details of what normal teeth should look like, how to determine a horse's age, how to identify various problems, and what to do about them.

Valentine, Beth A. and Michael J. Wildenstein. *Draft Horses: an Owner's Manual*. Cedar Rapids, IA: Rural Heritage Magazine, 2000. Cowritten by a veterinary pathologist and Cornell's resident farrier, this volume fills a significant void in maintenance health care information for the draft horse owner.

# DVDS AND VIDEOS

Ainsworth, Brandt. *Driving Draft Horses*. Rural Heritage Magazine, 2005. This DVD demonstrates how to attach and correctly handle driving lines, how to drive with one hand or both, the use of voice commands in conjunction with the lines, driving a single horse or a team, and ground driving.

———. *Harnessing Work Horses*. Rural Heritage Magazine, 2005. A demonstration of variations in work collars and pads, bridles and bits, harness parts and fit, driving lines, eveners and neck yokes, and a wide array of harness styles and their uses.

———. *Logging with Horses, Oxen and Mules*. Rural Heritage Magazine, 2004. A professional horse logger discusses safety gear, proper felling techniques, logging harness, log carts, ground skidding, scaling, and decking saw logs.

Bowers, Steve. *Harnessing, Hitching, and Driving the Draft and Driving Horse*. Bowers Farm, 2006. An in-depth look at harnessing for various disciplines and the dynamics of hitching a draft team or a single horse for pleasure driving.

Cuffey, Robyn. *Horse & Cart*. Rural Heritage Magazine, 2006. This DVD companion to the book *The Essential Guide to Carriage Driving* offers helpful tips on selecting affordable and safe combinations of beginner's cart and lightweight breast-collar pleasure driving harness for a single horse.

Edwards, Steve. *Communicating with Mules*. Queen Valley Mule Ranch, 2006. How to communicate through the bit, handle driving lines, initiate halter work, train a mule to pick up its feet on demand, and determine when your mule is, or is not, in a trainable mood.

———. *Foundations of Driving: Part 1, Mule Training*. Rural Heritage, 2004. A clear explanation of harness parts, how to harness a mule, the proper way to handle driving lines, and how to hitch up a mule team to a wagon for safe driving.

Gaton, Doris. *Breaking and Training the Driving Horse*. Wilshire Book Company, 1988. This DVD companion to the book with the same title offers a step-by-step introduction to training the novice driver and young horse, including harnessing with a breast collar and driving bridle, practicing handling the lines, ground driving, and driving a single horse to a cart.

Hammill, Doug. *Fundamentals Vols. I–IV: Gentle, Effective Techniques for Driving and Working Horses in Harness*. Doc Hammill's Horsemanship Workshops, 2003-2004. This instructional series starts out with the basics and gradually introduces more advanced, gentle but effective techniques for training, hitching, driving, and working equines in harness.

———. *Teaching Horses to Drive: A Ten Step Method*. Doc Hammill's Horsemanship Workshops, 2006. This ten-step method divides lessons into small, easy steps that allow the equine to master one step at a time to produce a calm, comfortable, willing, and reliable driving horse or mule.

Wildenstein, Michael. *Fundamentals of Draft Horse Shoeing*. Rural Heritage Magazine, 2006. Cornell's resident farrier shows how to care for the work horse hoof, with emphasis on trimming the hoof and shaping the shoe, how to pull a shoe, and what to do if your horse throws a shoe.

Zahm, Cathy. *Play to Win Series: Grooming Tips for Show and Sale*. Cathy Zahm Training, 1996. A professional draft horse groomer shows how to: trim the mane, tail, and legs; tie the tail; braid or roll the mane; groom the face and ears; and clip the coat.

## PERIODICALS

*Draft Horse Connection*
Addison, Ontario
613-924-9354
www.drafthorseconnection.ca
Quarterly, with emphasis on preserving the Canadian tradition of horse farming

*Draft Horse Journal*
Waverly, Iowa
319-352-4046
www.drafthorsejournal.net
Quarterly, focusing on show hitches, breeds, and bloodlines

*Rural Heritage*
Cedar Rapids, Iowa
931-268-0655
www.ruralheritage.com
Bimonthly, with emphasis on horses, mules and other draft animals harnessed for work

*Small Farmer's Journal*
Sisters, Oregon
800-876-2893
www.smallfarmersjournal.com
Quarterly, geared toward independent animal-powered family farms

## BREED ASSOCIATIONS

The best way to find current contact information for the following organizations is via their Web sites. In addition to these breed associations, many states have local draft organizations.

**American Brabant Association**
Liverpool, Texas
832-654-3537
www.theamericanbrabantassociation.com

**American Cream Draft Horse Association**
802-447-7612
Bennington, Vermont
www.acdha.org

**American Donkey and Mule Association**
Lewisville, Texas
972-219-0781
www.lovelongears.com

**American Draft Pony Association and Registry**
Fredericktown, Ohio
740-694-7913
www.americandraftpony.org

**American Haflinger Registry**
Akron, Ohio
330-784-0000
www.haflingerhorse.com

**American Shire Horse Association**
Effingham, South Carolina
843-629-0072
www.shirehorse.org

**American Suffolk Horse Association**
Ledbetter, Texas
979-249-5795
www.suffolkpunch.com

**Belgian Draft Horse Corporation of America**
Wabash, Indiana
260-563-3205
www.belgiancorp.com

**Clydesdale Breeders of the U.S.A.**
Pecatonica, Illinois
815-247-8780
www.clydesusa.com

**North American Spotted Draft Horse Association**
Goshen, Indiana
574-825-1924
www.nasdha.net

**North American Spotted Haflinger Association**
Cuyahoga Falls, Ohio
330-620-5358
www.nash2006.com

**Norwegian Fjord Horse Registry**
Webster, New York
585-872-4114
www.nfhr.com

**Percheron Horse Association of America**
Fredericktown, Ohio
740-694-3602
www.percheronhorse.org

**Pinto Draft Registry**
Estancia, New Mexico
505-384-1000
www.pinto-draft-registry.com

## ONLINE RESOURCES

Using an online search engine (such as www.google.com) you can find information on just about anything to do with draft horses and mules. The Internet is an extremely fluid medium, with new sites constantly being added and old sites disappearing or moving. Here are just a few of the many helpful sites available at the time this book was published.

**American Association of Equine Practitioners**
www.aaep.org
Guidelines for equine vaccinations

**Axwood Farm Library**
www.axwoodfarm.com/Library.php
Articles on longlining, hitching, driving, and evaluating an equine's movement

**Draft Dreams**
www.angelfire.com/ab/draftdreams
An international resource covering the broad spectrum of heavy horses and their uses

**Drafts For Sale**
www.draftsforsale.com
Classified ads for draft horses, draft crosses, and mules for sale

**Equine Law and Horsemanship Safety**
http://asci.uvm.edu/equine/law
Equine liability laws

**Horsepullresults.com**
www.horsepullresults.com
Horse pulling schedules, results, clubs, and puller profiles

*Rural Heritage*
www.ruralheritage.com
Devoted to equines worked in harness, this site offers apprenticeships, a calendar of events, teams and equipment for sale, forums, and loads of information

# Index

Page numbers in *italics* indicate illustrations or photographs; those in **bold** indicate tables.

## A

ability of animal, finding the ideal equine, 30, *30*
*Acer rubrum,* **145**
Achenbach grip for lines, 206, *206*
*Acroptilon repens* syn. *Centaurea repens,* **145**
actions and consequences, 104
advantages of drafts, 8–9, *8–15,* 11–13
age of equines
    finding the ideal equine, 4, 30, 37–38, *38,* 40, *40*
    housing and, 112
*Ageratina altissima* syn. *Eupatorium rugosum,* **146**
agricultural self-sufficiency, 4, 8–10, *9*
alertness of equine, 35
alfalfa hay, 135, 136, 148, 236
Allen, Cathie (Williams Lake, British Columbia), 24–25, *25*
American Association of Equine Practitioners, 163, 176
American Cream Drafts, 54, 64, 66, *69,* 69–70, 75, 216
American Livestock Breeds Conservancy, 69
Amish of Ohio, Iowa, and Indiana, 53
*Amsinckia* sp., **142, 143**
anesthesia, 81
animal-human relationship, 14–15, 16, *16,* 17, 18
apprenticeships, 21, 23, *23,* 24–26
approaching an equine, 100
Arbogast, Sam and Susan (Hillsboro, West Virginia), 129, *129*
ascarids (roundworms), 161, **162**
aspartate aminotransferase (AST), 184
attachment to owner, mule, 84, 91
attack pose, mule, 92, *92*
auctions for finding equines, 53
azoturia (EPSM), 41, 77, 156, 160, 183–84

## B

back at the knee (calf knee), 47
"Back" command, **205**
back conformation, *40,* 49, *49,* 76
backing harness components, 190, *190*
barbed wire fencing, 122
barefoot vs. shoeing, 166, 168
barns, 111, *111–12,* 112–18, *114–15, 117,* 121
barrel conformation, *40,* 49
bars of hoof, 52, *52*
base narrow/base wide, 46, *46,* 50, *51*
beans, 171
beauty considerations, 30, *30,* 31
beaver slide, 239, *239,* 240
bedding, 117
beet pulp, **139**
Belgian Horse Corporation of America, 66
Belgian mules, 39
Belgians, 6, 9, *9,* 30, 31, 39, 54, 58, *58,* 61, *61–62,* 62, 64, 65, *65,* 67, 71, 76, **76,** 79, 111, 129, 192, *192,* 216, 221, *221*
binocular vs. monocular vision, 94–95
birth and reproduction, mule, 81, 85
biting by equines, 97
black locust *(Robinia* spp.), **142**
black Percherons, 67
black urine, 156
black walnut *(Juglans nigra),* 117, **142,** 180
black water (EPSM), 41, 77, 156, 160, 183–84
blanketing, 110
blinders (blinkers), 94–95, 189
blind spots, 43, *43,* 94, 100
blister beetles *(Epicauta pennsylvanica, E. malculata, E. immaculata, E. lemniscata),* 136, *136,* 182
BLM (Bureau of Land Management), 92
blowing (vocalization), 101–2
board and block to counterbalance weight of mower, *236,* 238
bobsled, 225, 227
body condition of equine, 37, 40, *40*
body condition score, **132–33,** 132–35, *134–35*
body conformation, *40,* 48–49, *49,* 76

body language of equines, 93–101. *See also* communication (equine)
    ears, 95–96, *95–96,* 104
    eyes, 94–95, 102, 104
    feet, 100
    finding the ideal equine and, 28, 35, 36
    head, 90, 92, *92,* 94, 104
    laminitis, 100
    learning team's, 6, 28, 35, 36, 89, 90, 93–101
    legs, 100–101, *101*
    mouth (lips), 96–97, *97,* 102, 104
    mules learning human's body language, 17, 29
    neck, 90, 92, *92,* 94
    nostrils, 97–98, *98,* 104
    tail, 98–99, *99,* 104
body position for driving, 205, 215
bog spavin, 177, *177*
bolted-on tongue end, 222, *222*
bone spavin, 177, *177*
borium on shoes, 168, *168*
*Borrelia burgdorferi,* 164
Borsato, Rob (Williams Lake, British Columbia), 24–25, *25*
bot flies, 161, **162**
bottom of plow, 229, *230*
bow legs (out at the knee), 47, *47,* 51, *51*
Bowling, Ann T., 70
box breeching or western-style work harness, 199–200, *200,* 201
box stalls, 111, 112, *112,* 115, 116–17, *117*
box walking (vice), 125
Brabant Belgians, 58, *58,* 65, *65,* 66, 72, 72–73
brackenfern *(Pteridium acquilinum),* **142**
breast-collar harness, 189, *190,* 197–98, *198*
breeching placement, 194, *194*
breed associations for finding equines, 53
breeding equines, 33, 34, 54, 75, 77
bridle, 189, *189–90,* 193
brown urine, 155–56
brushes (grooming), 171

252

bucked knee (forward at the knee), 47
buck rake, 238, 239, *239*
buckthorn *(Amsinckia* sp.), **142, 143**
Buckwheat Blossom Farm (Wiscasset, Maine), 9
bulbs of heel, 52
Bureau of Land Management (BLM), 92

# C

calcium and phosphorus, 147–48, 177
calf knee (back at the knee), 47
calories needed daily, 130–31
camped under, *46*, 47, 50, *51*
camping and packing, *13*, 13–14, 18, 60, *60*, 227–29, *228*
Canadian Belgian Horse Association, 66
canine teeth (tushes), 37–38
canker (necrotic pododermatitis), 117, 179
cannon conformation, *40*, *45*, 47, *48*
capillary refill time (CRT), 155, 174
carriage business, 4, 8, 10, 18, 56, *56*, 225
carts, 224
castor bean *(Ricinus communis),* **142**
catching equine, ease of, 35
Cecil, Pete (Bend, Oregon), 111, *111*
cecum, 130, *130*
*Centaurea solstitialis,* **146**
center of gravity of equine and harnessing, 194
*Cestrum diurum,* 147
challenges of drafts, 15–17, *16*
champagne gene, 70
cherry *(Prunus* spp.), **143**
chest conformation, *40*, 44, 76
chestnuts, 82, *82*
chestnut (sorrel), 65, 69, 72
"chewing disease," **145, 146**
chewing surfaces, teeth, 37, 38, *38*, 152
children and drafts, 16, *16*, 56, *56*
chronic progressive lymphadema (CPL), 66, 73
CK (creatine kinase), 184
cleats on shoes, 168, *168*
clevis, 220, *222*, 231
climate and shelter needed, 110
clinics, 21, *21*, 23, *26*, 26–27
*Clostridium botulinum,* 136
Clydesdales, 54, 57, *57*, 64, 66, 67, *67*, 68, **76**
coat genetics, 70
cocklebur *(Xanthium strumarium),* **143**
Coggins test, 159, 185
cold-blooded vs. hot-blooded equines, 90
cold weather feeding, 140, *140*
colic, 126, 136, 137, 138, 140, 150, 161, 181–83
colitis, 182
collar, fitting the, 194–96, *195–96*

collar sores, 174–75, 194
color of hay (green vs. brown), 136
comfort style grip for lines, 206, *206*
commitment (your), 2–3
communication (equine), 88–106. *See also* body language of equines
    actions and consequences, 104
    blind spots, 43, *43*, 94, 100
    consistency importance, 105–6, *106*
    dominant equines, 92–93, *93*, 96, 97
    feral horses, 92–93
    flight instinct, 4, 10, 83, 90
    hearing of equines, 95–96, *95–96*
    herd social structure, 92–93, *93*, 105, 120, *120*
    herd structure, 92–93, *93*, 105, 120, *120*
    learning basics, 28, *28*
    learning basics of, 28, *28*
    learning by equines, 104–6, *105–6*
    listening, watching, acting equines, 102, *102*, 104
    memories of equines, 104, 105, 106
    monocular vs. binocular vision, 94–95
    pecking order in herd, 92–93, *93*, 120
    pressure and release concept, 104–5, *105–6*, 106
    prey animals, 90
    repetition, importance of, 105–6, *106*
    rewarding behavior, 102, 104
    sensitivity of equine, 35, 90
    smell sense of equines, 97–98, *98*
    startle posture, 90, 92, *92*
    submissive equines, 92–93, *93*
    thinking like an equine, *89*, 89–90, 92–93, *92–93*
    threat assessment, 90, 92, *92*, 94, 95, 97
    vision of equines, 43, *43*, 94–95
    vocalizations, 80, 93, 96, 101–2
communication harness components, 189, *189–90*
communication skills of mentor, 22
community-supported agriculture (CSA), 9, 24–25
companion animals for drafts, 8, *8*, 10, 31, 84, 120, *120*, 120–21
companionship from drafts, 8, 14–15, 16, *16*, 17
concentrated feeds, 130, 133, 137–40, **139**, 147, 177
conformation of ideal equine, 40–53. *See also* finding the ideal equine
    body, *40*, 48–49, *49*, 76
    chest, *40*, 44, 76
    feet (hooves), *40*, *45*, 48, 52–53, *52–53*, 76

finding the ideal equine and, 30, *30*, 37, 40, *40*
    forelegs, *40*, 44–48, *45–48*
    head, *40*, 43, *43*, 76
    height, measuring, 45
    hind legs, *40*, 50–51, *50–51*, 76
    judging, 76, **76**
    lameness, 41, 42, *42*
    movement symmetry, 41
    muscling, 41
    neck, *40*, 43–44, *44*, 76
    parts of equine body, *40*, *45*, 52
    pastern/hoof angle, 52, 53, *53*
    symmetry of general appearance, 41
    weight-to-bone ratio, 48
*Conium maculatum,* **145**
consistency, importance of, 105–6, *106*
copper, 147, 177
corn harvester, 58, *58*
corn (unprocessed) caution, 140
coronet, *40*, 52
corrective shoeing, 166, 169
coulter of plow, 229, *230*
courting buggy, 224
cow hay, 136
cow hocks, 51, *51*, 177
CPL (chronic progressive lymphadema), 66, 73
crazyweed *(Astragalus* and *Ozytropis* spp.), **144**, 149
Cream Drafts, 54, 64, 66, *69*, 69–70, 75, 216
creatine kinase (CK), 184
creativity for learning, 85, 105, 106
cresty neck, 43, 44, *44*, 181
cribbing (vice), 126
crosses, 12–13, 75, *75*
*Crotalaria* spp., **143**
croup conformation, *40*, 50
CRT (capillary refill time), 155, 174
cryptorchid, 34
CSA (community-supported agriculture), 9, 24–25
cultivation, 233–34, *234*
currycombs, 171
cutter bar of mower, *235*, 236

# D

daily wormers, 162, **162**
Damerow, Gail, ix–x
dam of mule, horse, 80, 81, 82, 83
death camas *(Zigadenus* spp.), **143**
decision, making the big, 5
Decker packsaddle, 227, *228*, 228–29
dedication to learning by mentor, 23
DE (digestible energy), 130, 137, 140
deer tick *(Ixodes scapularis),* 164
defecation status, 153, 156

degenerative joint disease (DJD), 177, *177*
dentistry, 152, 164–66, *165*
depth perception, 94, 95
developmental orthopedic disease (DOD), 177
development (maturation) of mules, 85
dewormers, 159, 161–63, **162,** *162*
deworming home remedy, 173
digestible energy (DE), 130, 137, 140
digestive system sensitivity, 128, 130, *130*, 135, 136
digging (vice), 125
digital pulse, 154, *154*
dirt eating (vice), 126
disc harrows, 62, *62*, 231, *232*
discipline and mules, 86
DJD (degenerative joint disease), 177, *177*
docking tails, 98
doctor's buggy, 224
DOD (developmental orthopedic disease), 177
dominant equines, 92–93, *93*, 96, 97
donkey sire of mules, 80, 81, 82, 83
double rigged packsaddles, 228, *228*
double-rumped, 50
dozing, 118, *118*
Draft Club, 27
*Draft Horse Journal,* 53
*Draft Horses, an Owner's Manual* (Valentine), 160
draft horses and mules, ix–x. *See also* body language of equines; communication (equine); driving equines; feeding equines; finding the ideal equine; first aid and illness; harnessing equines; health care; horses (draft); housing equines; learning the basics; mules (draft); owning drafts; working with draft power
dream of ownership, realizing, 17–18, *18*
driving equines, 202–6, 214–17. *See also* harnessing equines
    body position, 205, 215
    driving lines and, 202, 203, *203*, 204, 205–6, *206*, 215, *215*, 217, *217*
    hitching configurations, 203, *203*, 214, *214*
    holding lines, 205–6, *206*
    practice exercises without a team, 215, *215*, 217, *217*
    practicing, learning by, 202, 204
    single vs. team driving, 203, *203*
    starting with lines only, 205
    stopping with lines only, 205
    tension on lines, 202, 203, 204, 205, 215, *215*, 217, *217*
    turning, learning, 217, *217*
    turning with lines only, 205
    voice commands, 202, 203, 204–5, **205**
driving lines
    driving with, 202, 203, *203*, 204, 205–6, *206*, 215, *215*, 217, *217*
    harnesses and, 189, *189–90*, 193
dump rake, 238, *238*
dun gene, 70, 74
dynamometer, 12

## E

ears of equines
    body language, 95–96, *95–96*, 104
    temperature, 154
eastern equine encephalitis (EEE), 157, 158
"Easy" command, **205**
edema, 47
EEE (eastern equine encephalitis), 157, 158
EHV (equine herpes virus), 157, 159
EIA (equine infectious anemia), 159, 185
elbow conformation, *40*, *45–46*, 46
electric fencing, 122, 123, *123*, 124
electrolytes, 149–50
emergencies
    first aid and illness, 186
    horses (draft) vs. mules (draft), 3, 10
encephalomyelitis, 157, 158
energy requirements and feeding, 128, *128*, 138, **139**
England, Gene (Winder, Georgia), 76, *76*
English grip for lines, 206, *206*
enteritis, 182
enterolith, 137
*Epicauta pennsylvanica, E. malculata, E. immaculata, E. lemniscata* (blister beetles), 136, *136*, 182
epidydimis, 34
EPM (equine protozoal myeloencephalitis), 113
equine ehrlichiosis, 164
equine herpes virus (EHV), 157, 159
equine infectious anemia (EIA), 159, 185
equine polysaccharide storage myopathy (EPSM), 41, 77, 156, 160, 183–84
equine protozoal myeloencephalitis (EPM), 113
equipment components, 220, 222–23, *222–23*
*Equisetum* spp., **143**
eruption patterns of teeth, 37
estrous cycle, 34, 81, 99, *99*
evener, 222, *222*, 231
*Evener Directory (Rural Heritage* directory), 27
ewe neck, 44, *44*
examples of drafts, 55–62, *55–62*
exercise, importance of, 152, *152*

experienced equines for novices, 4, 6–8, *7*, 9, *9*, 14, 20, 31, 32, 39
experienced teamster or muleteer, 20, 31, 35, 40
Extension Service, 141
external parasites, 163–64
eyes of equines
    body language, 94–95, 102, 104
    conformation, *40*, 43, *43*, 76

## F

false rig (proud cut) geldings, 34
farming with drafts, 4, 8–10, *9*, 18, *18*, 229–34, *230*, *232–33*
farm wagon, 225
farriers. *See also* feet of equines
    barefoot vs. shoeing, 166, 168
    corrective shoeing, 166, 169
    manners of equine and, 35
    pastern/shoulder angle, 48
    pulling, shoeing for, 168
    show, shoeing for, 169, *169*
    traction, shoeing for, 168, *168*
    trimming and shoeing, 166–69, *166–69*
fats and fatty acids, 130, 137
feathering, 47, 65, 67, 68, 72
feeding equines, 127–50
    body condition score, **132–33,** *132–35*, *134–35*
    calcium and phosphorus, 147–48, 177
    calories needed daily, 130–31
    cold weather requirements, 140, *140*
    colic and, 136, 137, 138, 140, 150
    concentrated feeds, 130, 133, 137–40, **139**, 147, 177
    corn (unprocessed) caution, 140
    digestible energy (DE), 130, 137, 140
    digestive system sensitivity, 128, 130, *130*, 135, 136
    electrolytes, 149–50
    energy requirements and, 128, *128*, 138, **139**
    fats and fatty acids, 130, 137
    feeders, 111, *112*, 116–17, *117*
    grain, 130, 133, 138, **139**, 140, 148
    hay, 130, 131, *131*, 135–36, *136*, 148
    laminitis and, 137, 138, 140, 154, 179, 180, 181, 182
    minerals, 147–50, 177
    moldy feed caution, 113, 114, *114*, 136, 183
    pasture forage, 119, 130, 131, *131*, 136, 141, **142–46,** 148
    questions to ask yourself, 137
    requirements for drafts, 16, 32
    sandy soil (feeding on) caution, 137
    selenium, *148*, 148–49

storage of feed, 110, *112*, 113–14, *114*, *181*
supplements, 147–50, *148*
toxic plants, 141, **142–46**
trace-mineral salt block, 149–50
vitamins, 147, 163
water, 150, *150*
feet of equines. *See also* farriers
   body language, 100
   canker (necrotic pododermatitis), 117, 179
   conformation, 40, 45, 48, 52–53, *52–53*, 76
   digital pulse, 154, *154*
   health care, 166–69, *166–69*
   hoof cracks (quarter cracks), 53, 178, *178*
   lameness, 178–79
   mules, 167, *167*
   parts of hoof, *52*
   sidebone, 178
   sole bruises, abscesses, 52, 178–79
   temperature of hoof, 154, *154*
   thrush, 117, 179
Feltenberger, Dave (New Braunfels, Texas), 39, *39*
Fenbendazole, **162**
fencing, 6–7, 8, 108, *108*, 109, 110, 122–25, *122–25*
feral horses, 92–93
*Festuca elatior* syn. *Festuca arundinacea*, **146**
fetlock conformation, 40, 45, 47, 48
fever (high) and laminitis, 180
fiddleneck *(Amsinckia* sp.), **142**, *143*
financial considerations, 8, 15, 31
finding
   apprenticeship, 26
   mentor, 21
finding the ideal equine, 29–54. *See also* conformation of ideal equine
   age of equine, 4, 30, 37–38, *38*, 40, *40*
   alertness of equine, 35
   beauty considerations, 30, *30*, 31
   body condition, 37, 40, *40*
   body language of equines and, 28, 35, 36
   breeding equines, 33, 34, 54, 75, 77
   conformation of ideal equine and, 30, *30*, 37, 40, *40*
   experienced equines for novices, 4, 6–8, *7*, 9, *9*, 14, 20, 31, 32, 39
   experienced teamster or muleteer, 20, 31, 35, 40
   geldings vs. mares, 33–34
   "ideal," concept of, 30
   mares vs. geldings, 33–34
   needs, evolving with your, 33
   number of equines needed, 31–33, *32*

prepurchase exams, 54
prices, 54
purchasing, 53–54
questions to ask yourself, 31
stallions caution, 33
studdy geldings, avoiding, 34
teammates, working with, 36
teeth as indicator of age, 37–38, *38*
temperament of equine, 30, 35–37, 79, 83–84
test driving, 36
trailering and, 36–37
fireweed *(Amsinckia* sp.), **142**, *143*
first aid and illness, 172–86. *See also* health care; lameness
   colic, 126, 136, 137, 138, 140, 150, 161, 181–83
   collar sores, 174–75, 194
   emergencies, 186
   equine polysaccharide storage myopathy (EPSM), 41, 77, 156, 160, 183–84
   first-aid kit, content of, 174
   home remedies, 173
   shock, signs of, 174
   travel precautions, 185
   veterinarian, working with, 173, *173*
fistula of the withers, 44
fit of harness, 193–96, **194**, *194–96*
Fjord mules, 74, *74*
Fjords, 54, 64, 70, 73–74, *73–74*
flehmen response, 97, *97*, 181
flexible rail fencing, 122, 124, *124*
flexion test, 175, *175*, 176
flight instinct, 4, 10, 83, 90
floating teeth, 165
flooring, 117
Flowers, Tommy (Blackville, South Carolina), 72, *72–73*
forearm conformation, 40, 45–46, *46–47*
forecart, 223, *223*
forelegs conformation, 40, 44–48, *45–48*
fortified feed mix, 138, **139**
forward at the knee (bucked knee), 47
founder, 179
four-horse hitch, 62, *62*, 203
freezers for storing grain, 113
frog of hoof, 52, *52*
full-sweeney collar, 195
furrow wheel of plow, 229, *230*
Future Farmers of America, 221

## G

Galvayne's groove, 38
gang plow, 229
gas colic, 182
gaskin (second thigh) conformation, 40, *51*

gastric distention or rupture, 182
gastrointestinal (GI) tract, 156, 181, 182, 183
"Gee" command, **205**
geldings vs. mares, 33–34
genetics considerations, 41, 66, 71, 77
genitals, care of, 170–71
"Get up" command, **205**
GI (gastrointestinal) tract, 156, 181, 182, 183
glossary, 243–48
goals, setting, 4–5, *5*
goatweed *(Hypericum perforatum)*, **145**
Good Farming Apprenticeship Network, 26
grading lameness, 176
grain
   feeding, 130, 133, 138, **139**, 140, 148
   overload and laminitis, 137, 138, 179, 180, 181, 182
   storage, 110, 112, *112*, 113, 181
grass hay (timothy, orchard), 135, 136, 236
grass laminitis, 180, 181
groaning (vocalization), 102
grooming importance, 169–71
groundsel stinking willie *(Senecio* spp.), **146**
ground skidding, 240
grunting (vocalization), 102
gut sounds, 156

## H

Haflinger mules, 91, *91*
Haflingers, 8, 31, 54, 64, *71*, 71–72, 74, 76, **76**
hairless foal (JEB), 66, 77
half-breed, 228, *228*, 229
half-sweeney collar, 195
haltering equine, ease of, 35
hame-style harness, 189, *190*, 198–99, *199*, 216, *216*
hand-feeding treats caution, 96
handling vaccines yourself, 157
hand (measuring equine height), 45
harnessing equines, 187–202, 207–13. *See also* driving equines
   backing and stopping components, 190, *190*
   blinders (blinkers), 94–95, 189
   breast-collar harness, 189, *190*, 197–98, *198*
   breeching placement, 194, *194*
   bridle, 189, *189–90*, 193
   center of gravity of equine, 194
   collar, fitting the, 194–96, *195–96*
   communication components, 189, *189–90*

harnessing equines *(continued)*
   driving lines, 189, *189–90, 193*
   ease of, finding ideal equine, 35, 36
   fit importance, 193–96, **194,** *194–96*
   full-sweeney collar, 195
   half-sweeney collar, 195
   hame-style harness, 189, *190,* 198–99, *199,* 216, *216*
   hanging harness and collar, 115, *115,* 198, *198*
   how to, 188, *188,* 191, 207–13, *207–13*
   incomplete harnessing, danger, 202
   material types for, 197
   New England D-ring work harness, 199, *201*
   physical fitness (your) and, 16
   pleasure-draft-driving harness, 197–99, *198–99*
   pulling components, 189, *190*
   regular collar, 195
   safety, 6, 197, 199, 202
   show-draft-driving harness, 57, *57,* 197
   sizes (typical) of harnesses, **194**
   team harness components, *193*
   turning components, *190,* 191
   weight of harness, 16
   western-style or box breeching work harness, 199–200, *200,* 201
   work harness, *193,* 199–201, *200–201*
   Yankee breeching work harness, 199, *201*
harrowing, 231–32, *232*
harvesting crops, 234
harvesting hay, *238–39,* 238–40
Hatley, George (Deary, Idaho), 192, *192*
"Haw" command, **205**
hay
   feeding, 130, 131, *131,* 135–36, *136,* 148
   haymaking, 129, *129,* 234–36, 234–40, *238–39*
   moisture content, determining, 237
   storage, 110, 112, *112,* 113–14, *114*
head of equines
   body language, 90, 92, *92,* 94, 104
   conformation, 40, 43, *43,* 76
   lameness and head bobbing, 41, 42, *42*
health care, 151–71. *See also* first aid and illness
   capillary refill time (CRT), 155, 174
   Coggins test, 159, 185
   defecation status, 153, 156
   dewormers, 159, 161–63, **162,** *162*
   digital pulse, 154, *154*
   ear temperature, 154
   encephalomyelitis, 157, 158
   exercise, importance of, 152, *152*
   external parasites, 163–64

   feet (hooves), 166–69, *166–69*
   grooming importance, 169–71
   gut sounds, 156
   hydration, 153, *153*
   importance of, 7, 8, 15
   internal parasites, 161–63, **162**
   manure management, 141, 162, 163
   maxillary pulse, 154, *154*
   mucous membranes, 153, 155, *155*
   normal parameters, determining, 153–56, *153–56*
   parasite control, 159, 161–64, **162,** *162,* 183
   preventive health management, 152–53
   pulse, 153, 154, *154,* 155
   respiration rate, 153, 155
   teeth care (dentistry), 152, 164–66, *165*
   temperature, 153, 154, 155
   TPR (temperature, pulse, respiration), 155
   urination status, 153, 155–56
   vaccinations, 156–59
   weight management, 156, *156,* 181
health certificate, 185
"healthy as a mule," 82–83
hearing of equines, 95–96, *95–96*
hee-haw, 80, 101
heel of hoof, 52, *52*
height, measuring, 45
Henneke, D. R., 132
herd structure, 92–93, *93,* 105, 120, *120*
hindgut digestion, 130
hind legs conformation, 40, 50–51, *50–51,* 76
hindquarters conformation, 40, 50, *50–51*
hinny, 80
hitching configurations, 203, *203,* 214, *214*
hock conformation, 40, 51, *51,* 76, 177
holding driving lines, 205–6, *206*
home remedies, 173
hoof cracks, 53, 178, *178*
hoof picks, 171
hoof testers, 175, *175,* 176
hoof wall, 52, *52*
hooks on teeth, 165
hooves. *See* feet of equines
horn growth (hoof), 52
"horse apples," 156
horse dam of mules, 80, 81, 82, 83
Horse Progress Days, 27–28, 240
horses (draft). *See also* draft horses and mules; *specific breeds*
   breeds of, 63–77
   emergencies and, 3, 10
   genetics considerations, 41, 66, 71, 77

   grade horses, 74–75
   mistakes, forgiving your, 28, 39, 87
   preservation of breeds, 75, 77
horsetail *(Equisetum* spp.), **143**
Horse Team for Sale *(Rural Heritage* directory), 54
housing equines, 107–26
   barns, 111, *111–12,* 112–18, *114–15, 117,* 121
   bedding, 117
   blanketing, 110
   box stalls, 111, 112, *112,* 115, 116–17, *117*
   companion animals for drafts, 8, *8,* 10, 31, 84, *120,* 120–21
   feeders, 111, *112,* 116–17, *117*
   feed storage, 110, *112,* 113–14, *114,* 181
   fencing, 6–7, 8, 108, *108,* 109, 110, 122–25, *122–25*
   flooring, 117
   grain storage, 110, 112, *112,* 113, 181
   harness and collar, hanging, 115, *115,* 198, *198*
   hay storage, 110, 112, *112,* 113–14, *114*
   insect protection, 121
   legalities, 108–10, 226
   opossums caution, 110, 113
   pasture needs, 119–20, 119–21
   questions to ask yourself, 108, 115
   safety issues and liability, 17, 109, *109,* 110
   shelters, 110–19, *111–12, 114–15, 117–19*
   sleep of equines, 118, *118*
   space requirements for drafts, 16
   stalls, 111, 112, *112,* 115–17, *117*
   tack area, 112, *112,* 114–15, *115*
   three-sided shelters, 119, *119*
   tie rings, 111, 116, *117*
   tie stalls, 111, 115, 116, *117*
   turnout, 16, 119–20, *120*
   ventilation in barns, 112–13
   vices, dealing with, 125–26
   waterers, 106, *112,* 116, *117,* 150
   winter coat, 110
   zoning requirements, 108–10
Howe, Liza (Lena, Illinois), 216, *216*
hybrid vigor, 82–83
hydration, 153, *153*
*Hypericum perforatum,* **145**
hypersensitive equines, 103, *103*

Idaho Hall of Fame, 192
"ideal," concept of, 30
illness. *See* first aid and illness
impaction colic, 182
in at the knee (knock knee), 47, *47*

incisor cups, 38
incisors, 37, 38, *38*, 152
incomplete harnessing, danger of, 202
inflammation of small or large intestines, 182
inherent risks of equine activity, 109, *109*
injury and laminitis, 180
insecticide safety, 122
insect protection, 121
intelligence vs. stubbornness, mules, 79, 83–84, 86, 87
internal parasites, 161–63, **162**
Internet for finding learning opportunities, 21, 26, 28
internships, 21, 23, *23*, 24–26
intussusception, 182
iron, 147
*Isocoma* spp., **145**
Ivermectin, **162**
*Ixodes pacificus* (western black-legged tick), 164
*Ixodes scapularis* (deer tick), 164

## J

jack (male donkey), 80
jaw of equine, conformation, *40*, 43, 152
JEB (junctional epidermolysis bullosa), 66, 77
jennet (female donkey), 80
jimmyweed *(Isocoma* spp.), **145**
john mules, 81, 83–84
Johnson grass *(Sorghum* spp.), **146**
jointer of plow, 229
jointfir *(Equisetum* spp.), **143**
judging conformation, 76, **76**
*Juglans nigra*, 117, **142**, 180
junctional epidermolysis bullosa (JEB), 66, 77

## K

kicking (vice), 125
killed vaccines, 157
Klamath weed *(Hypericum perforatum)*, **145**
knee conformation, *40*, *45*, 47, *47*
knock knee (in at the knee), 47, *47*

## L

lab work diagnostics, 176, 177
Lakin, Harry, 69
lameness, 175–81. *See also* first aid and illness
    bog spavin, 177, *177*
    bone spavin, 177, *177*
    canker (necrotic pododermatitis), 117, 179
    degenerative joint disease (DJD), 177, *177*
    developmental orthopedic disease (DOD), 177
    diagnosing lameness, 176–77
    evaluating lameness, *175*, 175–76
    finding the ideal equine and, 41, 42, *42*
    grading lameness, 176
    hoof cracks (quarter cracks), 53, 178, *178*
    hooves and, 178–79
    sidebone, 178
    sole bruises, abscesses, 52, 178–79
    splints, 47, 178
    stocking up, 178
    thrush, 117, 179
    water and laminitis, 181
laminitis, 100, 137, 138, 140, 154, 179–81, 182
land acreage and number of equines needed, *32*, 32–33
*Lantana* spp., **144**
large intestine, 130, *130*
large strongyles, 161, **162**
larkspur, **144**
leader
    equine leader in herd, 92, 93
    mule wanting a, 84, 86, 87, 92
    teamster as, 6, 15, 17, 92, 93
leading manners, 35
"learning before operating," 219–20
learning by
    equines, 104–6, *105–6*
    mules, 84–85, *85*, 86, 87, *87*, 91, 104, 105–6
learning from equines, 2, *2*, 21, 28, 102, *102*
learning team's body language, 6, 28, 35, 36, 89, 90, 93–101
learning the basics, 19–28. *See also* communication (equine)
    apprenticeships (internships) for, 21, 23, *23*, 24–26
    clinics for, 21, *21*, 23, *26*, 26–27
    communication with equine, 28, *28*
    Draft Club for, 27
    equines, learning from, 2, *2*, 21, 28, 102, *102*
    experienced equines for novices, 4, 6–8, *7*, 9, *9*, 14, 20, 31, 32, 39
    experienced teamster or muleteer, 20, 31, 35, 40
    Horse Progress Days for, 27–28
    mentorship opportunities, 20–24, *22*
    mistakes, learning from your, 28, *28*
    public events for, 11, 27–28
    safety for, 20
    student's responsibility, 23
    workshops for, 21, 23, *26*, 26–27
leather harnesses, 197, 216

leaves vs. stems in hay, 136
left side, working from the, 101
legalities, 108–10, 226
legs body language, 100–101, *101*
liability, 17, 109, *109*, 110
lice, 163
light buggy, 224
lips body language, 96–97, *97*, 102, 104
listening, watching, acting equines, 102, *102*, 104
live vaccines, 157
lockjaw (tetanus), 156–57, 158
locoweed *(Astragalus* and *Oxytropis* spp.), **144**, 149
logging, 4, 11–12, 240, 241, *241*
logging forecart, 223, *223*
loin conformation, *40*, 49
long back, 49, *49*
long-distance hauling, 185
Lyme disease, 164

## M

Mader, Dave, 111
magnetic resonance imaging (MRI), 176
Mammoth Jacks, 39, 79, 81
mane, grooming, 170
mane chewing (vice), 126
manganese, 177
mange, 164
manners, equine, 30, 35–37, 79, 83–84
manure eating (vice), 126
manure management, 141, 162, 163
manure spreader, 14, 59, *59*
maple sap wagon, 56, *56*
mares vs. geldings, 33–34
mats for stalls, 117
Mautino, Vince (Colorado Springs, Colorado), 91, *91*
maxillary pulse, 154, *154*
mechanic skills, 220
medications and laminitis, 180
memories of equines, 104, 105, 106
mentorship opportunities, 20–24, *22*
metal pipe fencing, 122, 125, *125*
milk teeth, 37
milkvetch *(Astragalus* and *Oxytropis* spp.), **144**, 149
minerals, 147–50, 177
mistakes
    horses forgiving your, 28, 39, 87
    learning from your, 28, *28*
    mules exploiting your, 28, 37, 39, 84, 86, 87
mites, 164
modified live vaccines, 157
moisture content of hay, 237
molars, 37
moldboard of plow, 229, *230*

moldy feed caution, 113, 114, *114*, 136, 183
molly mules, 81, 83
Monday morning sickness (EPSM), 41, 77, 156, 160, 183–84
monocular vs. binocular vision, 94–95
monogastric digestive system, 130, *130*
motivation from mentor, 23
mouth (lips) body language, 96–97, *97*, 102, 104
movement symmetry, 41
mowers, 222
mowing hay, 235, *235–36*, 236–38
Moxidectin, **162**
MRI (magnetic resonance imaging), 176
mucous membranes, 153, 155, *155*
mules (draft), 78–87. *See also* draft horses and mules; *specific breeds*
   attachment to owner, 84, 91
   attack pose, 92, *92*
   birth and reproduction, 81, 85
   body language of humans, reading by, 17, 28
   connection to mules, 85, 87
   dam, horse, 80, 81, 82, 83
   development (maturation) of, 85
   discipline, 86
   donkey sire, 80, 81, 82, 83
   emergencies and, 3, 10
   examples of, 54, *54, 54–55*, 55, 58, *58*
   feet (hooves), 167, *167*
   "healthy as a mule," 82–83
   hee-haw, 80, 101
   hinny, 80
   horse dam, 80, 81, 82, 83
   hybrid vigor, 82–83
   intelligence vs. stubbornness, 79, 83–84, 86, 87
   jack (male donkey), 80
   jennet (female donkey), 80
   john mules, 81, 83–84
   leader, mule's wanting, 84, 86, 87, 92
   learning by, 84–85, *85*, 86, 87, *87*, 91, 104, 105–6
   mistakes (your), exploiting by, 28, 37, 39, 84, 86, 87
   molly mules, 81, 83
   origin of, 80–81, *81*
   pound-for-pound work, 83, *83*, 84
   registered mules, 80
   schedules and, 85, 86, 87, 100
   self-preservation sense, 82, 83, 92, *92*, 227
   sire, donkey, 80, 81, 82, 83
   United States Army mascot, 54, 87
   walking-on-by-pose, 92, *92*
Mule Team for Sale *(Rural Heritage* directory), 54

muscling and conformation, 41
mushroom feet, 52, 53, *53*
muzzle of equine, conformation, *40*, 43

## N
nasogastric tube, 153
navicular syndrome, 52
neck of equines
   body language, 90, 92, *92*, 94
   conformation, *40*, 43–44, *44*, 76
   grooming, 170
neck yoke, 222
necrotic pododermatitis (canker), 117, 179
needs (your), evolving, 33
Nelson Brothers Farm, 69
*Nerium oleander*, **145**
nerve and joint blocks, 176
New England D-ring work harness, 199, 201
nickering (vocalization), 93, 95, 102
nightshade *(Solanum* spp.), **144**
"no foot, no horse," 52
normal health parameters, determining, 153–56, *153–56*
North American Spotted Draft Horses, 54, 59, *59*, 64, *70*, 70–71
Norwegian Fjords, 54, 64, 70, 73–74, *73–74*
nostrils body language, 97–98, *98*, 104
Novak, Hal (McArthur, California), 86, *86*
Nowell, Justin (Monroe, Tennessee), 221, *221*
nuclear scanning (scintigraphy), 176
number of equines needed, 31–33, *32*
nylon webbing harnesses, 197, 216

## O
oat hay, 236
OCD (osteochondrosis dissecans), 147
offset knees, 47
oleander *(Nerium oleander)*, **145**
one-way plow, 229
opossums caution, 110, 113
Oregon State University, 160
ornamental plants caution, 141
osteochondrosis dissecans (OCD), 147
out at the knee (bow legs), 47, *47*, 51, *51*
ovariectomy (spaying) mares, 34
overbite (parrot mouth), 165, *165*
"Over" command, 35, 105, **205**
overo lethal white syndrome, 71
owning drafts, 1–18. *See also* draft horses and mules
   advantages, 8–9, 8–15, *11–13*
   animal-human relationship, 14–15, 16, *16*, 17, 18
   challenges of ownership, 15–17, *16*

   children and, 16, *16*, 56, *56*
   commitment (your), 2–3
   companionship from drafts, 8, 14–15, 16, *16*, 17
   decision, making the big, 5
   dream of, realizing, 17–18, *18*
   examples of, 55–62, *55–62*
   experienced equines for novices, 4, 6–8, *7*, 9, *9*, 14, 20, 31, 32, 39
   financial considerations, 8, 15, 31
   goals, setting, 4–5, *5*
   leader, teamster in position of, 6, 15, 17, 92, 93
   learning from equines, 2, *2*, 21, 28, 102, *102*
   physical fitness (your), 16–17, 31
   prevention vs. crisis management, 15
   professional, working with, 3
   public events and, 4, 10, 11, *11*, 18
   questions to ask yourself, 3, 5
   responsibility and liability, 17, 109, *109*, 110
   safety, importance of, 6
   skills required of teamsters, 2–3
   time commitment, 15
   trust-based relationship, 6, 17
   unexpected events, dealing with, 15
   work capability of drafts, 4, 8–10, *9*, 18, *18*
*Oxytropis* spp., **144**, 149

## P
pack boards, 228, *228*, 229
packing and camping, *13*, 13–14, 18, 60, *60*, 227–29, *228*
paddling, 178
palomino gene, 70
panniers, 227, 228, *228*
parasite control, 159, 161–64, **162**, *162*, 183
parrot mouth (overbite), 165, *165*
parts of equine body, *40, 45*, 52
pastern conformation, *40, 45*, 48
pastern/hoof angle, 52, 53, *53*
pastern/shoulder angle, 48
pasture
   forage, 130, 131, *131*, 136, 141, **142–46**, 148
   housing, *119–20*, 119–21
pasture brake *(Pteridium acquilinum)*, **142**
pawing (vice), 125
pecking order in herd, 92–93, *93*, 120
Percheron mules, 39, 58, *58*
Percherons, 6, 24, 25, *25*, 31, 39, 54, 58, *58*, 61, *61*, 64, *66*, 66–67, 71, **76**, 79, 216
personality of equine, 30, 35–37, 79, 83–84

physical fitness (your), 16–17, 31
physis, 85
"Pictou disease," **146**
Pinto Draft Registry, 71
pinworms, 161, **162**
planting your crop, 232–33
plastic-coated webbing harnesses, 197
pleasure-draft-driving harness, 197–99, *198–99*
pleasure driving drafts, 4, 5, *5*, 10–11, 18, 31–32, 192, *192*
Plowden, John, 9
plowing, 229–31, *230*
plowing bees, 11, 27
poison hemlock *(Conium maculatum)*, **145**
poisonous plants caution, 141, **142–46**
poisonvetch *(Astragalus* and *Ozytropis* spp.), **144**, 149
post-legged (straight behind), 50, *51*, 177
Potomac horse fever, 180
pound-for-pound work, 83, *83*, 84
power forecart, 223, *223*
power takeoff (PTO), 223, 237
practice exercises (driving) without a team, 215, *215*, 217, *217*
practicing, learning driving by, 202, 204
Praziquantel, **162**
premolars, 37
pre-purchase exams, 54
preservation of breeds, 75, 77
pressure and release concept, 104–5, *105–6*, 106
prevention vs. crisis management, 15
preventive health management, 152–53
prey animals, 90
prices of equines, 54
professional, working with, 3
proud cut (false rig) geldings, 34
*Prunus* spp., **143**
*Pteridium acquilinum,* **142**
PTO (power takeoff), 223, 237
publications for finding
    equines, 53–54
    learning opportunities, 21, 26, 28
public events, 4, 10, 11, *11*, 18, 27–28
pulling, shoeing for, 168
pulling competitions, 4, 12, *12*, 59, *59*
pulling harness components, 189, *190*
pulse, 153, 154, *154*, 155
purchasing equine, 53–54
purchasing equipment, 240
*Pyemotes tritici* (straw itch mite), 164
Pyrantel pamoate, **162**
Pyrantel tartrate, **162**
pyrocatechines, 155

## Q

qualities of a good mentor, 21–23, *22*

quality of hay, 136
quarter cracks, 53, 178, *178*
quarter of hoof, 52, *52*

## R

rabies vaccination, 157, 158
radiographs (X-rays), 176
ragwort *(Senecio* spp.), **146**
rakes, 55, *55*, 222, 238, *238–39*, 239
raking hay, *238–39*, 238–40
rapid-eye movement (REM), 118, *118*
rattlebox *(Crotalaria* spp.), **143**
rayless goldenrod *(Isocoma* spp.), **145**
rebellion, signs of, 104
red maple *(Acer rubrum),* **145**
red urine, 155
reel (side-delivery) rake, 55, *55*, 238, *238*
registered mules, 80
regular collar, 195
REM (rapid-eye movement), 118, *118*
repetition, importance of, 105–6, *106*
resources, 249–51
respiration rate, 153, 155
responsibility, 17, 109, *109*, 110
retained caps, 165
retained placental tissue, 180
rewarding behavior, 102, 104
rhabdomyolysis (EPSM), 41, 77, 156, 160, 183–84
rhinopneumonitis (EHV), 157, 159
Rice, Alina, ix–x
Rice, Jereld (Vale, Oregon), 186, *186*
*Ricinus communis* (castor bean), **142**
riding cultivator, *233*, 233–34
riding drafts, 12–14, *13*, 60, *60*
riding plow, 229–30, *230*, 231
ringbone, 52
roach back, 49
road stress and laminitis, 180
*Robinia* spp. (black locust), **142**
Roman nose, 43, 76, 81, 82
romantic side of working with draft power, 242, *242*
roundworms (ascarids), 161, **162**
ruminant digestive system, 130
*Rural Heritage,* ix, x, 26, 27, 53, 54
Russian knapweed, Russian thistle *(Acroptilon repens* syn. *Centaurea repens),* **145**

## S

saddles for drafts, 13
safety
    harnessing, 6, 197, 199, 202
    horse-drawn vehicles, 226, *226*
    housing equines, 17, 109, *109*, 110
    learning the basics and, 20
    owning drafts and, 6
safety stop and neck yoke, 222, *222*

sand colic, 182
sandy soil (feeding on) caution, 137
*Sarcocystis neurona,* 113
sawbuck packsaddle, 227, 228, *228*
schedules and mules, 85, 86, 87, 100
scintigraphy (nuclear scanning), 176
scouring rush *(Equisetum* spp.), **143**
second thigh (gaskin) conformation, *40*, 51
seedbed preparation, 231–32, *232*
Seemar, Rachel (Kennebunk, Maine), 9, *9*
selenium, *148*, 148–49
self-preservation sense, mules, 82, 83, 92, *92*, 227
selling and buying horses, 33
*Senecio* spp., **146**
sensitivity of equines, 35, 90
setfast (EPSM), 41, 77, 156, 160, 183–84
share of plow, 229, *230*
shares of dairy cows, selling, 129
shelters, 110–19, *111–12, 114–15, 117–19*
Shires, 54, 64, 66, 67, 68, *68*, *76*
shivers (EPSM), 41, 77, 156, 160, 183–84
shock, signs of, 174
shoeing and trimming equines, 166–69, *166–69. See also* farriers
short back, 49, *49*
short-wave sleep (SWS), 118, *118*
shoulder of equine conformation, *40*, 45–46, *45–46*
show-draft-driving harness, 57, *57*, 197
showing
    docked tails and, 98
    hoof size and, 52
    shoeing for, 169, *169*
sidebone, 178
side-delivery (reel) rake, 55, *55*, 238, *238*
side effects from vaccinations, 157
single-bottom plow, 229–30
single rigged packsaddles, *228*, 229
singletree, 222, *222*
single vs. team driving, 203, *203*
sire of mules, donkey, 80, 81, 82, 83
six-row feeder, 58, *58*
size of equines, 31
sizes (typical) of harnesses, **194**
skills required of teamsters, 2–3
sleep of equines, 118, *118*
sleigh, 57, *57*, 227
*Small Farmer's Journal,* 53
small strongyles, 161, **162**
smell sense of equines, 97–98, *98*
snorting (vocalization), 102
soil testing pasture, 141
*Solanum* spp., **144**
sole bruises, abscesses, 52, 178–79
sole of hoof, 52, *52*

*Sorghum* spp., **146**
sorrel (chestnut), 65, 69, 72
space requirements for drafts, 16. *See also* housing equines
spaying (ovariectomy) mares, 34
splints, 47, 178
Sponenberg, D. Phillip, 70
Spotted Drafts, 54, 59, *59*, 64, *70*, 70–71
spring- and spike-tooth harrows, 231–32, *232*
spring wagon, 225
squealing (vocalization), 93, 102
St.-John's-wort *(Hypericum perforatum),* **145**
stallions caution, 33
stalls, 111, 112, *112*, 115–17, *117*
standard stop and neck yoke, 222
standing near an equine, 100
starting with lines only, 205
startle posture, 90, 92, *92*
steering harness components, 189, *189–90*
stifle conformation, *40*, 51
stocking up, 178
stopping harness components, 190, *190*
stopping with lines only, 205
straight behind (post-legged), 50, *51*, 177
straw itch mite *(Pyemotes tritici),* 164
strongyles (large and small), 161, **162**
stubbornness, mules, 79, 83–84, 86, 87
studdy geldings, avoiding, 34
student's responsibility, 23
submissive equines, 92–93, *93*
Sudan grass *(Sorghum* spp.), **146**
Suffolk Punches, 31, 54, 64, *68*, 68–69, 75, 76, **76**, 129, 216, *216*
sulky plow, 61, *61*, 222, 229, *230*
supplements, 147–50, *148*
surrey, 224
swamp maple *(Acer rubrum),* **145**
swayback, 49
sweat scrapers, 171
sweeney, 41, 195
sweet feed, 138, **139**
switching side of tongue, 36
SWS (short-wave sleep), 118, *118*
symmetry of general appearance, 41

# T

tack area, 112, *112*, 114–15, *115*
tail body language, 98–99, *99*, 104
tail chewing (vice), 126
tails, docking, 98
tall fescue grass *(Festuca elatior* syn. *Festuca arundinacea),* **146**
tansy ragwort *(Senecio* spp.), **146**
tapeworms, 161, **162**, 182
tarweed *(Amsinckia* sp.), **142, 143**
*Taxus* spp., **146**

teaching commitment of mentor, 22
team harness components, *193*
teammates, working with, 36
tedding hay, 238
teeth
  care, 152, 164–66, *165*
  indicator of age, 37–38, *38*
temperament of equine, 30, 35–37, 79, 83–84
temperature, 153, 154, 155
temporal fossa (recessed), 40, *40*
tension on lines, 202, 203, 204, 205, 215, *215*, 217, *217*
test driving equine, 36
tetanus (lockjaw), 156–57, 158
Texas A&M University, 132
thinking like an equine, *89*, 89–90, 92–93, *92–93*
threat assessment, 90, 92, *92*, 94, 95, 97
three-horse hitch, 62, *62*
three-sided shelters, 119, *119*
threshing bees, 11, 27
thrush, 117, 179
thumps (laminitis) home remedy, 173
ticks, 164
tie rings, 111, 116, *117*
tie stalls, 111, 115, 116, *117*
time commitment, 15
toed in/toed out, 44–45
toe of hoof, 52, *52*
tongue, function of, 222, *222*
torsion, 181
toxic plants, 141, **142–46**
T-posts, 122, *122*, 123
TPR (temperature, pulse, respiration), 155
trace-mineral salt block, 149–50
traces, function of, 222, *222*
traction, shoeing for, 168, *168*
tractor mower, 237
traffic, fear of, 36
trailering equine, ease of, 36–37
training mules, 84–85, *85*, 86, 87, *87*, 91, 104, 105–6
transportation business vehicles, 225
transportation services, 4, 8, 10, 18, 56, *56*, 225
transportation vehicles, 224, *224*
trashcans for storing grain, 113
travel precautions, 185
treats (hand-feeding) caution, 96
trimming and shoeing equines, 166–69, *166–69*. *See also* farriers
trust-based relationship, 6, 17
turning, learning, 217, *217*
turning harness components, *190*, 191
turning with lines only, 205
turnout, 16, 119–20, *120*

tushes (canine teeth), 37–38
twisting, torsion, volvulus, 182
two-way plow, 229
tying up (EPSM), 41, 77, 156, 160, 183–84

# U

ulcers, 126, 135
ultrasound, 176, 177
underbite, 165, *165*
unexpected events, dealing with, 15
United States Army mascot, mule, 54, 87
United States Forest Service (USFS), 13, 92, 111, 227
University of California, Davis, 66, 70
University of Iowa, 12
urination status, 153, 155–56
USDA Service Center, 141

# V

vaccinations, 156–59
Valentine, Beth (Corvalis, Oregon), 160, *160*
VEE (Venezuelan equine encephalitis), 158
vegetable oil, 138, **139**, 140
vehicle, selecting right, 224–27, *224–27*
Venezuelan equine encephalitis (VEE), 158
ventilation in barns, 112–13
veterinarian, working with, 173, *173*
vices, dealing with, 125–26
vinyl fencing, 122, 124, *124*
Virginia Polytechnic Institute, 70
vis-à-vis, 225
vision of equines, 43, *43*, 94–95
vitamins, 147, 163
vocalizations, 80, 93, 96, 101–2
voice commands, 202, 203, 204–5, **205**

# W

wagons, 222, 225, *225*
wagon trail drives, 11, *11*, 27
walking cultivator, 233, *233*
walking-on-by-pose, mule, 92, *92*
walking plow, 230, *230*, 231
watching, equine learning by, 104
water
  colic and, 181, 182, 183
  feeding and, 150, *150*
  laminitis and, 181
  waterers, 106, *112*, 116, *117*, 150
weaving (vice), 125
weed management, 233–34, *234*
weeds in hay, 136
WEE (western equine encephalitis), 157, 158
weight management, 156, *156*, 181

weight of drafts, 80, 81
weight of harness, 16
weight-to-bone ratio, 48
western black-legged tick *(Ixodes pacificus)*, 164
western equine encephalitis (WEE), 157, 158
western-style or box breeching work harness, 199–200, *200*, 201
West Nile virus (WNV), 157, 158–59, 180
West Point Vet Clinic (New York), 87
wheel-driven mower, 237
whinnying (vocalization), 96, 102
white snakeroot *(Ageratina altissima* syn. *Eupatorium rugosum)*, **146**
"Whoa" command, **205**
wild jasmine *(Cestrum diurum)*, 147
windsucking (vice), 126
winter coat, 110
winter travel vehicles, 57, *57*, 225, 227, *227*, *242*
wire mesh fencing, 122, 123, 124–25, *125*
withers conformation, *40*, 44, 45–46, *46*
WNV (West Nile virus), 157, 158–59, 180
wolf teeth, 37, 165
wood chewing (vice), 125–26
wood fencing, 122, 124, *124*
work capability, drafts, 4, 8–10, *9*, 18, *18*
work harness, *193*, 199–201, *200–201*
working with draft power, 218–42. *See also* draft horses and mules
  cultivation, 233–34, *234*
  equipment components, 220, 222–23, *222–23*
  farming, 4, 8–10, *9*, 18, *18*, 229–34, *230*, *232–33*
  forecart, 223, *223*
  harrowing, 231–32, *232*
  harvesting crops, 234
  harvesting hay, *238–39*, 238–40
  haymaking, *234–36*, 234–40, *238–39*
  "learning before operating" mantra, 219–20
  legalities, 108–10, 226
  logging, 4, 11–12, 240, 241, *241*
  mowing hay, 235, *235–36*, 236–38
  packing and camping, *13*, 13–14, 18, 60, *60*, 227–29, *228*
  planting your crop, 232–33
  pleasure driving, 4, 5, *5*, 10–11, 18, 31–32, 192, *192*
  plowing, 229–31, *230*
  purchasing equipment, 240
  raking hay, *238–39*, 238–40
  riding drafts, 12–14, *13*, 60, *60*
  romantic side of, 242, *242*
  seedbed preparation, 231–32, *232*
  tedding hay, 238
  tongue, function of, 222, *222*
  traces, function of, 222, *222*
  transportation services, 4, 8, 10, 18, 56, *56*, 225
  vehicle, selecting, 224–27, *224–27*
  weed management, 233–34, *234*
workshops, 21, 23, *26*, 26–27
work vehicles, 225, *225*

## X
*Xanthium strumarium*, **143**
X-rays (radiographs), 176

## Y
Yankee breeching work harness, 199, *201*
yellow burr weed *(Amsinckia* sp.), **142, 143**
yellow star thistle *(Centaurea solstitialis)*, **146**
yew *(Taxus* spp.), **146**

## Z
*Zigadenus* spp., **143**
zinc, 177
zoning requirements, 108–10

## PHOTOGRAPHY CREDITS

p. 2 © Gail Damerow
p. 9 © Shuva Rahim/The Salt Institute for Documentary Studies
p. 18 © Tim J. Kaffenbarger
p. 21 © Sam Moore
p. 25 © Cathie Allen
p. 30 © Bob Langrish
p. 39 © Judi Feltenberger
p. 54 © Dusty Perin
p. 55 © Joseph Mischka
p. 56 © Bud Henderson top, © Dusty Perin bottom
pp. 57, 58 © Joseph Mischka
p. 59 © Fred Newman top, © Joseph Mischka bottom
p. 60 © Dusty Perin top, © Joseph Mischka bottom
p. 61 © Joseph Mischka top, © Bud Henderson bottom left, © Dusty Perin bottom right
p. 62 © Joseph Mischka
p. 65 © Bob Langrish top, © Eline Spek/iStockphoto bottom
pp. 66–69 © Bob Langrish
p. 70 © Lisa Bickford, Triple B Drafters
p. 71 © Dusty Perin
p. 72 © Joseph Mischka
p. 73 © Bob Langrish
p. 74 © Mary Hauser, Harmony Mules
p. 76 © Ellen Day
p. 79 © Dusty Perin
p. 86 Mary Novay
p. 87 © Joseph Mischka
p. 89 © Dusty Perin
p. 91 © Vince Mautino
p. 102 © Joseph Mischka
pp. 106, 108 © Dusty Perin
p. 109 © Gail Damerow
p. 111 © John Cecil
p. 128 © Rick and Linda Conley/Conley's Horse Photos, www.DraftHorsePhotos.com
p. 131 © Susan Arbogast
p. 150 © Dusty Perin
p. 152 © Andrea Burns
pp. 166, 171, 173 © Dusty Perin
p. 178 © Jeneè Franklin
p. 186 © Becky Rice
p. 188 © Gail Damerow
pp. 207–214 © Dusty Perin
pp. 216, 219 © Joseph Mischka
p. 221 © Mark Nowell
p. 224 © Gail Damerow
p. 227 © Dusty Perin
p. 230 © Sam Moore
p. 234 © Joseph Mischka
p. 242 © Bud Henderson

# Other Storey Titles You Will Enjoy

***Horsekeeping on a Small Acreage,* by Cherry Hill.**
A thoroughly updated, full-color edition of the author's best-selling classic about how to have efficient operations and happy horses.
320 pages. Paper. ISBN 978-1-58017-535-7.
Hardcover. ISBN 978-1-58017-603-3.

***How to Think Like a Horse,* by Cherry Hill.**
Detailed discussions of how horses think, learn, respond to stimuli, and interpret human behavior — in short, a light on the equine mind.
192 pages. Paper. ISBN 978-1-58017-835-8.
Hardcover. ISBN 978-1-58017-836-5.

***Oxen: A Teamster's Guide,* by Drew Conroy.**
The definitive guide to selecting, training, and caring for the mighty ox.
304 pages. Paper. ISBN 978-1-58017-692-7.
Hardcover. ISBN 978-1-58017-693-4.

***Storey's Barn Guide to Horse Handling and Grooming,* by Charni Lewis.**
Visual cues and instruction for every handling and grooming procedure in a convenient, hang-it-up format.
128 pages. Paper with concealed wire-o binding. ISBN 978-1-58017-657-8.

***Storey's Barn Guide to Horse Health Care + First Aid.***
Essential techniques for every horse owner, including dental and hoof care, nutrition, and treating wounds and lameness.
128 pages. Paper with concealed wire-o binding. ISBN 978-1-58017-639-2.

***Storey's Guide to Feeding Horses,* by Melyni Worth.**
A complete guide to designing a balanced feeding program according to the individual needs of every horse.
256 pages. Paper. ISBN 978-1-58017-492-3.

***Storey's Guide to Raising Horses,* by Heather Smith Thomas.**
The complete guide to intelligent horsekeeping: how to keep a horse healthy in body and spirit.
512 pages. Paper. ISBN 978-1-58017-127-4.

***Storey's Illustrated Guide to 96 Horse Breeds of North America,* by Judith Dutson.**
A comprehensive encyclopedia filled with full-color photography and in-depth profiles on the 96 horse breeds that call North America home.
416 pages. Paper. ISBN 978-1-58017-612-5.
Hardcover with jacket. ISBN 978-1-58017-613-2.

These and other books from Storey Publishing are available wherever quality books are sold or by calling 1-800-441-5700.
Visit us at *www.storey.com*.